Springer Series in Optical Sciences

Volume 231

Springer Series in Optical Sciences is led by Editor-in-Chief William T. Rhodes, Florida Atlantic University, USA, and provides an expanding selection of research monographs in all major areas of optics:

- lasers and quantum optics
- ultrafast phenomena
- optical spectroscopy techniques
- optoelectronics
- information optics
- applied laser technology
- industrial applications and
- other topics of contemporary interest.

With this broad coverage of topics the series is useful to research scientists and engineers who need up-to-date reference books.

More information about this series at http://www.springer.com/series/624

Feng Chen · Hiroshi Amekura · Yuechen Jia

Ion Irradiation of Dielectrics for Photonic Applications

Springer

Feng Chen
School of Physics, State Key
Laboratory of Crystal Materials
Shandong University
Jinan, Shandong, China

Yuechen Jia
School of Physics, State Key
Laboratory of Crystal Materials
Shandong University
Jinan, Shandong, China

Hiroshi Amekura
Hydrogen Materials Engineering Group,
Cryogenic Center for Liquid Hydrogen
and Materials Science
Center for Green Research on Energy
and Environmental Materials
National Institute for Materials Science
Tsukuba, Ibaraki, Japan

ISSN 0342-4111 ISSN 1556-1534 (electronic)
Springer Series in Optical Sciences
ISBN 978-981-15-4606-8 ISBN 978-981-15-4607-5 (eBook)
https://doi.org/10.1007/978-981-15-4607-5

This Springer imprint is published by the registered company Springer Nature Singapore Pte Ltd.
The registered company address is: 152 Beach Road, #21-01/04 Gateway East, Singapore 189721, Singapore

To our loved ones

Preface

Ion beam technology has been proved as a unique and powerful tool to modify the properties of various materials. As a series of more than 60 years old technologies, continuously growing interests in the interaction of energetic ion beams and matters bring out new applications of ion beam techniques in areas of materials science, chemistry, physics, biology, information science and technology, industry, and health science. There are quite a number of laboratories world widely devoting to fundamental and applied research on the ion-beam property modifications of diverse materials. We may use *Ion irradiation* or *Ion implantation* to describe the main technique of the ion beam technology. In *Optics* and *Photonics*, since 1960s, researchers have applied ion beam technology to fabricate versatile structures and devices towards a variety of intriguing applications. For dielectrics, such as glasses and crystals, one of the major ion-beam induced effects is to tailor the refractive indices of the target materials, forming diverse waveguiding structures, in which tight confinement of optical fields and diffraction-free light propagation can be achieved. Waveguides based on dielectric platforms could be utilized as key elements in compact photonic devices, achieving on-chip integration. Also, according to the nature of materials, practical applications, such as electrooptic modulators, frequency/wavelength converters, waveguide lasers, and amplifiers, and even monolithic devices with multi-functions have been realized by ion beam processing of distinguishing dielectrics. Particularly, the recent development of high-quality lithium niobate thin films may bring the dielectric materials processed by ion beam techniques to a more promising era. The so-called LNOI (lithium niobate on insulators) technique combines the advantages of low-loss membrane and excellent features of materials into a wafer-scale platform, enabling construction of groundbreaking on-chip photonic devices towards applications in telecommunications, light manipulations, and quantum science & technology. In addition to ion beam produced optical waveguides and membranes, nanostructures (e.g., nanotracks and nanoparticles) induced by ion beams also receive intensive attention from the researchers in the communities. The research on ion beam synthesis of nanoparticles in dielectrics started in 1970s. So far, the synthesis and reshaping of embedded nanoparticles has opened a new avenue to tailor the optical properties of the dielectric materials (glasses, crystals), which may offer alternative options for specific applications in nonlinear optics and lasers.

In 1994, Prof. P. D. Townsend, who was called the *father* of ion implanted optical waveguides, and his colleagues in Sussex published a famous professional book titled *Optical Effects of Ion Implantation* (by Cambridge University Press), which summarizes the research progress on ion implanted optical waveguides in dielectrics and ion-implantation induced luminescence to that date. Several review papers and book chapters were published on diverse aspects of this field by Chen and co-workers from Shandong or Jena, and by Olivares and co-workers from Madrid. In addition, the nanoparticles synthesized by diverse ion beams and their properties due to the surface plasmonic resonance effect have been studied widely on the fundamentals, and a couple of book chapters demonstrate the related works. Very recently, Chen and co-workers reveal that the nonlinear optical properties of dielectrics by embedded nanoparticles can be tailored and applied to generate ultrafast laser pulses. There has been a number of intriguing topics of ion beam processed dielectrics towards new photonic applications.

The primary purpose of this book is to introduce the fundamentals, applied techniques, and related properties of ion irradiation/implantation of dielectric materials (e.g., glasses, crystals) towards various photonic applications. This book is mainly aimed at a collection of optical or photonic research, and featured with the combination of ion beam solutions or scenarios for photonic devices and practical applications. This book was written by F. Chen (Shandong University), H. Amekura (NIMS), and Y. C. Jia (Shandong University). The structure of this book is as follows. Chapter 1 (by Chen and Jia) introduces the fundamentals of waveguides and nanoparticle systems, including the basic theories, facilities, and experimental techniques. Chapter 2 (by Chen and Jia) gives comprehensive overview of ion beam produced dielectric waveguides. In Chap. 3 (by Jia), the fabrication of photonic structures and devices based on thin films produced by ion beams are demonstrated. Chapters 4 and 5 (by Amekura) review the ion beam synthesis and elongation of nanoparticles embedded in dielectrics, which present the fundamentals on how the nanocomposites are produced and engineered by diverse ion beam techniques. Chapters 6–9 (by Chen) overview the research progress of ion beam irradiated waveguides and thin film devices on the applications in electrooptics, photoluminescence, nonlinear optics, and lasers, respectively. Chapter 10 (by Chen) is devoted to the tailoring of optical properties and its applications in laser technology. Chapter 11 (by Chen) summarizes the book contents and gives outlook for future research on the related topics. This book contains interdisciplinary research results, which could be particularly useful to readers who are interested in research on integrated optics, materials science, plasmonics, lasers, nonlinear optics, and ion beam technology.

We would like to thank Dr. S. Q. Zhou, Prof. D. Jaque, Prof. A. A. Bettiol, Dr. Y. Takeda, Dr. L. L. Pang, Prof. F. Ren, Prof. L. Wang, Prof. Y. Tan, Dr. W. J. Nie, Mr. R. Li, Mr. Z. Q. Li, Ms. C. Pang and Ms. S. X. Wang for their help on the preparation of some figures/tables in this book and the careful check of the draft. Prof. H. Hu gives his suggestion on the LNOI related chapters. We also thank Dr. J. Li and his colleague Dr. M. C. Huang of *Springer Nature* for the invitation and help on details of preparing this book. Finally, special acknowledgement is forwarded to National Natural Science Foundation of China (NSFC) for the long-term financial support, which enables us to engage in research on this interesting topic. I also would like to dedicate this book to Prof. K. M. Wang (1939–2013), who pioneered the research on ion implanted optical waveguides in China.

Jinan, China
Feng Chen

Contents

About the Authors

Feng Chen is currently a Professor at School of Physics, State Key Laboratory of Crystal Materials, Shandong University, China. He received his B.Sc. (1997) and Ph.D. (2002) degrees in Physics at Shandong Normal University and Shandong University, China, respectively. He spent 2 years at Technische Universität Clausthal (2003–2005), Germany, as an Alexander von Humboldt Research Fellow. His research interests include fabrication of micro-nanoscale photonic structures, optical waveguides, ion beam modification of nanomaterials, laser writing, plasmonics, lasers, nonlinear optics, etc. He has more than 300 papers and 4 chapters published in peer-reviewed journals or books and holds 7 patents. He was awarded the Fund for Distinguished Young Scholars (2009) from National Natural Science Foundation of China (NSFC). Feng Chen is a Fellow of Institute of Physics (IOP), UK, a Senior Member of Optical Society of America (OSA), a Senior Member of SPIE, and a Director Board Member of Chinese Physical Society (CPS). He serves as the Deputy Editor of *Chinese Optics Letters*, an Editorial Board Member of *Optical Engineering*, *Scientific Reports*, and *Physics* (in Chinese), and an active reviewer for more than 50 peer-reviewed journals. He has given keynote/invited talks in more than 60 international conferences/workshops and co-organized some conferences related to Optics (CLEO, CLEO-Europe, CIOP, etc.) or Ion Beam (IBMM, REI, etc.) societies.

Hiroshi Amekura is currently a Chief Researcher in National Institute of Materials Science (NIMS), Japan. He received B.Sc. (1989), M.Sc. (1999) from the University of Tokyo, and Ph.D. degree (1999) from Tsukuba University, Japan. His current interests include ion implantation/irradiation, synthesis of nanoparticles by ion beams, and their optical properties. He has published more than 150 papers in peer-reviewed journals and several book chapters. In 1998, he received Young Scientist Award in 12th International Conference on Ion Implantation Technology. He has been involved in the organization of major conferences in ion beam technology and materials society, such as IUMRS, MRS-J, and Swift Heavy Ion in Matter (SHIM).

Yuechen Jia is currently a Professor at School of Physics, Shandong University, China. He received degrees of B.Sc. (2010) and Ph.D. (2015) from Shandong University, China. His Ph.D. work focused on waveguide-based solid-state lasers and nonlinear optical frequency converters realized by ion beam technology and femtosecond laser micromachining. From 2016 to 2018, he worked on crystalline whispering gallery resonators and epitaxial ferroelectric thin films for applications in nonlinear optics at Albert-Ludwigs-Universität Freiburg, Germany, as an Alexander von Humboldt Research Fellow. In 2019, he researched on resonant all-dielectric nanostructures for strong light-matter interactions at Institut National de la Recherche Scientifique (INRS), Canada, as a Postdoctoral Research Fellow. In 2020, he was appointed Professor of Optics at Shandong University supported by "Taishan Scholars Youth Expert Program" of Shandong Province and "Qilu Young Scholar Program" of Shandong University. He has more than 40 papers published in peer-reviewed journals.

Chapter 1
Fundamentals of Ion Beam Technology, Waveguides, and Nanoparticle Systems

Abstract Ion beam technology is capable of engineering the properties of dielectrics in a flexible manner. Several techniques have been used to fabricate optical waveguides with diverse geometries and synthesize nanoparticles that are embedded in various materials, which enables broad applications in optics and photonics. In this chapter, the ion beam technology is briefly introduced in Sect. 1.1, giving an overview of the ion beam facilities and the applied techniques for waveguide fabrication and nanoparticle synthesis in dielectric materials. In Sect. 1.2, the fundamentals of optical waveguide theory based on dielectrics are briefly introduced. Section 1.3 describes the systems consisting of nanoparticles and their dielectric circumstances, and focuses on the general optical effects. The well-established techniques for characterization of optical waveguides and nanoparticle systems are overviewed in Sect. 1.4.

1.1 Ion Beam Technology Applied for Optical Dielectrics

1.1.1 The Ion Beam Facilities

The ion beam technology can contain a series of different techniques in terms of the used ion beam parameters, the ion-induced effects, and the main purpose of the processing. The basic process is to accelerate the ions of selected elements, and then bombard them onto a solid target. The bombardment of accelerated ions on the targets is accompanied by modification of mechanical, electrical, optical, and magnetic properties of the materials, which is closely correlated to the ion-matter interactions [1–4]. One mostly used term of the process is "*ion implantation*", which sometimes highlight the implantation effect of ions in solids, e.g., doping of specific elements in semiconductors. The modified properties of the materials by ion implantation could be mainly originated from or even be dominated by the presence of implanted foreign atoms. Another typical term is "*ion irradiation*", which generally underlines the ion-induced effects, e.g., defects and damages, rather than the nature of the used ions. Nevertheless, the research community well accepts both

F. Chen et al., *Ion Irradiation of Dielectrics for Photonic Applications*, Springer Series in Optical Sciences 231, https://doi.org/10.1007/978-981-15-4607-5_1

implantation and irradiation for ion-beam processing of dielectric materials, and uses traditional terms to demonstrate the different techniques according to ion beam parameters and application purposes. And in some cases, there is no fine line between ion implantation and irradiation.

In this section, the ion beam facilities and ion beams techniques used for dielectric materials processing will be surveyed. For research of materials science, the energy of ions is in the eV to tens of MeV region, whilst in some cases heavy ions with much high energy up to a few GeV may be also used to realize special applications. For processing dielectrics, the low-energy heavy ions of tens of keV are used to pattern the surface through the sputtering effect of ion interaction on the target. For example, 30–50 keV Ga^+ ion beams can be focused down to diameters less than 10 nm by focused ion beam (FIB) facility, which has been widely applied in nanoscale fabrication and engineering of solid materials. However, when we care more about the implantation or irradiation effect of energetic ions, the ion beam facilities are expected to generate higher-energy beams. So far, the machines for this purpose refer to various implanters, tandem accelerators, and ring-shaped accelerators. A schematic illustration of a typical ion implantation system is shown in Fig. 1.1

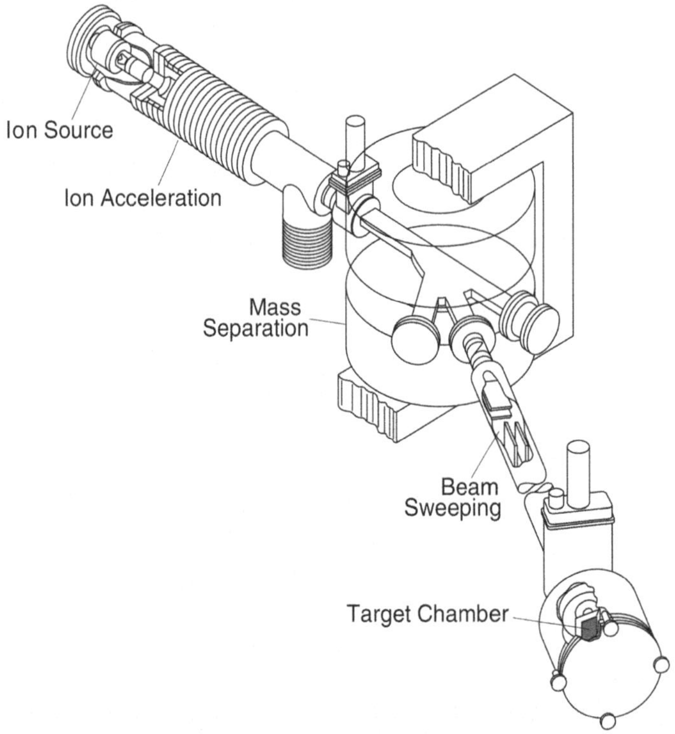

Fig. 1.1 Schematic illustration of an ion implantation system. Reproduced by permission from [1], ©2006 Springer

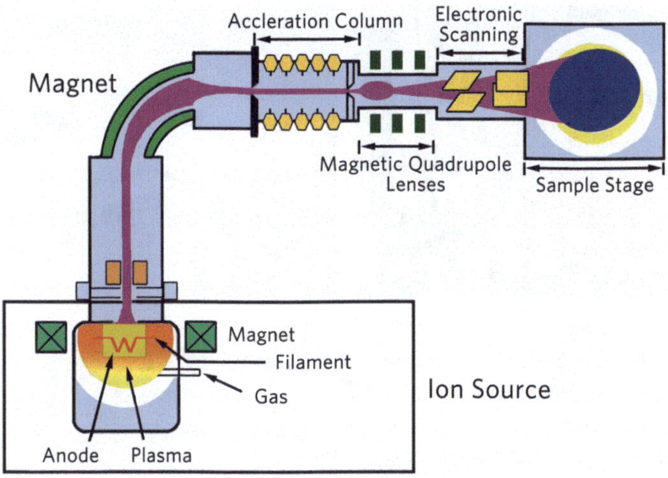

Fig. 1.2 The schematic of the 200-kV ion implanter at Wuhan University, China. Image courtesy of Prof. F. Ren (Wuhan University)

[1]. The basic components in the system include an ion source, an acceleration column, a mass-separator, and a target chamber. The mass-separating magnet is utilized to select the ion species (elements and isotopes) to be implanted. Beam-sweeping facilities are required for large-area uniform implantations. Generally, high energy accelerators require much more complex design and systems, and require a large room for operation.

Ion implanters (typically up to 500 kV) are utilized to achieve implantation or irradiation of various ions, ranging from light ions, like protons or He ions, to heavy ions like Au or Ag ions. Since the energy of ions is low, the ion current can be as high as 10 μA–1 mA, which is suitable for high-fluence implants. Figure 1.2 shows the schematic of a typical 200 kV implanter that is installed in Wuhan University, China. For dielectrics, such machines are usually used for doping of selected elements, nanoparticle synthesis, ion cut, and waveguide fabrication.

For ions at energies of several MeV, tandem accelerators or Van de Graafs are often used. Typical energy ranges from 1 MeV to tens of MeV, depending on the accelerating voltage and the ions' valence of the charge. For a 2 × 1.7 MV tandem, the highest energy for He^+ ions is 3.4 MeV, although in practice the energy used is below this value for safety and beam quality. Highly charged ion implantation could be used for nanoscale material processing and fabrication due to the strong electrical field effects. By choosing high charges of ions and high applied voltages, the ions (particularly heavy ions can be accelerated to 200 MeV). Figure 1.3 shows the photographs of Van de Graaf and tandem accelerators installed in the Institute of Ion Beam Physics and Materials Research, Helmholtz-Zentrum Dresden-Rossendorf (HZDR), Germany. MeV ion implantation or irradiation can be used to fabricate optical waveguides with diverse refractive index profiles in crystals or glasses, or to modify the geometries of nanoparticles in dielectrics.

Fig. 1.3 Photographs of **a** the 2 MV Van de Graaf, **b** the 3 MV tandem accelerator, and **c** the 6 MV tandem accelerator in Institute of Ion Beam Physics and Materials Research, HZDR, Germany. Image courtesy of Dr. S.Q. Zhou (HZDR)

Focusing Cyclotrons are used to generate high-energy ion beams with energy up to hundreds of MeV–a few GeV for heavy-ion research. Such huge machines require large rooms and complicated systems for operation. By using some blocks (e.g., aluminum foils) to reduce the speed of the ions before the chamber, the energy of ions can be reduced to a certain value. Figure 1.4 shows the full-view image of the Heavy Ion Research Facility at Lanzhou (HIRFL), which is a major research facility focusing on nuclear physics, atomic physics, heavy ion applications and interdisciplinary research. The high-energy ions can be used to form thick waveguide structures or elongate the nanospheres to nanorods in dielectrics. In general, high energy accelerators have smaller ion flux than implanters.

By using focused proton beams, the resolution for materials processing can be of sub-10 nm at MeV energy level. The MeV proton beam writing allows the production of high density and high aspect ratio 3D nanostructures [5–8]. Figure 1.5 shows the layout of the 2nd-generation proton beam writing facility installed at Centre for Ion Beam Applications (CIBA) of the National University of Singapore, Singapore. An accelerator is used to generate protons or He ions. The MeV proton beam or He ion beam writing have been successfully applied to fabricate buried waveguides in crystals and glasses. It has also been used to produce some small devices such as nanofluidic structures [9].

Fig. 1.4 The full-view image of the Heavy Ion Research Facility at Lanzhou (HIRFL), which is at Institute of Modern Physics, Chinese Academy of Sciences. Image courtesy of Dr. L.L. Pang (Institute of Modern Physics, Chinese Academy of Sciences)

Fig. 1.5 The layout of the 2nd-generation proton beam writing facility installed at Centre for Ion Beam Applications (CIBA) of the National University of Singapore. Reproduced by permission from [5], ©2012 American Institute of Physics

1.1.2 The Ion Beam Techniques

As ion beams are used to bombard on the dielectric materials, the induced modification of properties is based on the interaction of energetic ions with solids, which leads to energy loss to both atomic nuclei and electrons in materials. The nuclear and electronic energy losses are correlated to nuclear stopping through the ballistic collisions (between ions and nuclei of target atoms) and electronic stopping by inelastic collisions (between ions and electrons of target atoms), respectively [4]. The nuclear energy deposition is correlated to nuclear stopping power S_n, whilst the ion-induced ionization is determined by the electronic stopping power S_e. For the fabrication of dielectric waveguides, the lattice damage induced by S_n and S_e plays crucial roles on the changes of physical properties of materials and, accordingly, the alternation of the refractive indices in ion-irradiated regions [10]. For ion processing of dielectric materials towards optical and photonic applications (e.g., the waveguide fabrication or nanoparticle synthesis), one can simply classify the diverse ion beam techniques into a few regimes in terms of the ion species, energies, fluences, and beam sizes.

Light ion implantation usually refers to bombardment of protons or He ions. Typical energy regime of protons for waveguide fabrication is 300 keV–1 MeV, whilst for He ions, the typical energy ranges from 500 keV to 3 MeV. To form freestanding crystalline thin films of $LiNbO_3$, implantation of He ions at 2–3 MeV has been utilized. For wafer bonded $LiNbO_3$ thin films, the He ion implantation is performed at energy of a few hundred keV to obtain 300–900 nm thick membranes. For the above purpose, the ion fluences are of ~10^{16}–10^{17} cm^{-2}.

Heavy ions for waveguide fabrication in dielectric refer to those with atomic number $A \geq 8$ and ion mass heavier than that of light ions [11]. Sometimes they are called "medium-mass" ions, which is used to avoid misunderstanding with the "real" heavy ions such as Au, Ag, or Xe. The heavy ions at energies from a few MeV to ten MeV may be used to form optical waveguides in glasses or crystals. The typical ion fluence is of ~10^{14}–10^{15} cm^{-2}. Metallic ions such as Au, Ag, Zn, or Cu at 150–500 keV are implanted to dielectrics to synthesize nanoparticles, and the typical ion fluence ranges from ~10^{16} to 10^{17} cm^{-2} because the concentration of the implanted metallic ions is required to exceed the solubility in materials so that the ions can aggregate to nanoparticles [4].

Swift heavy ions refer to the large mass ions with energy no less than 1 MeV/u. The velocity effect of the swift heavy ion irradiation becomes more obvious during the process. For the waveguide formation in dielectrics, in this regime, the refractive index changes depend on the synergetic effect of several ions' impact or single ion's damage, which is correlated to the S_e values. In the low-energy cases, the S_e is very low and electronic damage could be removed by thermal annealing treatments. When the energy increases, the electronic damage could result in overlapping of several highly disordered nanotracks. As the energy reaches a certain threshold, the impact of a single ion will form an amorphous track, which allows to create refractive-index modifications for waveguides [10, 11].

1.2 Optical Dielectric Waveguides

Optical waveguides are critical components in modern integrated photonics [12, 13]. They are conventionally regarded as merely passive connecting elements within a photonic integrated circuit (PIC). As a matter of fact, however, optical waveguides are capable of providing far more than simply optical waveguiding. Active and passive integrated photonic devices based on waveguide technology play significant roles in construction and multi-functionalization of both current and future PICs. The greatest interest in such devices largely stems from the compact geometries of waveguide structures, which allows for good optical confinement within extremely compressed volumes with dimensions of micro/submicron scale, reaching considerably high local optical intensities in contrast to that in bulk counterparts. Thus, monolithically integrated waveguide-based photonic devices are expected to prevail against bulk setups for various applications in multiple areas such as information technology, life science, and quantum computing, just to name a few [14–16].

Optical waveguides defined in optical dielectric materials (e.g., single crystals, polycrystalline ceramics, and amorphous glasses), being compatible with robustness and multifunctionality endowed by a broad range of dielectric hosts, are expected to show superior performance in terms of tailoring light properties over that in semiconductors (e.g., Si) as active and functional miniature devices, such as high-speed optical modulators (further discussed in Chap. 6), high-efficiency nonlinear frequency converters (Chap. 8), and low-threshold lasers (Chap. 9). In this book, our discussion will focus more on optical waveguides defined in dielectrics (including optical crystals, ceramics, and glasses) rather than semiconductors.

Dielectric waveguides can be generally classified into two typical configurations according to their geometries: slab-like guiding layers (planar waveguides) offer one-dimensional light-field confinement and tunnel-like guiding cores (channel waveguides) provide an additional dimension of confinement (optical fibers and photonic crystal waveguides are beyond the scope of this book). The core concept of waveguide fabrication with a view to effectively implementing optical waveguiding based on total internal reflection is the construction of refractive index contrast between waveguide core and cladding/substrate regions. And this is fairly straightforward in dielectrics thanks to a broad range of material modification and deposition/bonding technologies [14–16]. Figure 1.6 schematically illustrates the typical waveguide geometries that usually defined in dielectrics by employing either individual or hybrid fabrication techniques. For example, planar waveguides in asymmetric slab (Fig. 1.6a) and symmetric slab/sandwich (Fig. 1.6b) configurations can be fabricated by employing surface-modification, layer-deposition, and wafer-bonding techniques. Channel waveguides in embedded (Fig. 1.6c), ridge (Fig. 1.6d), and buried (Fig. 1.6e) configurations can be defined based on etching/ablation of planar waveguides and/or on localized-modification techniques. The basic electromagnetic theory of optical waveguiding can be found elsewhere [13], here we only introduce briefly the number of modes that a commonly-used waveguide structure can support.

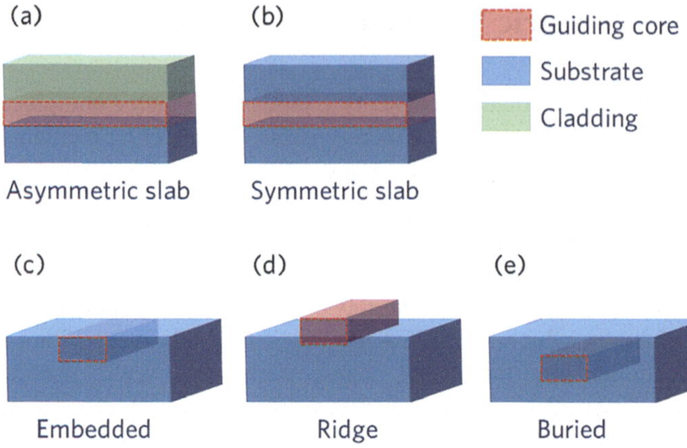

Fig. 1.6 Schematic illustrations of different types of dielectric waveguides

Asymmetric slab waveguide (Fig. 1.6a), with its cladding and substrate having different refractive indices of n_{clad} and n_{sub} and both of which are lower than that of the core n_{core}, is one of the most commonly used waveguide geometries in practice. Assuming $n_{core} > n_{sub} \geq n_{clad}$, the number of guided modes at certain wavelength λ that a slab waveguide with a core thickness of d can support for TE and TM polarizations are given respectively by [13].

$$m_{TE} = \left\{ \frac{2d}{\lambda} \cdot \mathrm{NA} - \frac{1}{\pi} \cdot \tan^{-1}\left[\left(\frac{n_{sub}^2 - n_{clad}^2}{n_{core}^2 - n_{sub}^2} \right)^{1/2} \right] \right\}_{\mathrm{int}} \tag{1.1}$$

and

$$m_{TM} = \left\{ \frac{2d}{\lambda} \cdot \mathrm{NA} - \frac{1}{\pi} \cdot \tan^{-1}\left[\left(\frac{n_{core}}{n_{clad}} \right)^2 \left(\frac{n_{sub}^2 - n_{clad}^2}{n_{core}^2 - n_{sub}^2} \right)^{1/2} \right] \right\}_{\mathrm{int}} \tag{1.2}$$

where, int denotes the next largest integer. The numerical aperture (NA) of a waveguide is defined as the sine of the maximum acceptance angle $\theta_{\max}$ of an incident ray with respect to the waveguide axis, so that the transmitted beam is guided in the core. In another word, the NA essentially determines the coupling efficiency from a light source, which is given by

$$\sin\theta_{\max} = \frac{1}{n_0}\sqrt{n_{core}^2 - n_{sub}^2} = \mathrm{NA} \tag{1.3}$$

where, n_0 is the refractive index of the medium around the waveguide, which equals to 1 in case of air. NA value is in principal related to the contrast between the refractive index of the waveguide core and that of the upper or lower cladding, whichever is higher. So, Eq. (1.3) is valid under the assumption of $n_{sub} \geq n_{clad}$.

Equations (1.1) and (1.2) would be more concise and explicit in case of a symmetric slab waveguide structure (Fig. 1.6b) with $n_{sub} = n_{clad}$, in which the number m is simplified to

$$m = \left(\frac{2d}{\lambda} \cdot \mathrm{NA}\right)_{\mathrm{int}} \tag{1.4}$$

for both TE and TM modes [13]. Thus, the waveguide is characterized as single-mode at a certain wavelength when $m = 1$. Clearly, this occurs when the slab is sufficiently thin or the NA is sufficiently small. Such a preliminary conclusion also applies to the case of a channel waveguide (Fig. 1.6e) buried in a dielectric with lower refractive index, giving

$$m \approx \left(\frac{d}{\lambda}\right)^2 \cdot (\mathrm{NA})^2 \tag{1.5}$$

where d refers to the cross-sectional width of a channel waveguide. It is worth noting that light can be effectively guided through a waveguide in one, two, or many modes. The actual number of modes that carry optical power depends strongly on the source of optical excitation, but the maximum number is m. In a dielectric waveguide, there is no forbidden band for guiding modes, i.e., the fundamental mode $m = 0$ is always allowed. However, each mode with different mode number (m = 1, 2, ...) has its own cut-off wavelength/frequency. The condition of optical excitation frequency ν for single-mode waveguide operation is given by [13]:

$$\nu > \nu_c = \frac{c}{2d \cdot \mathrm{NA}} \tag{1.6}$$

where c refers to the speed of light in vacuum.

Single-mode dielectric waveguides are more desired in many cases in photonic applications, but not all. According to the abovementioned formulas, the quantity of the waveguide modes can be readily reduced by simply adopting of lower NA and sufficiently small core size, which however, has the obvious disadvantage of weakening the light confinement/guidance. On the other hand, in waveguides with a high NA the thickness required for maintaining single-mode propagation is smaller. In fact, controlling light propagation and mode coupling in dielectric waveguides is a very active research topic partly driven by diverse emerging photonic nanostructures [17].

Benefiting from the merits of the waveguide geometry and the associated optical confinement, several prominent features of optical dielectric waveguides can be summarized [14–16]:

i. The original optical features and functionalities of dielectric hosts can be easily well preserved and even further enhanced within waveguide volumes, enabling active photonic devices with miniature footprints.

ii. Waveguide geometries allow for easier access to the active dielectric materials, thereby simplifying the incorporation of additional functionalities, such as optical modulators and gratings, to the waveguide devices.
iii. The development of multifunctional dielectric waveguide devices has facilitated their penetration into different photonic applications, many of which cannot be addressed by other classes of devices, offering immense scope for on-chip hybrid integration in compact and robust packages.

The advances achieved in the field of dielectric waveguides have gone hand-in-hand with the tremendous development of a broad range of fabrication technologies, including layer deposition on existing substrates (e.g., epitaxial growth and sputtering), material modification in existing substrates (e.g., ion exchange, ion implantation, and laser direct writing), and surface structuring techniques (e.g., wet-/dry-etching and lithography) [14–16]. These fabrication techniques have played a critical role in refining several key parameters in order to reach the optimal device performance that these integrated photonic devices need to progress from proof-of-concept demonstrations to practical applications [14–16]. Discussion of fabrication techniques in this book focuses mainly on ion beam related technology, which is of high applicability and flexibility in terms of material modification of optical waveguide structures thanks to its multiple adjustable processing parameters, such as ion energies, species and fluences, beam size and current, and some more detailed experimental conditions (e.g., temperature and masking) [10]. Alternatively, in combination with other techniques, in some cases, the real strength/functionality of ion beam technology in terms of material modification and micro-/nano-structuring can be fully exploited.

1.3 Optical Properties of Nanoparticle Systems

Nanoparticles (NPs) play increasingly significant roles in various areas. The nanoparticles in dielectric circumstance construct systems that are related to a term of "*Plasmonics*", which is a topic of *Plasmons* [18–21]. Numerous papers and books on Plasmonics reflect the fast development of the research in this area. Basically, plasmons are elementary excitations in a metallic solid. In dielectrics, the external field from the excited/illuminated light (light is a kind of electromagnetic wave) allows elementary charges movable in metallic NPs to form an intrinsic resonance. Plasmonics deals with the interaction of light with matter, focusing mainly on two phenomena: surface plasmon polaritons (SPPs, propagating modes) by surface electromagnetic excitations at a dielectric-metal interface, and localized surface plasmons that are confined in the vicinity of metallic NPs. The optical effect of nanoparticles is closely correlated to the localized surface plasmon resonance (LSPR) of metallic NPs in dielectric media [19]. In this book, we focus on the LSPR effect of the ion beam synthesized metallic NPs instead of the SPPs that can propagate along the interface of metals and dielectrics.

In such metallic NPs embedded dielectric systems, the light illumination enables strong enhancement of electromagnetic field by excitation of a plasmon [20]. Under the action of electric field of the light, the electrons move to form electron gas due to polarization, by which the arising restoring force in turn forms a plasmonic oscillation. Consequently, the metallic NPs act like oscillators and the correlated LSPR effect determines the optical properties of the systems. Figure 1.7 illustrates the excitation of plasmons through the polarization of metallic nanoparticles [20].

The most obvious optical effect of nanoparticle systems is the linear optical absorption. In fact, before the research on plasmon-related phenomena started from more than 100 years ago, the nanotechnology in ancient Rome period helps to create works of art that contain elements of metallic NPs. The Lycurgus Cup is one of the examples for applications of metallic NPs [22]. The cup was made by special type of ruby glass that was doped by a certain amount of Au and Ag NPs. It appears in green or red colors as light is reflected (e.g., by daylight) or transmitted through it. In addition, the stained glasses with silver and gold nanoparticles in some church windows exhibit colorful views: adding gold chloride to molten glass resulted in a red tint, and adding silver nitrate turned the glass into yellow [23]. For metallic NPs embedded in dielectrics produced by ion beams, the similar effect on the linear optical absorption has been observed [4]. Figure 1.8 shows the different colors of the metallic NPs in some glass and crystals. As one can see, different systems containing different metallic NPs in dielectrics appear different colors, which are determined by the dielectric constants, sizes, shapes, and concentrations of the metallic NPs and the dielectric media. The synthesis of these NPs in dielectrics will be demonstrated in Chaps. 2 and 3. And the detailed introduction of the optical absorption related to the metallic NPs in dielectrics will be given in Chap. 10.

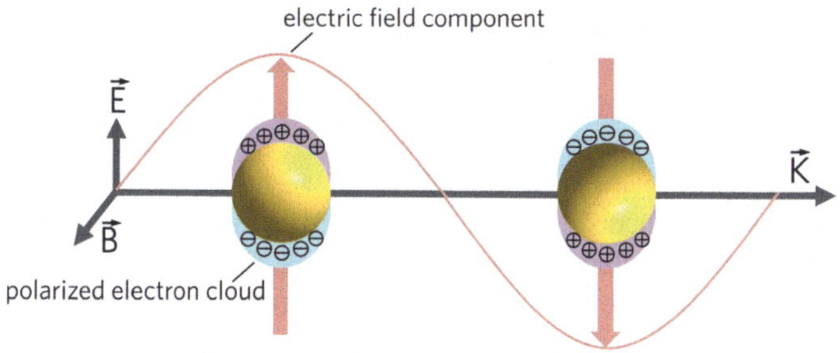

Fig. 1.7 The schematic plot of the excitation of plasmons through the polarization of metallic nanoparticles by light illumination. Reproduced by permission from [20], ©2016 Springer

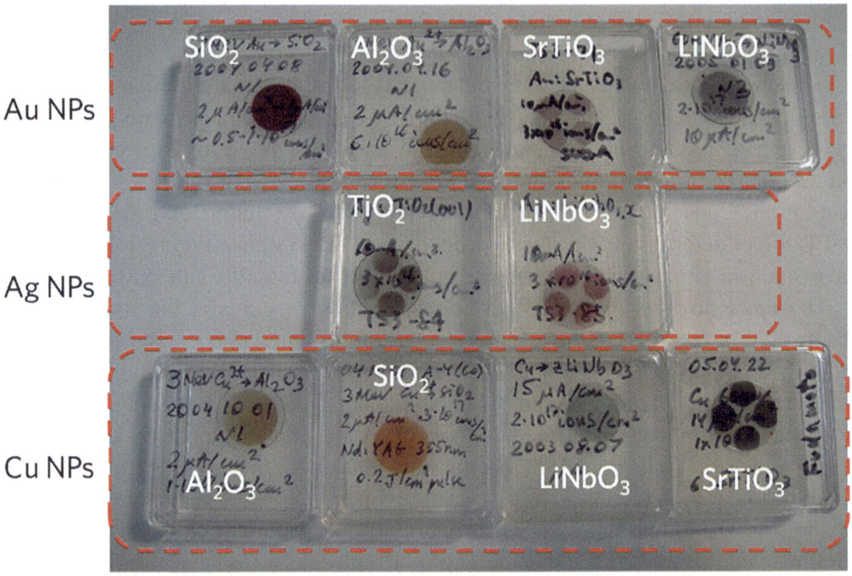

Fig. 1.8 The photographs of the different colors of the metallic NPs embedded in some glass and crystals produced by ion implantation. Image courtesy of Dr. Y. Takeda (National Institute for Materials Science)

1.4 Techniques for Investigation of Photonic Devices

1.4.1 Optical Coupling of Waveguides

Light propagates through optical waveguides in the form of guided modes. Therefore, to optically excite optical modes in an efficient way is of great significance, not only for studying the basic properties of fabricated waveguides via optical characterizations, but also for practical use of waveguide-based optical components in multi-functional PICs via high-quality integration.

In a very straightforward way, incident light can be coupled into a waveguide through one of its end-facets usually by using a focus lens, as schematically illustrated in Fig. 1.9a. This method is generally called *end-fire coupling*, which provides collinear input and output optical beams as well as plane output phase fronts. The achievable coupling efficiency η_c in such a coupling scheme can be very high, largely depending on how well the transverse distribution of the incident light $E_i(x)$ match to that of the guided mode $E_g(x)$, as described by [24].

$$\eta_c = \frac{\left[\int E_i(x) E_g^*(x) dx\right]^2}{\int E_i(x) E_i^*(x) dx \int E_g(x) E_g^*(x) dx} \tag{1.7}$$

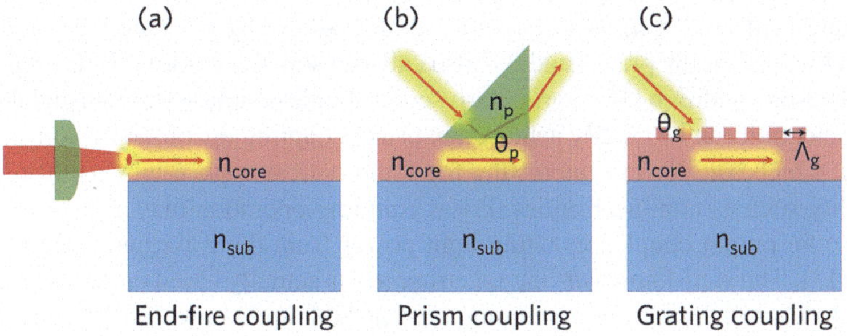

Fig. 1.9 Schematic illustrations of three commonly-used optical waveguide coupling methods: **a** end-fire coupling, **b** prism coupling, and **c** grating coupling

In case of optical coupling of single-mode laser beam into single-mode waveguide, assuming that the mode fields can be well approximated by Gaussian functions, Eq. (1.7) can be further simplified to be [24].

$$\eta_c = \frac{4w_i^2 w_g^2}{\left(w_i^2 + w_g^2\right)^2} \cdot \exp\left(-\frac{2(\Delta x)^2}{w_i^2 + w_g^2}\right) \tag{1.8}$$

where w_i/w_g is the effective incident/guided mode radius and Δx refers to a transverse offset between the two mode cores. Ideally, η_c can be close to unified if the modal distribution of the incident and guided modes are identical. However, this is generally quite hard in practice due to many constraints, such as imperfect optical alignment, waveguide end-facet roughness and flatness, mode mismatch, and Fresnel reflections at waveguide end-facets, which have to be well considered when experimentally studying the optical properties of the waveguides. Nevertheless, end-fire coupling is a widely practical method for optical excitation of guided modes since it is not limited on the waveguide configurations or geometries, which is a remarkable advantage over that of longitudinal coupling method as discussed below [16].

An alternative end-fire coupling configuration, even without any in-coupling lens, simply align the end of laser source (usually a light-emitting diode or a laser diode) and the waveguide, leaving a small space for optimizing coupling condition (the distance between the two ends can be usually adjustable). This configuration has been widely used for high-power diode-pumped solid-state waveguide lasers [14–16]. In addition, end-fire coupling, as a transversal coupling strategy, can be performed using free-space light from a laser source as abovementioned or even guided light from an optical fiber or an additional waveguide.

In contrast to that in the transversal coupling scheme, longitudinal coupling method excites guided modes by using a source wave (usually via evanescence field) entering from the surface at some angle, as schematically illustrated in Fig. 1.9b (*prism coupling*) and Fig. 1.9c (*grating coupling*). When a prism with refractive index $n_p > n_{core}$ placed at a distance d_p (air gap in general) close to the waveguide surface, the incident light at an angle θ_p can be coupled into the waveguide core via evanescent field if d_p is sufficiently small (usually by pressing the waveguide against the base of the prism) and θ_p (also called resonance angle) fulfills total internal reflection condition [13]. Particularly, controllable coupling strength and flexible mode selection can be readily achieved in prism coupling operation by adjusting d_p and θ_p, which can be very interesting for some optical applications requiring high tunability such as nonlinear optics. Prism coupling operation may also be reversed to make an output coupler, exacting light power from the waveguide into the free space [13]. The well-known *m*-line spectroscopy is actually based on prism coupling method, detailed discussion concerning this can be found elsewhere [24]. Just for an example, Fig. 1.10 shows the typical mode lines for ion implanted planar waveguides in quartz obtained by prism coupling technique [25].

Alternatively, when an incident light is diffracted by periodically distributed surface gratings, a portion of diffracted light components may appear in the longitudinal direction, which can be used for optical excitation of waveguide modes [24]. This method is called grating coupling, as schematically illustrated in Fig. 1.9c. The coupling efficiency in this operation can be adjusted by the incident light angle θ_g, the grating period Λ_g and the grating height, which can be optimized for a maximum coupling strength [24]. Furthermore, similar to the abovementioned end-fire and prism coupling methods, the incident beam profile also has a major impact on the grating coupling condition. However, the grating structures in this method have to be fabricated on the surface of the waveguide by advanced lithography and micro-/nano-structuring technology, which complicates significantly the entire waveguide structures and the resulting PICs. In practice, both prism and grating coupling requires surface-type waveguide configurations, e.g., asymmetric (Fig. 1.6a), embedded (Fig. 1.6c), and ridge waveguides (Fig. 1.6d). So there is no surprise why the end-fire coupling method is of most interest currently for basic optical waveguide characterizations.

1.4.2 Microscopic and Spectroscopic Investigations of Waveguides

For dielectric waveguides fabricated by material-modification-based techniques, e.g., ion implantation/irradiation technology, a variety of microscopic and spectroscopic techniques, such as polarizing light microscopy, Raman spectroscopy, fluorescence spectroscopy, second harmonic (SH) spectroscopy, etc. can be well employed for detailed optical investigations of the damaged/modified regions, where in most cases either waveguide cores or waveguide boundaries are located. The most suitable applied investigation technique can be chosen according to the target specific properties of the bulk, which have been locally modified in/around waveguide regions to a certain extent by the incident ion beam. In this section, we will briefly overview some of the commonly-used techniques.

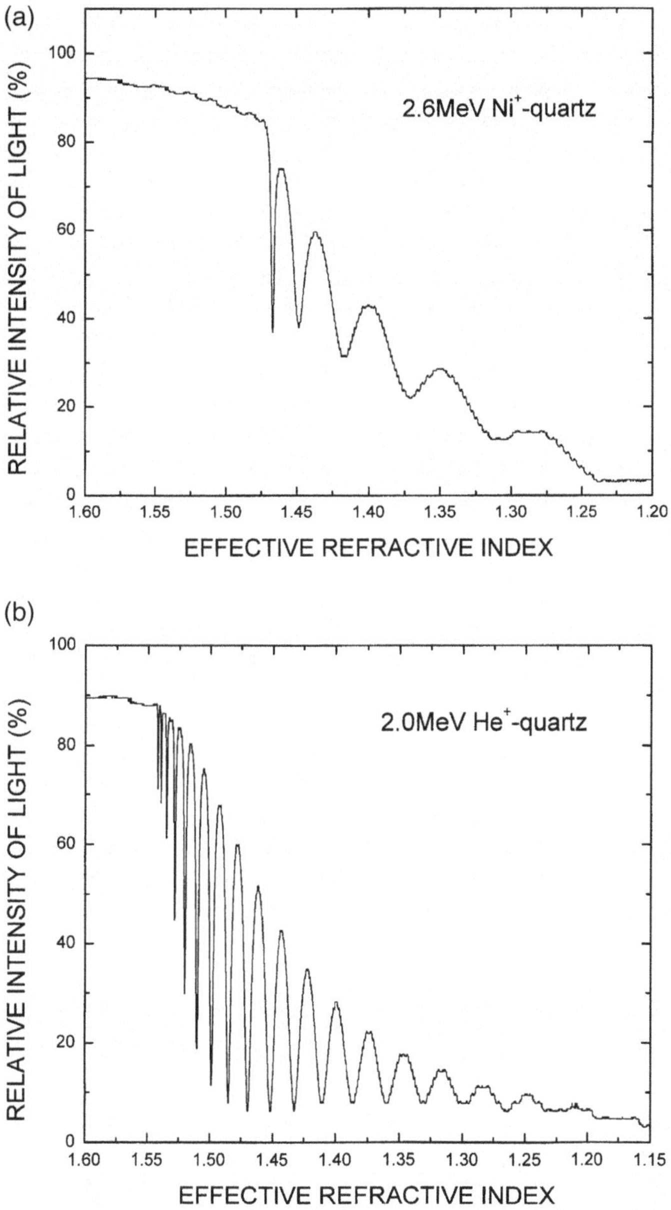

Fig. 1.10 Measured mode lines for ion implanted quartz waveguides by **a** 2.6 MeV Ni and **b** 2.0 MeV He ion implantation. Each dip in the spectra represents a possible mode of waveguides. Reproduced by permission from [25], ©2001 Elsevier

Once ion-implanted, a very first impression on the geometrical/structural information of the fabricated waveguides can be rapidly collected by using standard polarizing light microscopy, which is a contrast-enhancing imaging technique allowing for a direct, but superficial, evaluation of composition and structure of anisotropic samples. It uses polarizing filters to make use of polarized light, usually in two observation modes: reflection and transmission modes, to highlight the local features modified by the ion-beam-induced lattice damages. For example, Fig. 1.11a, b are microscopic images of a ZnSe embedded waveguide (fabricated by ion implantation and standard photolithography, the trapezoidal shape of the waveguide cross section is due to the wedged geometry of the photoresist masks at their edges) [26] and a Nd:GGG ridge waveguide (fabricated by a combination of ion irradiation and femtosecond laser ablation) [27] cross sections obtained by employing polarizing light microscopy in reflection and transmission modes, respectively. Clearly, these figures highlight the geometrical, positional and dimensional information of the ion-beam-damaged areas with good selectivity and enhanced contrast due to the local material modifications.

To further study more detailed information, e.g., photophysical and photochemical mechanisms responsible for the formation of ion-implanted optical waveguides, of the localized material modifications induced by ion beams, confocal microscopy has been proved to be a useful and reliable tool [10]. Figure 1.12 schematically illustrates a standard confocal microscopy system. Confocal microscopy was proposed originally to minimize the interference impacts from the abundant scattered light, i.e., out-of-focus light, associated with the illuminated sample used in conventional wide-field microscopy by preventing it from entering the objective of the microscope. For this, a pinhole aperture suited in front of the light source and a second one positioned in front of the detector are used, collecting only limited light rays emitted from points on the sample and blocking the unwanted scattered light rays. By using coherent laser systems as excitation sources, the confocal principle has been well-developed in Raman, fluorescence and SH spectroscopy. These techniques, which are non-destructive, can provide point-by-point local μ-Raman, μ-photoluminescence

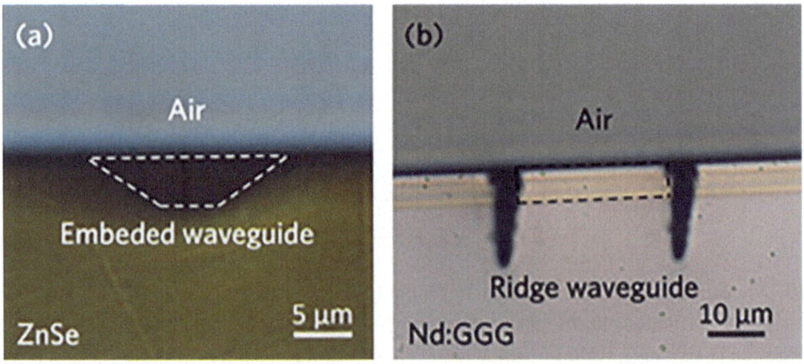

Fig. 1.11 Cross-sectional images of **a** a ZnSe embedded waveguide and **b** a Nd:GGG ridge waveguide obtained by using polarizing light microscopy in reflection and transmission modes, respectively. Reproduced by permission from [26, 27], ©2012 Optical Society of America

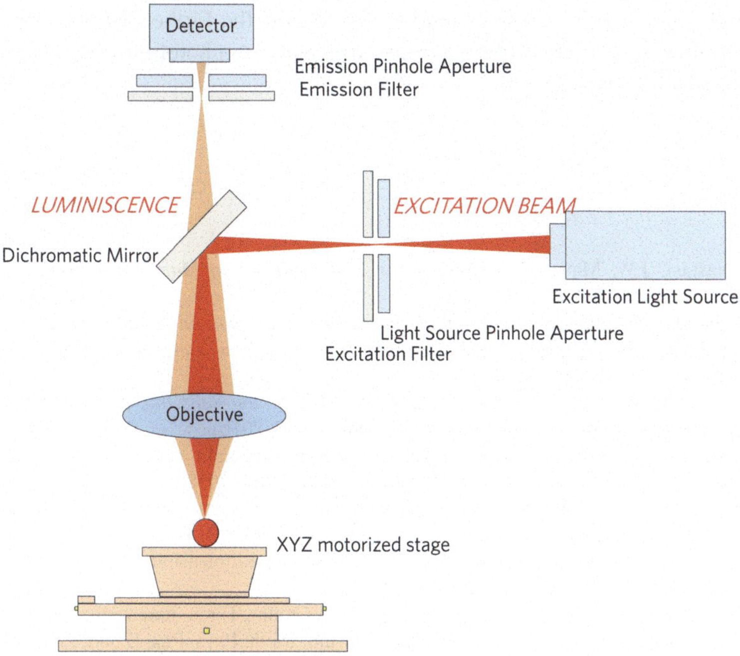

Fig. 1.12 A schematic illustration of a typical confocal fluorescence microscopy system. Image courtesy of Prof. D. Jaque (Universidad Autónoma de Madrid)

(if the sample in study is doped with fluorescence-active ions) and/or μ-SH information by which to analytically characterize, at a micron/submicron resolution, the material-modification mechanisms, such as localized lattice damage and distortion, in ion-damaged regions [10]. For example, Fig. 1.13 shows the confocal μ-photoluminescence cross-sectional images of Nd:YAG waveguides fabricated by focused proton beam writing [28], in which the local lattice modification at the

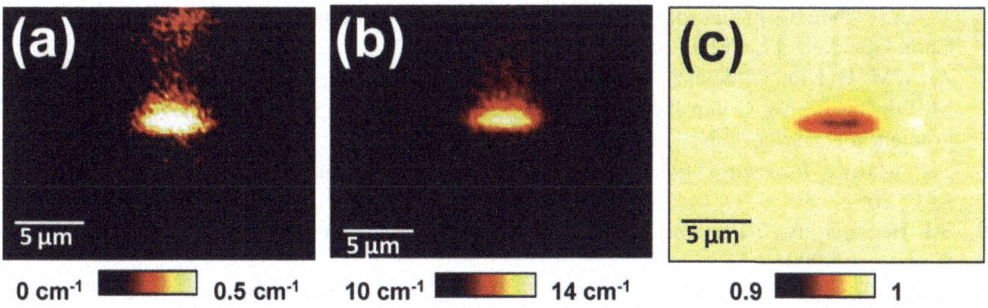

Fig. 1.13 The μ-photoluminescence images of the cross section of a buried Nd:YAG waveguide fabricated by proton-beam writing: Spatial dependence of the **a** spectral shift, **b** bandwidth, and **c** integrated intensity of the hypersensitive $^4F_{3/2} \rightarrow {}^4I_{9/2}$ Nd^{3+} emission line at around 938 nm. Reproduced by permission from [28], ©2010 Optical Society of America

waveguide cores can be clearly identified. Additionally, further details concerning confocal microscopy-based spectroscopic, especially μ-photoluminescence, results are discussed in Chap. 7.

References

1. M. Nastasi, J.W. Mayer, *Ion Implantation and Synthesis of Materials* (Springer, Berlin, Heidelberg, 2006)
2. B. Schmidt, K. Wetzig, *Ion Beams in Materials Processing and Analysis* (Springer, 2013)
3. R. Hellborg, H.J. Whitlow, Y. Zhang (eds.), *Ion Beams in Nanoscience and Technology* (Springer, Berlin Heidelberg, 2009)
4. W. Wesch, E. Wendler, *Ion Beam Modification of Solids: ion-Solid Interaction and Radiation Damage* (Springer International Publishing Switzerland, 2016)
5. J.A. van Kan, P. Malar, A. Baysic de Vera, The second generation Singapore high resolution proton beam writing facility. Rev. Sci. Instrum. **83**, 02B902 (2012)
6. Y. Yao, J.A. van Kan, Automatic beam focusing in the 2nd generation PBW line at sub-10 nm line resolution. Nucl. Instrum. Meth. Phys. Res. B **348**, 203 (2015)
7. S.K. Vanga, A.A. Bettiol, Proton beam writing of three-dimensional microcavities. Nucl. Instrum. Meth. Phys. Res. B **306**, 28 (2013)
8. A.A. Bettiol, S.P. Turaga, Y. Yan, S.K. Vanga, S.Y. Chiam, Three-dimensional metamaterials fabricated using proton beam writing. Nucl. Instrum. Meth. Phys. Res. B **306**, 271 (2013)
9. J.A. van Kan, C. Zhang, P.P. Malar, J.R.C. van der Maarel, High throughput fabrication of disposable nanofluidic lab-on-chip devices for single molecule studies. Biomicrofluidics **6**, 036502 (2012)
10. F. Chen, Micro- and submicrometric waveguiding structures in optical crystals produced by ion beams for photonic applications. Laser Photonics Rev. **6**, 622 (2012)
11. A. Rivera, J. Olivares, G. García, J.M. Cabrera, F. Agulló-Rueda, and F. Agulló-López, Giant enhancement of material damage associated to electronic excitation during ion irradiation: The case of $LiNbO_3$, Phys. Stat. Solidi (a) **206**, 1109 (2009)
12. G. Lifante, *Integrated Photonics: fundamentals* (Wiley, Hoboken, 2003)
13. B.E.A. Saleh, M.C. Teich, *Fundamentals of Photonics*, 3rd edn. (Wiley, Hoboken, 2019)
14. C. Grivas, Optically pumped planar waveguide lasers, Part I: fundamentals and fabrication techniques. Prog. Quant. Electron. **35**, 159 (2011)
15. C. Grivas, Optically pumped planar waveguide lasers: Part II: gain media, laser systems, and applications. Prog. Quant. Electron. **45**, 3 (2016)
16. Y. Jia, F. Chen, Compact solid-state waveguide lasers operating in the pulsed regime: a review. Chin. Opt. Lett. **17**, 012302 (2019)
17. Z. Li, M.-H. Kim, C. Wang, Z. Han, S. Shrestha, A.C. Overvig, M. Lu, A. Stein, A.M. Agarwal, M. Lončar, N. Yu, Controlling propagation and coupling of waveguide modes using phase-gradient metasurfaces. Nat. Nanotechnol. **12**, 675 (2017)
18. S.A. Maier, *Plasmonics: fundamentals and Applications* (Springer Science+Business Media LLC, 2007)
19. S.I. Bozhevolnyi, *Plasmonic Nanoguides and Circuits* (Pan Stanford Publishing Pte. Ltd., Singapore, 2009)
20. A. Trügler, *Optical Properties of Metallic Nanoparticles: basic Principles and Simulation* (Springer International Publishing Switzerland, 2016)
21. M. Quinten, *Optical Properties of Nanoparticle Systems: Mie and Beyond* (Wiely-VCH, 2011)
22. U. Leonhardt, Invis. Cup. Nat. Photon. **1**, 207 (2007)
23. Y. Wei, H. Ebendorff-Heidepriem, J. Zhao, Recent advances in hybrid optical materials: integrating nanoparticles within a glass matrix. Adv. Opt. Mater. **7**, 1900702 (2019)

24. A. Boudrioua, *Photonic Waveguides* (Wiley, Hoboken, 2009)
25. F. Chen, H. Hu, F. Lu, B. Shi, J. Zhang, K. Wang, D. Shen, X. Wang, Comparison between MeV Ni^+ and He^+ ion-implanted planar optical waveguides in quartz. Opt. Commun. **190**, 153 (2001)
26. Y. Jia, F. Chen, Optical channel waveguides in ZnSe single crystal produced by proton implantation. Opt. Mater. Express **2**, 455 (2012)
27. Y. Jia, N. Dong, F. Chen, J.R. Vázquez de Aldana, Sh Akhmadaliev, S. Zhou, Ridge waveguide lasers in Nd:GGG crystals produced by swift carbon ion irradiation and femtosecond laser ablation. Opt. Express **20**, 9763 (2012)
28. A. Benayas, D. Jaque, Y. Yao, F. Chen, A.A. Bettiol, A. Rodenas, A.K. Kar, Microstructuring of Nd:YAG crystals by proton-beam writing. Opt. Lett. **35**, 3898 (2010)

Chapter 2
Overview of Ion Beam Produced Dielectric Waveguides

Abstract Optical waveguides have been fabricated in dielectric materials by using diverse ion beam techniques. In this chapter, an overview of dielectric materials, including glasses, crystals, and ceramics, for waveguide fabrication will be given in Sect. 2.1. Section 2.2 describes material modification of dielectrics under processing of ion beam techniques, which is the base for waveguide fabrication. Section 2.3 reviews the fabrication techniques of ion beam solutions and the geometries of ion beam fabricated waveguides. In Sect. 2.4, the basic waveguiding properties of ion irradiated structures are summarized.

2.1 Overview of Dielectric Materials for Waveguide Fabrication

Dielectrics are important materials for optical and photonic applications. The group of dielectric materials typically includes glasses (amorphous), single crystals (monocrystalline), and ceramics (polycrystalline). All these materials receive respective applications according to their properties in optics and other areas. In this section, the dielectrics for waveguide fabrication will be classified into these three families by highlighting their representative examples.

2.1.1 Glasses

Ion beam processed waveguides have been fabricated in various glasses, including fused silica, silicate, phosphate, fluoride, chalcogenide, germanate, soda-lime glasses, and filter glasses, etc. In this subsection, several families of glasses for ion beam fabricated waveguides are introduced.

Fused silica is a high-purity synthetic amorphous silicon dioxide (SiO_2), which is almost the cheapest glass for practical applications. It has been widely applied in almost all areas of human lives. The excellent properties, such as near-zero thermal expansion, good chemical inertness, low dielectric constant and losses, enable fused

F. Chen et al., *Ion Irradiation of Dielectrics for Photonic Applications*, Springer Series in Optical Sciences 231, https://doi.org/10.1007/978-981-15-4607-5_2

silica to be used for extensive applications in various fields [1]. Fused silica also serves as the major material for commercial optical fibers for construction of optical fiber communication networks and fiber lasers [2, 3]. In addition, waveguide chips based on fused silica substrates receive intriguing applications in on-chip quantum devices [4, 5]. The earliest waveguide in fused silica was reported by Schineller et al., who utilized proton implantation to form planar waveguide layer [6]. Subsequent study was further performed on extended investigation of He^+, N^+, O^+, Si^+, P^+, Ti^+, Ge^+ or other ions [7–11]. By using proton beam writing, in particular, the buried waveguides have been produced in fused silica as well [12].

Silicate glass is considered as excellent candidate for waveguide amplifiers and lasers because of their high solubility for rare-earth ions, such as Nd^{3+}, Er^{3+}, and Yb^{3+}, as well as their chemical robustness. Usually, these doped silicate glasses are used as gain media for solid-state lasers and amplifiers. Ion implantation has been proved effective for the refractive index modulation of silicate glasses. However, silicates are amorphous, particularly with no fixed composition; different silicates may have different ratio for the same composition. Earlier work on the ion-implanted waveguides in silicate glasses was performed by Kakarantzas et al. in 1992 [13]. Chen et al. reported on optical waveguides in a Nd^{3+}-doped silicate glass, which exhibited index increase in the surface region, resulting in fabrication of non-tunneling waveguide structures [14]. Because silicate is a mixture, refractive index behaviors may be different for diverse silicate glasses.

Doped by Er^{3+}/Yb^{3+} ions, phosphate glass is extensively used to form erbium-doped waveguide amplifiers (EDWA) in the area of modern telecommunications. The most common method for waveguide formation in these glasses is the ion exchange. A three-dimensional fabrication technique is the femtosecond laser writing/inscription, which is able to produce more complicated waveguides in this glass. Alternatively, implantation of light as well as heavy ions can be used to construct waveguides in phosphates. The earlier work by Kakarantzas et al. showed that similar refractive index behaviors were observed for the He^+ ion implanted waveguides in their phosphate and silicate glass samples [15]. In addition, Chen's group studied the planar and ridge waveguide properties in an Er^{3+}/Yb^{3+} co-doped phosphate glass formed by means of He or Si ion implantation [16, 17]. Sum et al. investigated the channel waveguides by focused proton beam writing technique [18] for amplification at wavelength of 1.55 μm [19].

Fluoride glass possesses wide transparent wavelength range from the deep-ultraviolet (UV) to the mid-infrared (MIR) regimes. The ZBLAN is the most stable fluoride glass and the most commonly used material for the MIR optical fiber. When doped by active ions such as Tm or Er, ZBLAN becomes an excellent gain medium for lasing and signal amplification. By using ion beam or femtosecond laser writing technology, waveguides in ZBLAN glass have been produced successfully. For example, Zhao et al. used He ion implantation to form single-layer and double-layer planar waveguides in ZBLAN glasses, achieving a refractive index change of ~0.0015 [20].

Chalcogenide glass features very promising properties, such as wider infrared transparent band than silica or fluoride glasses. It is composed of covalently bonded heavy elements. Chalcogenide glass is transparent for the MIR, which brings out considerable applications in MIR operating devices. Moreover, it also has some unique optical properties for nonlinear and waveguide optics, which can be applied in phase-change memories, solar cells, sensors and photonics [21]. Gallium lanthanum sulfide glass (GLS) is a major member of chalcogenide glass family and has significant advantages over other chalcogenide glasses: wider transmission window (0.5–10 μm), higher refractive index (n~2.2767), excellent thermal and mechanical stabilities, larger Kerr nonlinear coefficient, lower toxicity, and higher solubility of rare-earth ions. Consequently, GLS glass can be used as a bulk platform for optical devices, such as compact waveguide-based devices for trace gas sensing or infrared imaging and detection in space [22]. Waveguide structures have been achieved in GLS chalcogenide glass by proton implantation and proton beam writing [23, 24].

2.1.2 Crystals

Different from the amorphous matrix of glasses, single crystals have periodically arrayed lattice structures through the whole bulks. According to the symmetry of the lattice (on the basis of the unit cell), there are seven crystal systems (namely, the systems of Cubic, Tetragonal, Orthorhombic, Hexagonal, Trigonal, Monoclinic, and Triclinic) and 32 point-groups for all the crystals [25]. Figure 2.1 shows the schematic illustrations of the configurations of seven primary crystal systems. The crystals with cubic systems are optically isotropic, whilst those with other six systems are anisotropic, that is, of birefringence. For ion beam processed optical waveguides,

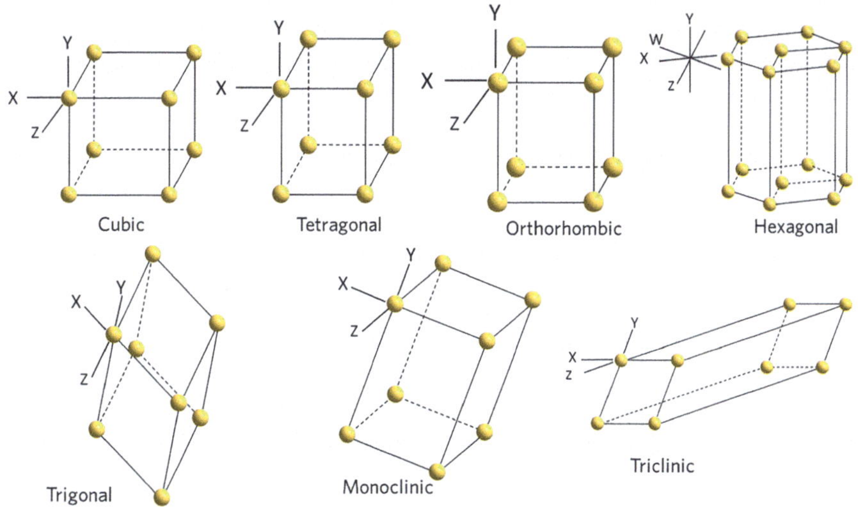

Fig. 2.1 Schematic plots of the configurations of seven primary crystal systems

the lattice structure as well as the birefringence of the dielectric crystals are extremely important for the refractive index changes induced by ion beams, which determines the waveguide properties. In addition, one can classify dielectric crystals by their properties and functionalities, such as laser crystals, nonlinear crystals, electrooptic crystals, acousto-optic crystals, photoluminescent crystals, etc. Some crystals possess more than one feature, e.g., $LiNbO_3$ is a multi-functional material for many applications. The optical devices based on ion beam produced waveguides will be introduced in following chapters, according to their features and functions. In this section, an overview of the dielectric crystals for waveguide fabrication based on symmetry (lattice crystal systems) will be given. In fact, it has been found that similar ion beam induced effect on the refractive index will be valid for those crystals with the same crystal system.

2.1.2.1 Cubic Crystals

In crystallography, the cubic crystal system refers to those with the unit cell in the geometry of a cube. This system has three crystallographic axes that are all perpendicular to each other and equal in length (see Fig. 2.1a), which results in the homogenous behaviors for the light propagation. This is one of the simplest shapes but with highest symmetry among all the dielectric crystals. The most investigated cubic crystals are garnets, e.g., YAG ($Y_3Al_5O_{12}$), GGG ($Gd_3Ga_5O_{12}$), and other similar materials. Doped YAG (e.g., Nd:YAG, Yb:YAG, Er:YAG, or Tm:YAG) is one of a most extensively used laser crystals, and the commercial solid-state lasers based on the bulk Nd:YAG are quite mature. Up to now, a large number of high-power lasers use this crystal as their working medium [26]. Due to the high chemical stability of YAG, it seems very difficult to form optical waveguide in this crystal by means of chemical techniques such as metal thermal in-diffusion or ion exchange. Successful techniques for YAG waveguide fabrication include femtosecond laser writing and ion irradiation/implantation. Implantation/irradiation of selected ions, such as protons, He, C, O, and Kr ions have been used to fabricate waveguides in Nd:YAG or Yb:YAG, and hence realizing waveguide lasers [27–35]. In addition, Nd:GGG is an excellent laser crystal due to its high thermal diffusivity. The H, He, C, and Kr ions with different parameters have been applied to fabricate planar and channel waveguides in this crystal [36–39], and based on the Nd:GGG waveguides, lasing in both continuous wave and pulsed regimes have been realized.

Other cubic crystals include BGO, diamond, and CaF_2, etc. BGO ($Bi_4Ge_3O_{12}$) is used as a scintillator in many areas such as particle physics and aerospace physics. The waveguides in this crystal have been realized by implantation of H, He and O ions [40, 41]. Diamond is also with cubic system, and proton beam writing has been applied to write buried channel waveguides in this material [42, 43]. CaF_2 waveguides were produced by N ion implantation [44].

2.1.2.2 Tetragonal Crystals

In tetragonal system, the crystal lattices are referred to three mutually perpendicular axes, and two of which are equal in length (see Fig. 2.1b). For optical applications, the crystals belonged to the vanadate family, such as Nd:YVO_4, Nd:$GdVO_4$, are most favorite ones for waveguide applications. The undoped vanadates can be used as birefringent devices due to the large contrast of ordinary and extraordinary indices. The Nd doped vanadates are favorite gain media for solid-state lasers which have been applied in a number of areas. The first planar waveguide in Nd:YVO_4 was produced by MeV Si ion implantation [45] in 2001, by Chen et al., and the mode tailoring was realized by thermal annealing at moderate temperatures by the same group [46]. Later investigation of Nd:YVO_4 planar waveguides was performed by using Raman technique in proton or He ion implanted samples [47]. In 2008, Chen et al. utilized mask-assisted MeV O^{3+} ion implantation to fabricate channel waveguides in Nd:YVO_4 [48]. Waveguide lasing in ion irradiated Nd:YVO_4 was also realized at 1064 nm [49]. Nd:$GdVO_4$ waveguides were produced by swift N ion irradiation [50].

Strontium-barium niobate (SBN, $Sr_xBa_{1-x}Nb_2O_6$) and potassium sodium strontium barium niobate (KNSBN) are excellent photorefractive crystals. They can be doped by Ce, Cr, Co or Fe ions to enhance the photorefractive responses. By using proton or He ion implantation, the planar and ridge waveguides in SBN and KNSBN crystals have been produced [51–54]. In KNSBN waveguide arrays, the optical band structures were investigated [55].

2.1.2.3 Orthorhombic Crystals

In the orthorhombic crystal system, the crystal lattices are characterized by three mutually perpendicular axes that are unequal in length. All three bases intersect at 90° angle, so the three lattice vectors remain mutually orthogonal (see Fig. 2.1c). In optics, the nonlinear crystals potassium titanyl phosphate ($KTiOPO_4$ or KTP) and potassium titanyl arsenate ($KTiOAsO_4$ or KTA) belong to this system. $KTiOPO_4$ crystal is usually used as frequency/wavelength conversion elements for laser wavelength from near infrared to visible (e.g., from 1064 to 532 nm) based on phase-matched second harmonic generation (SHG). $KTiOPO_4$ crystals can be periodically poled to be PPKTP with switched domain regions for various frequency conversion. By engineering the domain structures (mainly the period of the ferroelectric domains), PPKTP can be tailored for diverse quasi-phase-matched (QPM) nonlinear optical process, such as SHG, sum frequency generation (SFG), and difference frequency generation (DFG). $KTiOAsO_4$ crystal is also an important medium for nonlinear optical applications, similar to $KTiOPO_4$. It is mostly used for optical parametric oscillation (OPO) to achieve MIR light. $KTiOPO_4$ and $KTiOAsO_4$ waveguides have been produced by implantation/irradiation of various ions, such as H, He, and O [56–61]. At room temperature, $KNbO_3$ also belongs to orthorhombic system. This crystal possesses excellent EO, nonlinear optical and photorefractive properties. Ion implanted waveguides in $KNbO_3$ have been produced by implantation of protons and He ions [62–64]. Optical waveguide was also fabricated in LBO crystal by He ion implantation, which is also a nonlinear crystal belonging to this system [65, 66].

2.1.2.4 Trigonal and Hexagonal Crystals

Trigonal crystals have three axes of equal length intersecting at oblique angles (see Fig. 2.1e). It is a sub-division of the hexagonal crystal system (see Fig. 2.1d), which consists of three equivalent horizontal axes (their positive ends make an angle of 120°). In the trigonal family, some well-known crystals such as $LiNbO_3$, $LiTaO_3$, β-phase barium borate (β-BaB_2O_4 or β-BBO), quartz (crystalline SiO_2), have received broad applications in numerous areas. The 6H and 4H silicon carbide (SiC) is also with hexagonal structure.

Lithium niobate ($LiNbO_3$) is the most used single crystals for optical and photonic applications due to its excellent electrooptic, acousto-optical, and nonlinear optical properties [67]. It can also be used as gain media for lasers and amplifiers, when doped by active ions such as Er and Nd. Ion implanted waveguides in $LiNbO_3$ receive wide investigations by a number of groups [68–71], and ion slicing or thin film wafer bonding enable ion beam processed $LiNbO_3$ to be used as unique platforms for on-chip devices [72]. The ion beam effect on $LiNbO_3$ is not only a research topic, but also related to industrial products for various applications. Lithium tantalate ($LiTaO_3$) is with similar lattice structure to $LiNbO_3$. The ion implanted $LiTaO_3$ waveguides have been produced by various ion beams [73, 74]. Both $LiNbO_3$ and $LiTaO_3$ can be periodically poled to form ferroelectric domain engineered materials to achieve quasi-phase-matched nonlinear processes. For example, SHG based on PPLN waveguides has been realized [75]. β-BaB_2O_4 is a common nonlinear optical material for wavelength conversion and EO modulation. The broad transparent band (190–3300 nm) enables β-BaB_2O_4 to be useful for UV light generation. It is also used for spontaneous parametric down-conversion (SPDC), which is to generate single photons or entangled photon pairs for quantum photonic applications. The β-BaB_2O_4 waveguides were produced by ion implantation to realize EO modulation at UV band [76, 77]. In addition, planar waveguides were produced in quartz by He or Ni ion implantation [78]. For hexagonal crystals, Luan et al. reported planar and ridge waveguide fabrication in 4H-SiC by using implantation of C ions [79].

2.1.2.5 Monoclinic Crystals

In monoclinic system, the lengths of all axes are different, and two axes meet at 90° but the other one does not (see Fig. 2.1f). The low symmetry of this kind of crystal lattice is suitable for nonlinear optical applications. This crystal family includes tribismuth borate (BiB_3O_6, BIBO), gadolinium or yttrium calcium oxoborate (GdCOB or YCOB), etc. BiB_3O_6 is used for SHG, OPO and SPDC application. Planar waveguides in BiB_3O_6 crystals were firstly produced by He or P ion implantation [80], and later also by C ion implantation [81]. Oxoborate crystals are nonlinear optical crystals, and can be doped by Nd or Yb ions for lasing. Nd or Yb doped YCOB and GdCOB are self-frequency-doubled materials, which can generate visible light (green or yellow) by direct pump at ~800 nm by diode. Pure and doped YCOB and GdCOB waveguides have been produced by implantation of He or C ions [82–85].

2.1.3 Ceramics

In the 1990s, polycrystalline ceramics have been fabricated mainly by vacuum sintering [86]. The most popular ceramic is YAG. Nowadays, the rare earth ion doped YAG ceramics, such as Nd:YAG, Yb:YAG, Er:YAG, and Tm:YAG, have become excellent gain media for solid-state lasers that possess comparable performances with their single crystalline counterparts. For ion implanted ceramic waveguides, the first one was realized in Nd:YAG by Chen et al. using mask-assisted O ion implantation [87]. Later work was performed by proton implantation and proton/He ion writing in Nd:YAG [88, 89], and C ion implantation in Yb:YAG [90]. Lasing at 1064 nm based on ion beam irradiated Nd:YAG ridge waveguides has also been realized [91, 92].

2.2 Refractive Index Profiles

Ion beams change the local refractive indices of the materials based on material-modification mechanism, i.e., energy transfer from the ion beam to the local dielectric matrix, which is correlated to the ion-induced damages. In Sect. 1.1.2, it has been demonstrated that the incident energetic ions lose their energies through nuclear collisions and electronic excitation, which are characterized by S_n and S_e, respectively. The accumulation of damage effects leading to the local refractive index modifications mainly depends on ion beam parameters, including ion energy, ion fluence, and in some cases, also on single ion's impact or ion beam scales, and target material properties, such as crystal symmetry and optical birefringence. Accordingly, the refractive index profiles of ion-damaged waveguide structures can be generally subcategorized into five typical configurations, which will be briefly discussed in this section [93–95]. Particularly, characterizations of refractive index profiles of ion-implanted waveguides usually rely on refractive index reconstruction computation method (such as RCM, Reconstruction Calculation Method) [96] instead of direct experimental measurements. Nevertheless, first measurement of effective refractive index (n_{eff}) by prism coupling method (*m*-line spectroscopy) is necessary. Then, it is required to assume a conceivable profile that is characterized by a set of parameters, and followed by adjusting certain parameters, the calculated profile, with computed n_{eff}, may match the measured indices in the best way. Finally, with these optimized parameters, such obtained profile is considered as the refractive index distributions of the waveguides [96].

One of the most well-known and frequently-used refractive index profiles of ion-implanted waveguides is the so-called "*optical barrier*" type, which was firstly proposed by Townsend in 1976 and is particularly suitable for light-ion-implantation regime [93]. Briefly, as penetrating the target materials, the majority of the incident ions will stop at the end of ion trajectory, where the nuclear damage mainly occurs. Accompanying with the nuclear damage, there is usually an expansion of lattice matrix in this region, resulting in a buried layer with negative changes of refractive index. While in the region between the target surface and the barrier layer of the ions' trajectory, S_n is usually very low, forming a relatively-high-index layer experiencing no changes or very slight alternations, where the waveguide core sections are usually located [95]. Early works on ion-implanted waveguides were mostly performed by He ion implantation at several MeV-energy level. In these cases, S_n plays dominating roles whilst S_e introduces only negligible point defects that could be removed by thermal post-annealing treatment [95]. Figure 2.2a shows a typical S_n-barrier-type refractive index profile. This model has been successfully applied to many crystals and glasses engineered by light ions [93–95]. One shortcoming of the S_n-barrier type is that the low-index layer for single-energy implantation usually has a thickness <1 μm with a refractive-index modification (Δn) in the order of 10^{-3}, which is prone to cause optical field leakage from waveguide cores into substrates, resulting in unwanted waveguide losses. To this end, multiple-energy light ion implantation (e.g., 1.8 + 2.0 + 2.2 MeV He^+ ion implantation) can be used to further broaden the barrier region [95, 97–99]. In addition, multiple-energy implantation can also be utilized to fabricate multi-layer waveguides [95].

In contrast to that of light ion implantation, the impact of S_e cannot be ignored any more in heavy ion implantation case, which leads to an increment of refractive index in electronic damage region in some crystals [45, 100, 101]. Together with the refractive index reduction induced by S_n at the end of ion trajectory, an "*enhanced well*" (where $\Delta n > 0$) plus "*barrier*" (where $\Delta n < 0$) type refractive index profile (as shown in Fig. 2.2b) can be formed as a result of combining the respective contributions of S_e and S_n [95]. The cause of the "enhanced well" formation can be multiple, and is closely correlated to the material properties, e.g., the crystal birefringence and crystal symmetry. For example, for crystals with large optical birefringence, such as $LiNbO_3$ and YVO_4, it may be possible to obtain an increment of the relatively lower refractive index (n_e in case of $LiNbO_3$ while n_o in case of YVO_4) towards the relatively higher one (n_o in case of $LiNbO_3$ while n_e in case of YVO_4) by ion beam induced damages [95]. In addition, the refractive index modification depends on the ion beam parameters as well. Usually, higher ion fluence causes larger refractive index changes. Particularly, such a refractive index profile is also possible for light-ion-implanted crystals, in which the damage is associated with material modifications, e.g., alternation of spontaneous polarization [68].

In the regime of swift heavy ion irradiation, the single ion's impact or a synergetic effect of a few ions may create a thick barrier buried inside target materials as a result of electronic damages [102, 103]. The maximum index change in this case could be as large as 0.1, which is much larger than that in regimes of ion implantation.

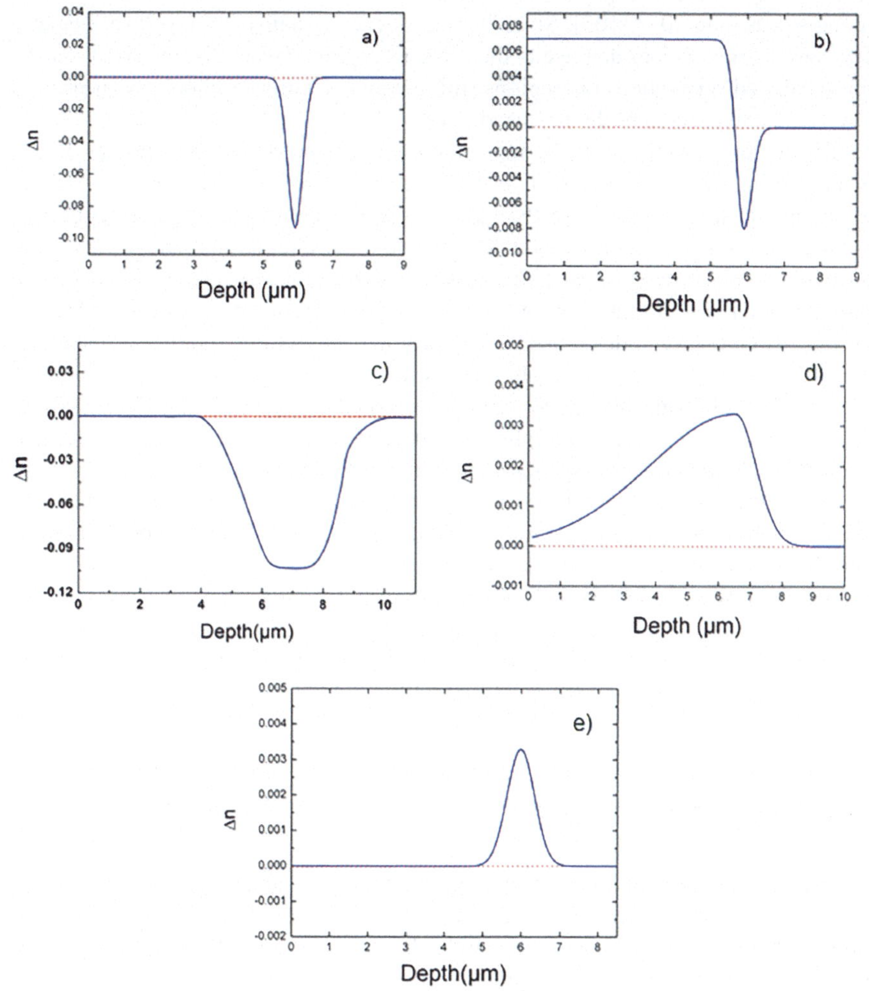

Fig. 2.2 Typical refractive-index profiles of ion-beam-produced waveguides: **a** optical barrier (S_n), **b** enhanced well (S_e) plus optical barrier (S_n), **c** optical barrier (S_e), **d** enhanced well (S_e), and **e** enhanced well (S_n) types

Figure 2.2c shows a typical profile of S_e-based barrier type refractive index. A well-studied example is swift-ion-irradiated $LiNbO_3$, in which the high index change of S_e barrier suggests that the electronic-damaged region experiences a phase transition from crystalline to amorphous [102, 103]. Similar effects have been confirmed in other birefringent crystals, such as in $KGd(WO_4)_2$ [104]. In contrast, for some cubic crystals like YAG and GGG, swift heavy ion irradiation at low fluences usually causes positive refractive index changes in most of the ion trajectory, forming a typical enhanced-well type profile induced by electronic excitation, as shown in Fig. 2.2d. In this case, the waveguide layer could be thick whilst the general index change remains less than 10^{-3} [105]. Since the ion fluence remains at a low level (usually 10^{12} cm^{-2}), the nuclear damage is therefore negligible. Generally, the swift-heavy-ion-irradiated waveguide core regions still remain crystalline without any overlap of ion tracks due to the low-fluence irradiation.

The enhanced well may be buried at the nuclear damage region in case of proton beam writing, the typical refractive index profile of which is shown in Fig. 2.2e. Since the electronic damage of the light ions at the end of ion range is negligible, the nuclear collisions are dominant. The proton fluence required in the proton beam written waveguides (for forming an enhance well) is usually lower than that of proton implantation (for forming a low-index barrier). For example, in case of Nd:YAG, the buried channel waveguides fabricated by proton beam writing requires a fluence of 5×10^{15}–2×10^{16} ions/cm^2, while this value will be raised to the order of 10^{17} ions/cm^2 for achieving an efficient proton-implanted waveguide [106]. Such a profile is quite attractive for practical applications because it offers symmetric confinement of light field and beam propagation in a channelized manner.

In a short summary, ion beam technology, based on ion damage mechanism, offers a flexible and practical approach to introduce localized material modifications and define waveguides with on-demand refractive index profiles in a given dielectric material by adjusting the ion implantation/irradiation parameters [95]. Furthermore, by properly combining ion beam technology and different fabrication/surface-structuring techniques, waveguides with flexible geometries can be produced, which will be discussed in the following section.

2.3 Fabrication Techniques and Waveguide Geometries

The direct ion implantation/irradiation (excluding proton beam writing) often forms a slab guiding layer (or multilayers in case of multi-energy ion implantation/irradiation), i.e., a planar waveguide that restricts light diffraction along the vertical direction, offering 1D optical confinement. However, for more practical applications in photonic chips, channel waveguides, which allows 2D optical confinement, are highly desired. Therefore, apart from proton beam writing which is able to directly define a channel waveguide, a lot of efforts have been made on developing some surface-patterning and surface-structuring techniques, such as mask-assisted implantation, laser ablation or diamond saw dicing, and selective optical illumination, to define on-demand channelized geometries in ion-implanted/irradiated waveguides. In this section, we will overview these fabrication techniques.

Mask-assisted ion implantation/irradiation is a commonly-used method in which a patterned mask on top surface of the target material is used to block the incident ions in the mask-covered region, where no impacts/damages in the substrate will be introduced, remaining perfectly the original properties of the bulk after the ion implantation/irradiation operation. In this way, the geometrical characteristics of the patterned mask can be well transferred to the waveguide embedded in the substrate surface, forming embedded waveguides possibly with straight or/and curved optical guiding channels. Generally, the commonly-used mask can be photoresist mask and metallic mask [95, 107], as schematically illustrated in Fig. 2.3. It is worth noting that the mask thickness, shape and material properties are of crucial importance for the embedded waveguide formation. For example, the cross section of a photoresist mask is typically with a trapezoidal shape, and the mask thickness should be >5 μm [108, 109]. Such a design is essential for completing the side wall of refractive index barrier without complicated processing, e.g., by multi-energy ion implantation, in case of rectangular cross-sectional masks. For swift heavy ion irradiation, metal masking, which has been commercially available and can be produced by micromachining techniques, is more suitable because the penetration depth of the high-energy ions is usually >10 μm [107]. In addition, one can use ion implantation with on-demand mask to fabricate branched waveguides, such as Y branch or 1 × 4 splitters. Figure 2.4 shows the schematic, optical images (surface) and output mode profiles of the 1 × 4 branch waveguide by using masked O ion implantation [110].

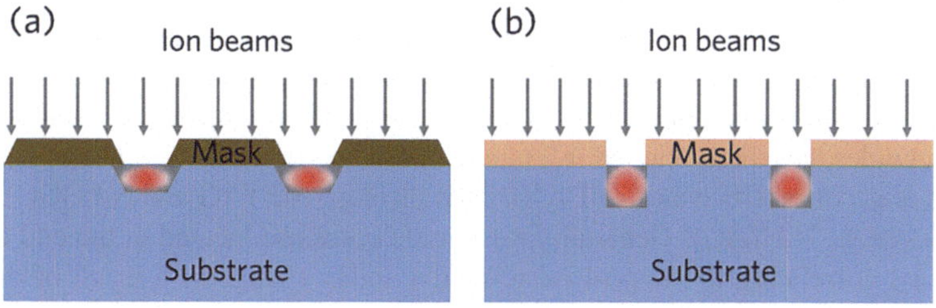

Fig. 2.3 Schematic illustrations of mask-assisted ion implantation/irradiation for embedded waveguide fabrication by using **a** photoresist masking and **b** metal masking

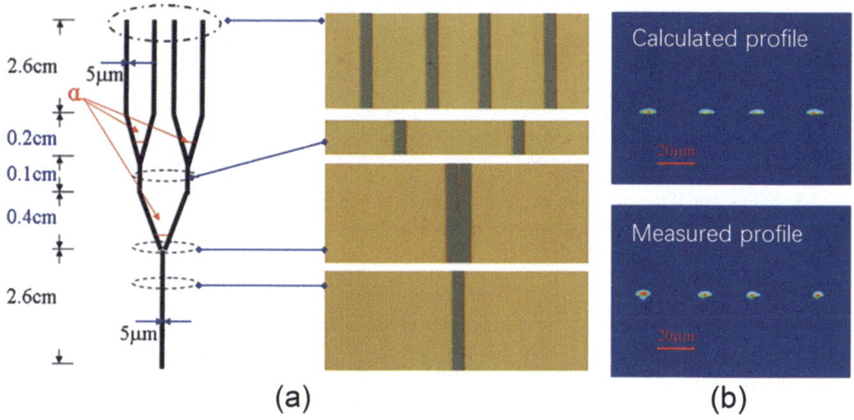

Fig. 2.4 **a** Schematic, optical images (surface) and **b** output mode profiles of the 1 × 4 branch waveguide by using masked O ion implantation. Reproduced by permission from [110], ©2010 Elsevier

In addition to embedded waveguides, ridge waveguides are able to realize 2D confinement of light propagation as well. One of the most intriguing features of ridge waveguides is that the lateral optical confinement is often achieved by air, allowing a high refractive index contrast and easy integration of additional elements, such as electrodes, to the waveguides. On the basis of ion-implanted/irradiated planar waveguides, so far, a few surface-structuring techniques, such as ion beam etching, femtosecond laser ablation, and diamond saw dicing, have been utilized to introduce air grooves and to define the ridge waveguide side walls [32, 38, 53, 95], as schematically illustrated in Fig. 2.5. In particular, the air grooves/side walls defined by femtosecond laser ablation are usually with relatively high roughness (>0.5 μm), whilst the roughness could be as low as 1 nm in case of diamond saw dicing [111]. Nevertheless, one clear advantage of femtosecond laser ablation over that of diamond blade dicing is its 3D micromachining capability, enabling curved waveguide channels, e.g., Y-branches [112]. Reduction of high side-wall roughness can be expected by applying post-ablation processing, such as chemical etching and ion beam milling [113].

For direct fabrication of buried channel waveguides in dielectric materials, focused beam writing using protons or He ion beams has proven to be a powerful tool, as schematically illustrated in Fig. 2.6. So far, focused proton and He^+ ion beam writing of waveguides have been applied in a number of dielectrics [12, 114, 115], including glasses (e.g., fused silica and ZBLAN), crystals (e.g., Nd:YAG and Nd:GGG) , and ceramics (e.g., Nd:YAG). Generally, waveguide cores are located at the end of ion trajectory, correlating to the positive refractive index change as a result of the ion

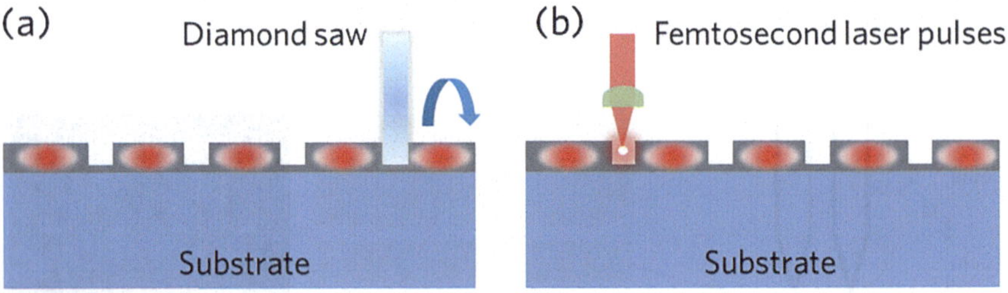

Fig. 2.5 Schematic illustrations of ridge waveguide fabrication based on ion-implanted/irradiated planar waveguide layer using subsequent, **a** precise diamond saw dicing or **b** femtosecond laser ablation

Fig. 2.6 Schematic illustration of buried channel waveguides produced by proton beam writing

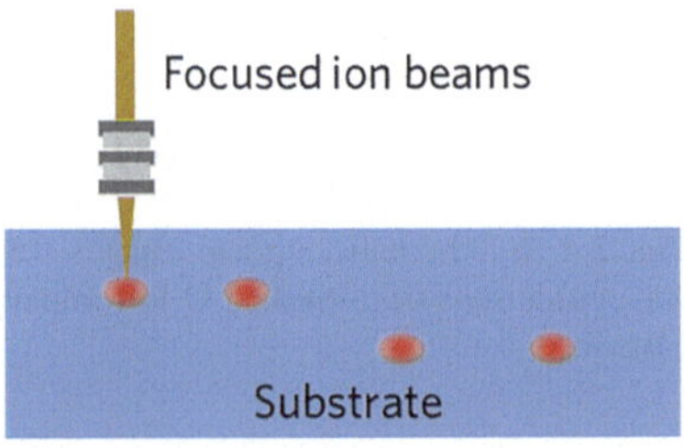

beam induced damage. In this way, the depth of the waveguide cores can be tailored by the energy of the protons of He ions. Figure 2.7 shows the mode profiles of the graded depth waveguides in proton beam writing of fused silica that illuminated by white light [12].

Furthermore, for active ion doped dielectrics, the μ-photoluminescence mapping of the waveguide cross sections can be also conducted to confirm micro-modifications at the nuclear damage regions. In contrast to buried waveguides, particularly, surface-type embedded channel waveguides are also achievable by using proton beam writing, which has been demonstrated in GLS glasses [24]. Therefore, similar to all the other ion beam techniques, the target material matrix also plays an important role in formation of waveguide structures in focused beam writing operation.

For dielectric crystals with unique photoactive properties, e.g., for $LiNbO_3$ crystals with excellent photorefractive properties [116], reconfigurable waveguides can be produced by selective light illumination. As a result of the photorefractive effect under light illumination, the refractive index of the irradiated region in $LiNbO_3$ may be reduced, forming high-index guiding cores in the neighboring areas. A typical design for this kind of waveguide uses photomasked illumination (with stripes or patterns) on an ion-implanted $LiNbO_3$ planar waveguide, as schematically illustrated in Fig. 2.8. The selective illumination can be performed by either coherent light (e.g.,

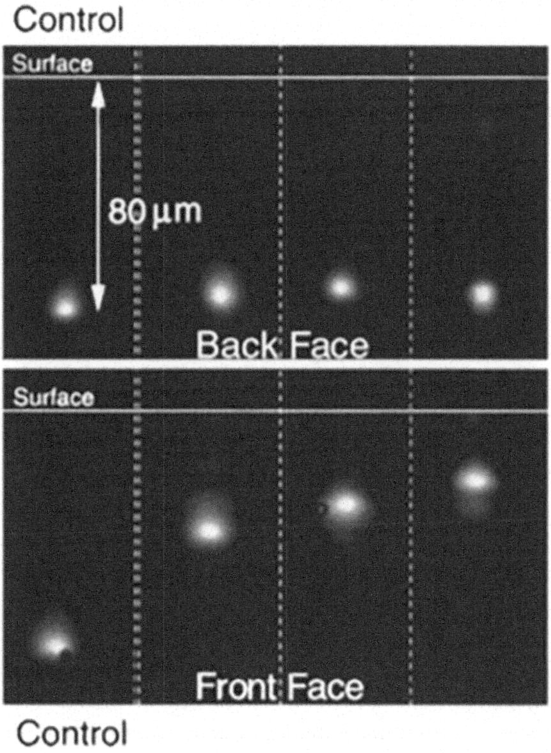

Fig. 2.7 Images of graded-depth fused-silica waveguides illuminated with fiber coupled white light. The white light mode images match the control mode shape at both the front and back face. Reproduced by permission from [12], ©2000 Elsevier

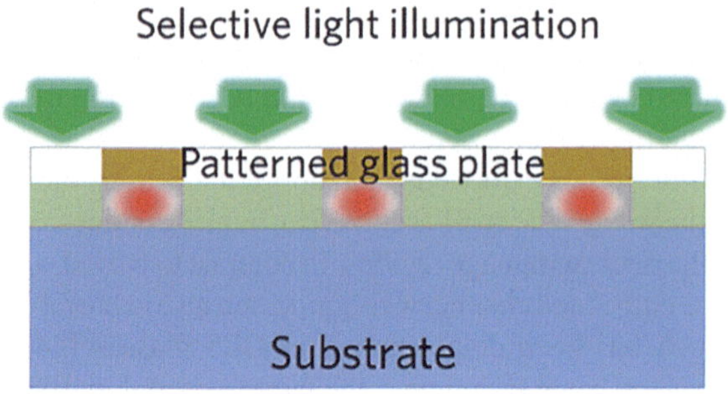

Fig. 2.8 Schematic illustration of reconfigurable waveguide fabrication based on ion-implanted/irradiated planar waveguide layer using subsequent selective illumination of coherent or incoherence light through a glass plate with stripe patterns

green laser) or incoherent light (e.g., white light) [117, 118], inducing only negative index changes in the open areas. The waveguide cores are therefore located in the ion-implanted/-irradiated regions covered by mask stripes, where refractive index remains unaffected. When the illuminated regions are under homogenous white light illumination again, the space charge field will be redistributed due to electrons' movement, resulting in a refractive index recovery, i.e., erasing process of waveguide channels. Thus, $LiNbO_3$ channel waveguides can be constructed, erased and even reconstructed by steps of selective light illumination, forming the so-called reconfigured channel waveguides. Clearly, this technique is only useful for limited crystals with reconfigurable photoactive properties.

2.4 Basic Waveguiding Properties

The basic optical waveguiding properties of ion-implanted/-irradiated waveguides can be characterized based on the optical coupling methods introduced in Sect. 1.4.1. To have a good understanding of the loss information of the fabricated waveguides, a suitable loss measurement method has to be chosen/developed on the basis of fully considering the type of waveguide in terms of geometry and optical/material characteristics. In waveguide loss measurements, generally, the total waveguide loss α_{total} consists of waveguide propagation loss α_p and coupling loss α_c. The former one is often said the "real" waveguide loss, which is mainly caused by imperfect fabrication and processing, inefficient optical confinement, and material absorption. While the latter one α_c is introduced because unified in-/out-coupling efficiency is generally very hard to achieve in practice using any kind of coupling system, resulting optical coupling efficiency $\eta_c < 1$. And normally, it is very hard to directly determine the exact value of α_c. Therefore, many methods have been developed to circumvent direct α_c measurement.

A very straightforward method, namely *cut-back method*, relies on successive single-pass optical transmission measurements through waveguides fabricated under same conditions but with different lengths. Usually, the in-coupled optical power P_{in} is kept constant in all the measurements, then both the out-coupled power P_{out} and the total waveguide loss α_{total} are related to the waveguide length L_w, as described by [119].

$$-\alpha_{\text{total}} \cdot L_w = 10 \cdot \log\left(\frac{P_{\text{out}}}{P_{\text{in}}}\right) \quad (2.1)$$

In this way, the difference in α_{total} between measurements depends only on α_p because the coupling condition, i.e., η_c or α_c, remains the same in all cases. Therefore, α_p can be easily extracted from simple addition and subtraction calculations.

An alternative approach based on end-fire coupling system is called *Fabry-Pérot method*, in which α_p is estimated under different effective L_w [120]. In this case, the two end-facets of the waveguide are regarded as two parallel reflecting surfaces, forming a low finesse planar Fabry-Pérot resonator. Then, by either tuning the excitation wavelength or adjusting the waveguide temperature, the effective L_w can be modulated, leading to variations of the internal phase difference φ_d and the transmitted P_{out}, as decried by [120, 121]

$$\frac{P_{\text{out}}}{P_{\text{in}}} = \frac{(1-R)^2 \cdot \exp(\alpha_p L_w)}{[1 - R \cdot \exp(\alpha_p L_w)]^2 + 4R \cdot \exp(\alpha_p L_w) \sin^2(\varphi_d/2)} \quad (2.2)$$

where $R = [(n_{\text{eff}} - 1)/(n_{\text{eff}} + 1)]^2$ is the effective reflectivity. Therefore, the α_p can be derived by measuring the periodically modulated output power maxima ($P_{\text{out,max}}$) and minima ($P_{\text{out,min}}$), without considering α_c either, according to the following expression [120, 121].

$$\alpha_{\text{total}} \cdot L_w = \ln\left(\frac{RK}{1 - \sqrt{1 - K^2}}\right) \quad (2.3)$$

with

$$K = \frac{P_{\text{out,max}} - P_{\text{out,min}}}{P_{\text{out,max}} + P_{\text{out,min}}} \quad (2.4)$$

For example, Fig. 2.9a depicts a power measurement of Fabry-Pérot transmission resonances/fringes from an ion-implanted $LiNbO_3$ channel waveguide by heating slowly the sample [122]. And α_p as low as 0.17 dB/cm at 1.55 μm can be determined by comparing the intensity contrast and considering the waveguide properties.

In case of prism coupling, *streak imaging method*, by which loss is determined by measuring the longitudinal distribution of the scattered light from the waveguide surface at different positions along the guiding direction (as shown in Fig. 2.9b), is usually used [123, 124]. This method is only valid for surface-type waveguides, but it has been proven to be a very simple and useful method for directly determining α_p, according to [120]

$$P_{\text{out}}(z) = P_{\text{in}} \cdot \exp(-\alpha \cdot z) \quad (2.5)$$

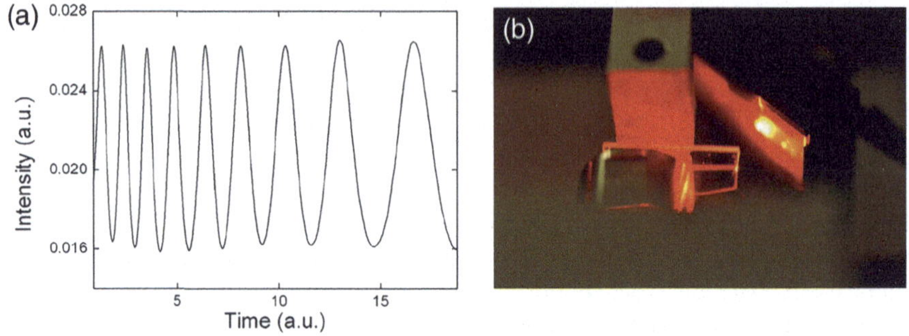

Fig. 2.9 **a** Measured Fabry-Pérot transmission resonances/fringes from an ion-implanted $LiNbO_3$ channel waveguide by heating slowly the sample. **b** A photograph of scattered light from an ion-implanted $LiNbO_3$ planar waveguide in streak imaging measurement based on prism coupling. Reproduced by permission from [122], ©2012 Optical Society of America

where z refers to the position along the guiding direction. Notably, this method is usually less precise when applied for measuring high-quality waveguides, in which case no much light is scattered from the waveguide surface at all [121].

Detailed discussion of these and some more loss measurement methods can be found elsewhere [121]. In the following, we will briefly overview the basic waveguiding properties of waveguides fabricated by ion beam technology, using $LiNbO_3$ waveguide as an example.

The factors determining the total loss of the ion-beam-processed $LiNbO_3$ waveguides are: absorption (absorption loss, L_{abs}), refractive index profiles (mode leakage loss, L_{leak}), and end-facet polishing quality (coupling loss, L_c). The total L_{abs} of $LiNbO_3$ waveguides are contributed by two factors: material absorption and ion-beam-enhanced absorption [68]. Material absorption occurs due to the original optical properties of $LiNbO_3$ crystals, while ion-beam-enhanced absorption is a result of ion-induced point defects (color centers). Therefore, the crystal lattice structures in waveguide regions are not as perfect as the original crystals. The introduced color centers enlarge the absorption of light penetrating the waveguide, increasing largely the propagation loss. For ion implantation with a high fluence (around 10^{16} ions/cm^2), e.g., light ion implantation, significant degradation of the optical transmission properties in the as-implanted guiding regions is often observed [68]. Fortunately, thermal post-annealing treatment is proven to be an effective and reliable approach for either completely removing or largely minimizing the induced lattice defects. The commonly-used temperature is around 300–450 °C in air, depending on the ion fluence used in practice. In contrast, heavy-ion-implanted $LiNbO_3$ waveguides, fabricated at a relatively low ion fluence (around 10^{14} ions/cm^2), only slightly influence the optical absorption of the waveguide structures. However, in this case, thermal annealing (with temperatures of 200–300 °C) is still necessary for further improving the guiding properties [68].

The refractive index profiles of $LiNbO_3$ waveguides depend strongly on the used ion beam techniques, as discussed in Sect. 2.2. For those waveguiding confined only by negative barriers, the barrier layer thickness is actually a major parameter for waveguide loss due to tunneling effect, which causes light energy penetration through the barrier, leaking into the substrate. To this end, the barrier can be made broader by using multi-energy ion implantation or swift heavy ion implantation, minimizing the tunneling effect [122, 125]. For enhanced-well-confined $LiNbO_3$ waveguides, in contrast, it is not necessary to broaden the barrier because these guiding structures are nonleaky. In addition, single-mode waveguides in $LiNbO_3$, which usually have a lower propagation loss than that of multimode ones, can be obtained by controlling the refractive index profiles through an optimization of the implantation parameters and thermal post-annealing conditions.

The typical loss values are 0.5–4 dB/cm for light ion-implanted $LiNbO_3$ waveguides and 0.2–10 dB/cm for heavy ion-implanted ones. It is worth nothing that the guiding properties in $LiNbO_3$ are usually strongly dependent on the polarization of the excitation laser due to its original optical birefringence [68, 95].

References

1. R. Kitamura, L. Pilon, M. Jonasz, Optical constants of silica glass from extreme ultraviolet to far infrared at near room temperature. Appl. Opt. **46**, 8118 (2007)
2. L.L. Blyler, A.C. Hart, R.E. Jaeger, P. Kaiser, J.P. Luongo, J.R. Simpson, Low-loss silicone-clad fused silica fibers for optical communications. Am. Ceram. Soc. Bull. **55**, 455 (1976)
3. P.D. Dragic, M. Cavillon, J. Ballato, Materials for optical fiber lasers: a review. Appl. Phys. Rev. **5**, 041301 (2018)
4. J.B. Spring, B.J. Metcalf, P.C. Humphreys1, W.S. Kolthammer, X. Jin, M. Barbieri, A. Datta1, N. Thomas-Peter, N.K. Langford, D. Kundys, J.C. Gates, B.J. Smith, P.G.R. Smith, I.A. Walmsley, Boson sampling on a photonic chip. Science **339**, 798 (2013)
5. C.Y. Wang, J. Gao, X.M. Jin, On-chip rotated polarization directional coupler fabricated by femtosecond laser direct writing. Opt. Lett. **44**, 102 (2019)
6. E.R. Schineller, R.P. Flam, D.W. Wilmot, Optical waveguides formed by proton irradiation of fused silica. J. Opt. Soc. Am. **58**, 1171 (1968)
7. A.P. Webb, A.J. Houghton, P.D. Townsend, Changes in the chemical stability of ion-implanted silica glass. Radiat. Eff. **30**, 177 (1976)
8. C.M. Johnson, M.C. Ridgway, P.W. Leech, Thermal annealing of implantation-induced compaction for improved silica waveguide performance. Appl. Phys. Lett. **69**, 984 (1996)
9. X.L. Wang, K.M. Wang, G. Fu, S.L. Li, D.Y. Shen, H.J. Ma, R. Nie, Low propagation loss of the waveguides in fused quartz by oxygen ion implantation. Opt. Express **12**, 4675 (2004)
10. P.W. Leech, M. Ridgway, M. Faith, Channel waveguides formed in fused silica and silica-on-silicon by Si, P and Ge ion implantation. IEE Proc. Optoelectron. **143**, 281 (1996)
11. E.V.K. Rao, D. Moutonnet, Buried optical waveguides in fused silica by high-energy oxygen ion implantation. J. Appl. Phys. **46**, 955 (1975)
12. M.L. von Bibra, A. Roberts, S.D. Dods, Ion beam energy attenuation for fabrication of buried, variable-depth, optical waveguides. Nucl. Instrum. Methods Phys. Res. B **168**, 47 (2000)
13. G. Kakarantzas, L. Zhang, P.D. Townsend, Modifications Induced by Irradiation in Glasses. E-MRS Symp. Proc. **29**, 97 (1992)
14. L. Wang, F. Chen, X.L. Wang, K.M. Wang, Y. Jiao, L.L. Wang, Low-loss planar and stripe waveguides in Nd^{3+}-doped silicate glass produced by oxygen-ion implantation. J. Appl. Phys. **101**, 053112 (2007)
15. G. Kakarantzas, P.D. Townsend, J. Wang, Very low loss ion implanted planar waveguides in lead germanate glass. Electron. Lett. **29**, 489 (1993)

16. F. Chen, X.L. Wang, X.S. Li, L.L. Hu, Q.M. Lu, K.M. Wang, B.R. Shi, D.Y. Shen, Ion-implanted waveguides in Nd^{3+}-doped silicate glass and Er^{3+}/Yb^{3+} co-doped phosphate glass. Appl. Surf. Sci. **193**, 92 (2002)
17. Y. Tan, F. Chen, L.L. Hu, P.F. Xing, Y.X. Chen, X.L. Wang, K.M. Wang, Ridge optical waveguide in an Er^{3+}/Yb^{3+} co-doped phosphate glass produced by He^{+} ion implantation combined with Ar^{+} ion beam etching. J. Phys. D **40**, 6545 (2007)
18. T.C. Sum, A.A. Bettiol, K. Liu, M.Q. Ren, E.Y.B. Pun, S.V. Rao, J.A. van Kan, F. Watt, Proton beam writing of erbium-doped waveguide amplifiers. Nucl. Instrum. Methods Phys. Res. B **231**, 394 (2005)
19. K. Liu, E.Y.B. Pun, T.C. Sum, A.A. Bettiol, J.A. van Kan, F, Erbium-doped waveguide amplifiers fabricated using focused proton beam writing. Appl. Phys. Lett. **84**, 684 (2004)
20. Y. Zhao, S. LaRochelle, É. J. Knystautas, N. Bélanger, A. Villeneuve, M. Saad, Planar waveguides in ZBLAN fabricated by He ion implantation. IPR paper Ith17 **45**, 174 (2000)
21. A.B. Seddon, Chalcogenide glasses: a review of their preparation, properties and applications. J. Non-Cryst. Solids **184**, 44 (1995)
22. B.J. Eggleton, B. Luther-Davies, K. Richardson, Chalcogenide photonics. Nat. Photon. **5**, 141 (2011)
23. F. Qiu, T. Narusawa, Proton-implanted planar waveguide in gallium lanthanum sulphide glass. Jpn. J. Appl. Phys. **49**, 092503 (2010)
24. Q. An, C. Cheng, S.K. Vanga, A.A. Bettiol, F. Chen, Proton beam writing of chalcogenide glass: a new approach for fabrication of channel waveguides at telecommunication O and C bands. J. Lightwave Technol. **32**, 3763 (2014)
25. H. Brown, R. Bülow, J. Neubüser, H. Wondratschek, H. Zassenhaus, *Crystallographic Groups of Four-Dimensional Space* (Wiley, New York, 1978)
26. W.C. Fricke, Fundamental mode YAG:Nd laser analysis. Appl. Opt. **9**, 2045 (1970)
27. P. Moretti, M.F. Joubert, S. Tascu, B. Jacquier, M. Kaczkan, M. Malinowskii, J. Samecki, Luminescence of Nd^{3+} in proton or helium-implanted channel waveguides in Nd:YAG crystals. Opt. Mater. **24**, 315 (2003)
28. G.V. Vazquez, J. Rickards, H. Marquez, G. Lifante, E. Cantelar, M. Domenech, Optical waveguides in Nd:YAG by proton implantation. Opt. Commun. **218**, 141 (2003)
29. P.J. Chandler, L. Zhang, P.D. Townsend, Ion-implanted wave-guides in laser host materials. Nucl. Instrum. Methods Phys. Res. B **59**, 1223 (1991)
30. G.V. Vázquez, J. Rickards, G. Lifante, M. Domenech, E. Cantelar, Low dose carbon implanted waveguides in Nd:YAG. Opt. Express **11**, 1291 (2013)
31. Y.C. Jia, Y. Tan, C. Cheng, J.R.V. de Aldana, F. Chen, Efficient lasing in continuous wave and graphene Q-switched regimes from Nd:YAG ridge waveguides produced by combination of swift heavy ion irradiation and femtosecond laser ablation. Opt. Express **22**, 12900 (2014)
32. Y.C. Jia, C.E. Rüter, S. Akhmadaliev, S.Q. Zhou, F. Chen, D. Kip, Ridge waveguide lasers in Nd:YAG crystals produced by combining swift heavy ion irradiation and precise diamond blade dicing. Opt. Mater. Express **3**, 433 (2013)
33. Y.C. Jia, N.N. Dong, F. Chen, J.R.V. de Aldana, S. Akhmadaliev, S.Q. Zhou, Continuous wave ridge waveguide lasers in femtosecond laser micromachined ion irradiated Nd:YAG single crystals. Opt. Mater. Express **2**, 657 (2012)
34. Y.Y. Ren, N.N. Dong, F. Chen, D. Jaque, Swift nitrogen ion irradiated waveguide lasers in Nd:YAG crystal. Opt. Express **19**, 5522 (2011)
35. A. Benayas, N.N. Dong, Y.C. Yao, F. Chen, A.A. Bettiol, D. Jaque, Thermal optimization and erasing of Nd:YAG proton beam written waveguides. Opt. Lett. **36**, 3278 (2011)
36. Y.Y. Ren, N.N. Dong, Y. Tan, J. Guan, F. Chen, Q.M. Lu, Continuous wave laser generation in proton implanted Nd:GGG planar waveguides. J. Lightwave Technol. **28**, 3578 (2010)
37. Y.Y. Ren, F. Chen, Q.M. Lu, H.J. Ma, Optical waveguides in Nd:GGG crystals produced by H^{+} or C^{3+} ion implantation. Appl. Opt. **49**, 2085 (2010)
38. Y.C. Jia, N.N. Dong, F. Chen, J.R.V. de Aldana, S. Akhmadaliev, S.Q. Zhou, Ridge waveguide lasers in Nd:GGG crystals produced by swift carbon ion irradiation and femtosecond laser ablation. Opt. Express **20**, 9763 (2012)

39. S.J. Field, D.C. Hanna, A.C. Large, D.P. Shepherd, A.C. Tropper, P.J. Chandler, P.D. Townsend, L. Zhang, Ion-implanted Nd:GGG channel waveguide laser. Opt. Lett. **17**, 52 (1992)
40. R.Y. He, S.K. Vanga, A.A. Bettiol, F. Chen, Focused ion-beam writing of channel waveguides in BGO crystal for telecommunication bands. Opt. Eng. **54**, 057108 (2015)
41. J. Yang, C. Zhang, F. Chen, S. Akhmadaliev, S.Q. Zhou, Planar optical waveguides in $Bi_4Ge_3O_{12}$ crystal fabricated by swift heavy ion irradiation. Appl. Opt. **50**, 6678 (2011)
42. S. Lagomarsino, P. Olivero, F. Bosia, M. Vannoni, S. Calusi, L. Giuntini, M. Massi, Evidence of light guiding in ion-implanted diamond. Phys. Rev. Lett. **105**, 233903 (2010)
43. H.N. Jin, S.P. Turaga, S.K. Vanga, A.A. Bettiol, Single-mode light guiding in diamond waveguides directly written by a focused proton beam. Opt. Lett. **43**, 2648 (2018)
44. I. Bányász, Z. Zolnai, M. Fried, S. Berneschi, S. Pelli, G. Nunzi-Conti, Leaky mode suppression in planar optical waveguides written in Er:TeO_2-WO_3 glass and CaF_2 crystal via double energy implantation with MeV N^+ ions. Nucl. Instrum. Methods Phys. Res. B **326**, 81 (2014)
45. F. Chen, X.L. Wang, K.M. Wang, Q.M. Lu, D.Y. Shen, Optical Waveguides Formed in Nd:YVO_4 by MeV Si^+ Implantation. Appl. Phys. Lett. **80**, 3473 (2002)
46. F. Chen, X.L. Wang, K.M. Wang, Q.M. Lu, D.Y. Shen, Ion-implanted Nd:YVO_4 planar waveguide: refractive index characterization and propagation mode reduction. Opt. Lett. **27**, 1111 (2002)
47. M.E. Sanchez-Morales, G.V. Vazquez, P. Moretti, H. Marquez, Optical waveguides in Nd:YVO_4 crystals by multi-implants with protons and helium ions. Opt. Mater. **29**, 840 (2007)
48. F. Chen, L. Wang, Y. Jiang, X.L. Wang, K.M. Wang, G. Fu, Q.M. Lu, C.E. Rüter, D. Kip, Optical channel waveguides in Nd:YVO_4 crystal produced by O^+ ion implantation. Appl. Phys. Lett. **88**, 071123 (2006)
49. Y.C. Yao, N.N. Dong, F. Chen, L.L. Pang, Z.G. Wang, Q.M. Lu, Continuous wave waveguide lasers of swift argon ion irradiated Nd:YVO_4 waveguides. Opt. Express **19**, 24252 (2011)
50. N.N. Dong, Y.C. Yao, F. Chen, Optical waveguides in Nd:$GdVO_4$ crystals fabricated by swift N^{3+} ion irradiation. Opt. Mater. **35**, 310 (2012)
51. D. Kip, E. Krätzig, V. Shandarov, P. Moretti, Thermally induced self-focusing and optical beam interactions in planar strontium barium niobate waveguides. Opt. Lett. **23**, 343 (1998)
52. D. Kip, M. Wesner, E. Krätzig, All-optical beam deflection and switching in strontium-barium-niobate waveguides. Appl. Phys. Lett. **72**, 1960 (1998)
53. Y. Tan, F. Chen, H.J. Zhang, Optical ridge waveguides in SBN crystal produced by low-dose carbon ion implantation followed by a sputter etching technique. Opt. Express **15**, 16696 (2007)
54. F. Chen, L. Wang, C.L. Jia, Q.M. Lu, X.L. Wang, K.M. Wang, Characterization of optical planar waveguide in Ce:KNSBN crystal formed by triple-energy helium ion implantation. Appl. Surf. Sci. **253**, 3589 (2007)
55. F. Chen, L. Wang, X.L. Wang, K.M. Wang, Q.M. Lu, Channel waveguide array in Ce-doped potassium sodium strontium barium niobate crystal fabricated by He^+ ion implantation. Appl. Phys. Lett. **86**, 191102 (2006)
56. P. Bindner, A. Boudrioua, J.C. Loulergue, P. Moretti, Formation of planar optical waveguides in potassium titanyl phosphate by double implantation of protons. Appl. Phys. Lett. **79**, 2558 (2011)
57. L. Zhang, P.J. Chandler, P.D. Townsend, Z.T. Alwahabi, S.L. Pityana, A.J. McCaffery, Frequency doubling in ion-implanted $KTiOPO_4$ planar waveguides with 25% conversion efficiency. J. Appl. Phys. **73**, 2695 (1993)
58. F. Chen, H. Hu, Q.M. Lu, K.M. Wang, F. Lu, B.R. Shi, D.Y. Shen, Refractive index profiles of MeV phosphor ion implanted planar waveguide in KTP. Appl. Surf. Sci. **183**, 39 (2001)
59. N.N. Dong, F. Chen, D. Jaque, Q.M. Lu, Micro second harmonic and Raman spectra of He^+ implanted $KTiOPO_4$ waveguides. Opt. Express **19**, 13934 (2011)
60. N.N. Dong, Y.C. Yao, Y.C. Jia, F. Chen, S.K. Vanga, A.A. Bettiol, Q.M. Lu, Buried channel waveguides in $KTiOPO_4$ nonlinear crystal fabricated by focused He^+ beam writing. Opt. Mater. **35**, 184 (2012)

61. F.X. Wang, F. Chen, X.L. Wang, Q.M. Lu, K.M. Wang, D.Y. Shen, H.J. Ma, R. Nie, Fabrication of optical waveguides in $KTiOAsO_4$ by He or Si ion implantation. Nucl. Instrum. Methods Phys. Res. B **215**, 389 (2004)
62. D. Fluck, P. Günter, Second-harmonic generation in potassium niobate waveguides. IEEE J. Sel. Top. Quant. Electron. **6**, 122 (2000)
63. T. Bremer, W. Heiland, B. Hellermann, P. Hertel, E. Krätzig, D. Kollewe, Waveguides in $KNbO_3$ by He^+ implantation. Ferroelectric Lett. **9**, 11114 (1988)
64. F.P. Strohkendl, D. Fluck, P. Günter, Nonleaky optical waveguides in $KNbO_3$ by ultralow dose MeV He ion implantation. Appl. Phys. Lett. **59**, 3354 (1991)
65. K.M. Wang, H. Hu, F. Lu, F. Chen, B.R. Shi, C.Q. Ma, Q.M. Lu, D.Y. Shen, Refractive index profiles of LiB_3O_5 waveguides formed by MeV He ion irradiation. J. Appl. Phys. **92**, 3551 (2002)
66. Y. Jiang, C.L. Jia, L. Wang, X.L. Wang, F. Chen, K.M. Wang, Q.M. Lu, H.J. Ma, D.Y. Shen, The optical properties of planar waveguides in LiB_3O_5 crystals formed by Cu^+ implantation. Appl. Surf. Sci. **253**, 2674 (2006)
67. L. Arizmendi, Photonic applications of lithium niobate crystals. Phys. Status Solidi A **201**, 253 (2004)
68. F. Chen, Photonic guiding structures in lithium niobate crystals produced by energetic ion beams. J. Appl. Phys. **106**, 081101 (2009)
69. J. Olivares, M.L. Crespillo, O. Caballero-Calero, M.D. Ynsa, A. García-Cabañes, M. Toulemonde, C. Trautmann, F. Agulló-López, Thick optical waveguides in lithium niobate induced by swift heavy ions (~10 MeV/amu) at ultralow fluences. Opt. Express **17**, 24175 (2009)
70. G.L. Destefanis, P.D. Townsend, J.P. Gailliard, Optical waveguides in $LiNbO_3$ formed by ion implantation of helium. Appl. Phys. Lett. **32**, 293 (1978)
71. G.G. Bentini, M. Bianconi, M. Chiarini, L. Correra, C. Sada, P. Mazzoldi, N. Argiolas, M. Bazzan, R. Guzzi, Effect of low dose high energy O^{3+} implantation on refractive index and linear electro-optic properties in X-cut $LiNbO_3$: planar optical waveguide formation and characterization. J. Appl. Phys. **92**, 6477 (2002)
72. G. Poberaj, H. Hu, W. Sohler, P. Gunter, Lithium niobate on insulator (LNOI) for microphotonic devices. Laser Photonics Rev. **6**, 488 (2012)
73. K.M. Wang, F. Chen, H. Hu, J.H. Zhang, F. Lu, B.R. Shi, Q.M. Lu, C.Q. Ma, Waveguide formation in $LiTaO_3$ and LiB_3O_5 by keV hydrogen ion implantation. Opt. Commun. **196**, 215 (2001)
74. V. Tormo-Marquez, M. Díaz-Hijar, M. Carrascosa, V.Ya. Shur, J. Olivares, Low loss optical waveguides fabricated in $LiTaO_3$ by swift heavy ion irradiation. Opt. Express **27**, 8696 (2019)
75. Q. Huang, P. Liu, T. Liu, L. Zhang, Y.F. Zhou, X.L. Wang, Second harmonic generation in periodically poled $LiNbO_3$ waveguides formed by oxygen-ion implantation. Phys. Status Solidi RRL **6**, 205 (2012)
76. X.L. Wang, F. Chen, F. Lu, G. Fu, S.L. Li, K.M. Wang, Q.M. Lu, D.Y. Shen, H.J. Ma, R. Nie, Refractive index profiles of planar optical waveguides in beta-BBO produced by silicon ion implantation. Opt. Mater. **27**, 459 (2004)
77. R. Degl'Innocenti, A. Majkic, P. Vorburger, G. Poberaj, P. Günter, M. Döbeli, Ultra-violet electro-optic amplitude modulation in β-BaB_2O_4 waveguides. Appl. Phys. Lett. **91**, 051105 (2007)
78. F. Chen, H. Hu, F. Lu, B.R. Shi, J.H. Zhang, K.M. Wang, D.Y. Shen, X.M. Wang, Comparison between MeV Ni^+ and He^+ ion-implanted planar optical waveguides in quartz. Opt. Commun. **190**, 153 (2001)
79. Q.F. Luan, Y.C. Jia, Y.T. Wang, S. Akhmadaliev, S.Q. Zhou, J.R.V. de Aldana, Y. Tan, F. Chen, Optical ridge waveguides in 4H–SiC single crystal produced by combination of carbon ion irradiation and femtosecond laser ablation. Opt. Mater. Express **4**, 1166 (2014)
80. F. Chen, H. Hu, K.M. Wang, B. Teng, J.Y. Wang, Q.M. Lu, D.Y. Shen, Formation of planar optical waveguide by MeV He^+ and P^+ ion-implanted into BiB_3O_6 crystal. Opt. Lett. **26**, 1993 (2001)

81. L. Wang, F. Chen, X.L. Wang, K.M. Wang, Q.M. Lu, H.J. Ma, Formation of planar waveguide in BiB_3O_6 crystal by MeV carbon implantation. Nucl. Instrum. Methods Phys. Res. B **266**, 899 (2008)
82. A. Boudrioua, J.C. Loulergue, P. Moretti, B. Jacquier, G. Aka, D. Vivien, Second-harmonic generation in He^+-implanted gadolinium calcium oxoborate planar waveguides. Opt. Lett. **24**, 1299 (1999)
83. R. Li, W.J. Nie, Z. Shang, C. Cheng, S. Akhmadaliev, S.Q. Zhou, Q.M. Lu, F. Chen, Guided-wave second harmonics in Nd:YCOB ridge waveguides produced by combination of carbon ion irradiation and precise diamond blade dicing. Opt. Mater. **57**, 153 (2016)
84. Y.Y. Ren, N.N. Dong, Y.C. Jia, L.L. Pang, Z.G. Wang, Q.M. Lu, F. Chen, Efficient laser emissions at 1.06 μm of swift heavy ion irradiated Nd:YCOB waveguides. Opt. Lett. **36**, 4521 (2011)
85. Y.Y. Ren, Y.C. Jia, N.N. Dong, L.L. Pang, Z.G. Wang, Q.M. Lu, F. Chen, Guided-wave second harmonics in Nd:YCOB optical waveguides for integrated green lasers. Opt. Lett. **37**, 244 (2012)
86. A. Ikesue, Y.L. Aung, Ceramic laser materials. Nat. Photonics **2**, 721 (2008)
87. F. Chen, Y. Tan, D. Jaque, Ion implanted optical channel waveguides in neodymium doped yttrium aluminum garnet transparent ceramics. Opt. Lett. **34**, 28 (2009)
88. Y. Tan, F. Chen, Proton implanted optical channel waveguides in Nd:YAG laser ceramics. J. Phys. D **43**, 075105 (2010)
89. Y.C. Yao, C. Zhang, S.K. Vanga, A.A. Bettiol, F. Chen, Proton or helium ion beam written channel waveguides in Nd:YAG ceramics. Opt. Mater. **35**, 2557 (2013)
90. N.N. Dong, A. Benayas, D. Jaque, Y. Tan, F. Chen, Fluorescence-quenching free channel waveguides in Yb:YAG ceramics produced by carbon ion implantation. J. Lightwave Technol. **29**, 1460 (2011)
91. Y. Tan, S. Akhmadaliev, S.Q. Zhou, S.Q. Sun, F. Chen, Guided continuous-wave and graphene-based Q-switched lasers in carbon ion irradiated Nd:YAG ceramic channel waveguide. Opt. Express **22**, 3572 (2014)
92. Y.C. Jia, J.R.V. de Aldana, S. Akhmadaliev, S.Q. Zhou, F. Chen, Femtosecond laser micromachined ridge waveguide lasers in Nd:YAG ceramics. Opt. Mater. **36**, 228 (2013)
93. P.D. Townsend, P.J. Chandler, L. Zhang, *Optical Effects of Ion Implantation* (Cambridge University Press, 1994)
94. F. Chen, X. Wang, K. Wang, Developments of ion implanted optical waveguides in optical materials: a review. Opt. Mater. **29**, 1523 (2007)
95. F. Chen, Micro-and submicrometric waveguiding structures in optical crystals produced by ion beams for photonic applications. Laser Photonics Rev. **6**, 622 (2012)
96. P.J. Chandler, F.L. Lama, A new approach to the determination of planar waveguide profiles by means of a non-stationary mode index calculation. Opt. Acta **33**, 127 (1986)
97. K. Wang, F. Chen, H. Hu, X. Wang, B. Shi, Q. Lu, C. Ma, D. Shen, Double waveguide in $NdLiP_4O_{12}$ laser crystal formed by MeV He^+ ion implantation. Nucl. Instrum. Methods Phys. Res. B **227**, 271 (2005)
98. K. Wang, F. Lu, M. Meng, W. Li, X. Wang, B. Shi, Z. Zhou, X. Gao, Double barrier structures in $LiNbO_3$ waveguide created by MeV He^+ implantation. Opt. Commun. **141**, 141 (1997)
99. F. Chen, X. Wang, K. Wang, Q. Lu, F. Lu, D. Shen, H. Ma, Bi-layer optical waveguide formed by double boron ion implantation into strontium barium niobate crystal. Nucl. Instrum. Methods Phys. Res. B **234**, 467 (2005)
100. H. Hu, F. Lu, F. Chen, B. Shi, K. Wang, D. Shen, Extraordinary refractive-index increase in lithium niobate caused by low-dose ion implantation. Appl. Opt. **40**, 3759 (2001)
101. H. Hu, F. Lu, F. Chen, B. Shi, K. Wang, D. Shen, Monomode optical waveguide in lithium niobate formed by MeV Si^+ ion implantation. J. Appl. Phys. **89**, 5224 (2001)
102. J. Olivares, G. Garcia, A. Garcia-Navarro, F. Agullo-Lopez, O. Caballero, A. Garcia-Cabanes, Generation of high-confinement step-like optical waveguides in $LiNbO_3$ by swift heavy ion-beam irradiation. Appl. Phys. Lett. **86**, 183501 (2005)

103. J. Olivares, A. Garcia-Navarro, G. Garcia, A. Mendez, F. Agullo-Lopez, A. Garcia-Cabanes, M. Carrascosa, O. Caballero, Nonlinear optical waveguides generated in lithium niobate by swift-ion irradiation at ultralow fluences. Opt. Lett. **32**, 2587 (2007)
104. A. Garcia-Navarro, J. Olivares, G. Garcia, F. Agullo-Lopez, S. Garcia-Blanco, C. Merchant, J.S. Aitchison, Fabrication of optical waveguides in KGW by swift heavy ion beam irradiation. Nucl. Instrum. Methods Phys. Res. B **249**, 177 (2006)
105. Y. Ren, N. Dong, F. Chen, A. Benayas, D. Jaque, F. Qiu, T. Narusawa, Swift heavy-ion irradiated active waveguides in Nd:YAG crystals: fabrication and laser generation. Opt. Lett. **35**, 3276 (2010)
106. A. Benayas, D. Jaque, Y. Yao, F. Chen, A.A. Bettiol, A. Rodenas, A.K. Kar, Micro-structuring of Nd:YAG crystals by proton-beam writing. Opt. Lett. **35**, 3898 (2010)
107. Y. Tan, Q. Luan, F. Liu, S. Akhmadaliev, S. Zhou, F. Chen, Swift carbon ion irradiated Nd:YAG ceramic optical waveguide amplifier. Opt. Express **21**, 13992 (2013)
108. D. Fluck, T. Pliska, P. Günter, S. Bauer, L. Beckers, C. Buchal, Blue-light second-harmonic generation in ion-implanted $KNbO_3$ channel waveguides of new design. Appl. Phys. Lett. **69**, 4133 (1996)
109. Y. Tan, F. Chen, L. Wang, X.L. Wang, K.M. Wang, Q.M. Lu, Optical channel waveguides in $KTiOPO_4$ crystal produced by proton implantation. J. Lightwave Technol. **26**, 1304 (2008)
110. J.-H. Zhao, X.-L. Wang, F. Chen, 1×4-Branch waveguide power splitters in lithium niobate by means of multi-energy O ion implantation. Opt. Mater. **32**, 1441 (2010)
111. M.F. Volk, S. Suntsov, C.E. Rüter, D. Kip, Low loss ridge waveguides in lithium niobate thin films by optical grade diamond blade dicing. Opt. Express **24**, 1386 (2016)
112. C. Chen, S. Akhmadaliev, C. Romero, J.R.V. de Aldana, S. Zhou, F. Chen, Ridge waveguides and Y-branch beam splitters in $KTiOPO_4$ crystal by 15 MeV oxygen ion implantation and femtosecond laser ablation. J. Lightwave Technol. **35**, 225 (2017)
113. R. Degl'Innocenti, S. Reidt, A. Guarino, D. Rezzonico, G. Poberaj, P. Günter, Micromachining of ridge optical waveguides on top of He^+-implanted β-BaB_2O_4 crystals by femtosecond laser ablation. J. Appl. Phys. **100**, 113121 (2006)
114. A.A. Bettiol, S.V. Rao, E.J. Teo, J.A. Van Kan, F. Watt, Fabrication of buried channel waveguides in photosensitive glass using proton beam writing. Appl. Phys. Lett. **88**, 171106 (2006)
115. Y. Yao, N. Dong, F. Chen, S. Vanga, A.A. Bettiol, Proton beam writing of Nd:GGG crystals as new waveguide laser sources. Opt. Lett. **36**, 4173 (2011)
116. D. Kip, Photorefractive waveguides in oxide crystals: fabrication, properties, and applications. Appl. Phys. B **67**, 131 (1998)
117. Y. Tan, F. Chen, X.L. Wang, L. Wang, V.M. Shandarov, D. Kip, Formation of re-configurable optical channel waveguides and beam splitters on top of proton-implanted lithium niobate crystals using spatial dark soliton-like structures. J. Phys. D **41**, 102001 (2008)
118. Y. Tan, F. Chen, M. Stepić, V. Shandarov, D. Kip, Reconfigurable optical channel waveguides in lithium niobate crystals produced by combination of low-dose O^{3+} ion implantation and selective white light illumination. Opt. Express **16**, 10465 (2008)
119. B.E.A. Saleh, M.C. Teich, *Fundamentals of Photonics* (Wiley, Hoboken, 2019)

120. R. Regener, W. Sohler, Loss in low-finesse Ti:$LiNbO_3$ optical waveguide resonators. Appl. Phys. B **36**, 143 (1985)
121. C. Grivas, Optically pumped planar waveguide lasers, Part I: fundamentals and fabrication techniques. Prog. Quant. Electron. **35**, 159 (2011)
122. H. Hu, F. Lu, X.L. Wang, F. Chen, K.M. Wang, Low-loss optical waveguides and Y-branch splitters in lithium niobate fabricated by MeV oxygen ions with low dose. Opt. Express **20**, 21114 (2012)
123. Y. Okamura, S. Yoshinaka, S. Yamamoto, Measuring mode propagation losses of integrated optical waveguides: a simple method. Appl. Opt. **22**, 3892 (1983)
124. N. Nourshargh, E.M. Starr, N.I. Fox, S.G. Jones, Simple technique for measuring attenuation of integrated optical waveguides. Electron. Lett. **21**, 818 (1985)
125. M. Jubera, J. Villarroel, A. García-Cabañes, M. Carrascosa, J. Olivares, F. Agullo-López, A. Méndez, J.B. Ramiro, Analysis and optimization of propagation losses in $LiNbO_3$ optical waveguides produced by swift heavy-ion irradiation. Appl. Phys. B **107**, 11 (2012)

Chapter 3
Photonic Structures Based on Thin Films Produced by Ion Beams

Abstract By combining ion beam technology with a series of advanced micro-/nano-structuring methods such as photolithography, chemical and dry etching techniques, high-density single-crystal-based photonic integrated circuits consists of a number of on-chip optical components such as waveguides, micro-cavities, as well as photonic crystal structures can be fabricated. In this chapter, fabrication of crystalline thin films by crystal ion slicing (Sect. 3.1) and lithium-niobate-on-insulator technology (Sect. 3.2) as well as micro-/nano-structuring of on-chip optical waveguides (Sect. 3.3), whispering gallery resonators (Sect. 3.4) and photonic crystal cavities (Sect. 3.5) are overviewed.

The development of modern photonics has in fact called forth the urgent need for dense photonic integrated circuits incorporated with different optical materials, various optical devices and diverse optical structures. Therefore, a hybrid integration of dielectrics, e.g., ferroelectric crystals like $LiNbO_3$ that have been extensively used in a variety of applications based on bulk optics, in the form of micro- and submicron-thick thins films onto other, often non-compatible and low-refractive-index, heterogeneous platforms, is a very attractive prospect for integrated and nanoscale photonics. In fact, the $LiNbO_3$-based thin film technology and novel on-chip micro/nano-scale photonic devices are revolutionizing both the $LiNbO_3$ industry and the modern integrated photonics, enabling higher performance, lower cost and entirely new devices and applications.

3.1 Freestanding Crystalline Thin Films by Crystal Ion Slicing

Crystal ion slicing (CIS) is a versatile epitaxial lift-off technique allowing for exfoliation of freestanding single-crystalline thin films with thicknesses of several microns from original donor substrates. The detached thin films can be subsequently bonded onto desired heterogeneous substrates, which has been proven useful for a great variety of crystals [1–3]. In contrast to thin films fabricated by epitaxial-growth-based

F. Chen et al., *Ion Irradiation of Dielectrics for Photonic Applications*, Springer Series in Optical Sciences 231, https://doi.org/10.1007/978-981-15-4607-5_3

deposition techniques, such as pulsed laser deposition, chemical vapor deposition, and RF sputtering processing, CIS-based single-crystalline thin films is free of grain boundaries, minimizing greatly the volume scattering losses as well as retaining near-perfectly the original bulk features [4–6]. Due to its broad material applicability and unique capability of epitaxial lift-off of high-quality thin films, CIS is becoming of central importance in integrated photonics as a preliminary step for hybrid integration of functional PICs on a CMOS-compatible substrate [1–3, 7].

CIS was initially developed for silicon-on-insulators (SOI) technology and later imitated to ferroelectric crystals [1–3]. Depending on the material platforms and the desired device applications, CIS has diversified into several variants since it was demonstrated in 1995 [1, 8]. Even so, the core concept remains the same. In the following, without loss of generality, CIS of lithium niobate ($LiNbO_3$) thin films, firstly proposed by Osgood's group [1, 9], is used as an example to describe this technique in detail.

Light ions, e.g., hydrogen (H or proton) and helium (He) ions, are conventionally chosen as the implantation species because of their small atomic mass, thus yielding a deeply buried damage layer [1–3]. The lattice damage mechanism can be briefly described as following: At high ionic energies, the implanted ion along its trajectory loses its energy (E_{ion}) dominantly caused by electronic scattering (adequately described by the Lindhard-Scharff-Schiott theory [10], which predicts an electric stopping power $S_{\mathrm{e}} \propto E_{\mathrm{ion}}^{1/2}$), which has minor/negligible lattice damage over most of the implantation depth. While at low energies, the energy loss is primarily due to Rutherford scattering with the host nuclei (with nuclear stopping power $S_{\mathrm{n}} \propto E_{\mathrm{ion}}^{2}$) [3]. This nonlinear dependence ensures the majority of implanted ions are eventually stationed in a relatively narrow spatial region (sacrificial layer) beneath the target substrate surface, where lattice defects are introduced by the transfer of ion energy to the target nuclei. These induced defects, usually in forms of platelets, bubbles and micro-cracks, are acting as nucleation centers for the agglomeration of diffusing He or H during subsequent annealing. The evolution of lattice defects filled with He or H is actually responsible for the eventual layer-splitting. Calculation based on SRIM code [11] for 3.8 MHz He ion implantation of $LiNbO_3$ shows a highly peaked implantation profile centered at a depth of 10 μm beneath the surface, as illustrated in Fig. 3.1a. A frequently-used ion implantation recipe for CIS of $LiNbO_3$ thin films (10 μm thickness with 400 nm sacrificial layer) is summarized as following: charged 3.8 MeV He ions (a total fluence of 5×10^{16} ions/cm^2) are implanted into the $LiNbO_3$ target wafer (implantation with an angle of 7° to the target surface normal to suppress channeling effects), which is mounted on a water-cooled target holder with temperature maintained <400 °C, with a low beam current (<0.25 μA/cm^2) [1, 9, 12–14]. Particularly, it has been found that blistering and even exfoliation is possible to occur directly during ion implantation if the temperature is too high [1].

Subsequently, thin film exfoliation or lift-off from the implanted $LiNbO_3$ donor crystal can be achieved either by wet-etching [1] or thermal treatment [2]. In the latter case, however, a controlled exfoliation is possible only after proper bonding of the implanted sample onto a supporting substrate [2]. We will come back to this in the next section. In this section, we focus more on the wet-etching method.

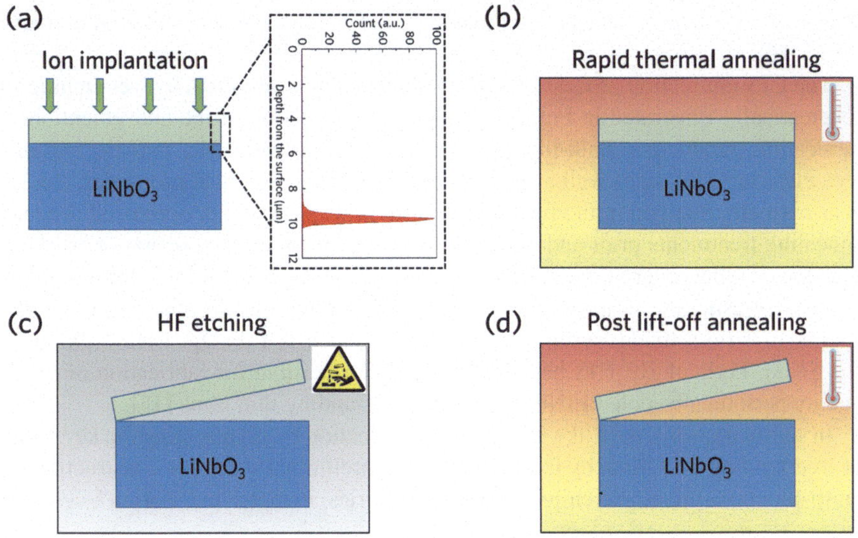

Fig. 3.1 CIS fabrication sequence of $LiNbO_3$ freestanding thin films by wet-etching: **a** He ion implantation; **b** rapid thermal annealing before. **c** HF acid etching; **d** post lift-off annealing. The inset in (**a**) illustrates the He ion distribution after implantation (in case of 3.8 MeV He ion implantation)

By immersing the whole implanted $LiNbO_3$ in a dilute HF acid solution for hours, the sacrificial layer can be etched away, forming a deep undercut (10 μm depth beneath the $LiNbO_3$ surface in case of the abovementioned recipe) or a full detachment of a $LiNbO_3$ freestanding thin film [1]. Particularly, the etching rate and selectivity can be further enhanced (by a factor of 10^4) by an additional rapid thermal annealing (RTA) treatment right after the ion implantation (prior to the chemical etching operation) [1, 9, 12–14]. In such a RTA process, the development of highly localized stress-induced high-density micro-domains/micro-cracks tightly confined within the sacrificial layer are observed and identified as the primary cause of accelerated etching. And due to the same reason, the subsequent etching rate is found to be nearly independent of the target crystal orientation nor the ion implantation energy [12]. Additionally, the RTA is able to further improve the quality of the crystal above the sacrificial layer by out-diffusing He and its attendant residual stress [13]. An optimized RTA and subsequent etching recipe (on the basis of the abovementioned ion implantation recipe) is: RTA in forming gas (5% H_2, 95% N_2) for 30 min at 200–250 °C and etching in 5% HF at room temperature, resulting in an etching rate of 3.2 ± 0.3 μm/min (background etching rate of only 0.1 nm/min) [1, 14]. Besides, to protect the exposed surface from the etchant, a suitable protective layer (such as an Apiezon black wax) can be used.

After exfoliation, a post lift-off annealing (PLA) treatment (400–800 °C for hours in a laboratory-atmosphere ambient) is proven to be useful and necessary with a view to removing residual stress and/or ion damage. In this way, the optical, e.g., electro-optic (EO), response of the detached $LiNbO_3$ freestanding thin films can be improved to a degree that is comparable to their single-crystal bulk counterparts [1, 15, 16].

The CIS fabrication process of $LiNbO_3$ freestanding thin films by wet-etching is schematically illustrated in Fig. 3.1. With the combination of the abovementioned fabrication parameters, bulk-like properties, including dielectric, pyroelectric and ferroelectric response as well as EO, linear, and nonlinear optical properties, can be near-perfectly retained in the freestanding thin films [1–3]. With optimized thermal annealing treatments prior and posterior to the chemical etching operation, surface roughness of the order of optical grade polished $LiNbO_3$ is achievable for thin films [1, 17]. By using the similar/modified recipes, CIS technique has been successfully applied to various oxide crystals beyond $LiNbO_3$, such as $Y_3F_5O_{12}$, $LiTaO_3$, $KTaO_3$, etc. [1–3]. Particularly, CIS has also been proven feasible for fabrication of high-quality periodically-poled $LiNbO_3$ (PPLN) freestanding thin films [18].

In summary, we would like to conclude this section by highlighting the key word of freestanding thin films by CIS, which is the preliminary step for construction of multi-functional photonic chips based on dielectrics, namely "exfoliation".

3.2 Direct Bonding of Crystalline Membranes on Insulators by LNOI Technology

Once exfoliated, the $LiNbO_3$ thin films fabricated by wet-etching method can be directly bonded onto heterogeneous planar surfaces [1]. This works fine for bonding of micron-thick thin films with small areas. For example, the pioneer study employed an epoxy resin for direct bonding of freestanding $LiNbO_3$ thin films (with a thickness of 10 μm and an area of 0.3 cm^2) onto silicon and GaAs substrates [9]. However, wet-etching is limited in fabrication of submicron-thick films with large areas and uniform thicknesses. Such films vary gradually in thickness, being thicker in the center and thinner towards the edges, where the exposure time to HF is longer [1]. In view of this limitation, thermally induced exfoliation method was developed [2, 19, 20]. Similar to the RTA process in wet-etching fabrication, thermal treatment induces growth of micro-cavities in the highly damaged layer, leading to creation of micro-domains/micro-cracks and finally to the desired detachment of the implanted layer [2]. Compared to the preliminary exfoliation, the subsequent bonding is actually one of the most critical steps in fabrication of large-area crystalline thin films with uniform thickness. Two key points that have to be seriously considered in order to achieve a high-quality wafer bonding: the optical quality, such as flatness, roughness and cleanness, of the involved surfaces and the thermal expansion coefficient mismatch between the supporting substrate and the implanted layer, either of which could have destructive impacts on the bonding strength and thus on the resulting thin film quality. A controlled layer exfoliation is possible only after a proper bonding of the implanted sample onto the supporting substrate [2].

A well-developed approach, firstly proposed by Günter's group [19], utilizes SiO_2 layer (with a thickness of 1–2 μm) by plasma enhanced chemical vapor deposition (PECVD) between the ion-sliced $LiNbO_3$ thin film and the supporting substrate. The stress between SiO_2 layer and $LiNbO_3$ substrate can be controlled by PECVD parameters. Alternatively, electrodes can be deposited before SiO_2 layer deposition if an electrode below the $LiNbO_3$ thin film is required [2]. With such a sandwich structure, i.e., $LiNbO_3$ film/SiO_2 layer/$LiNbO_3$ substrate, the large thermal expansion coefficient mismatch between $LiNbO_3$ and SiO_2 can be significantly alleviated. Prior to wafer bonding, the acceptor substrate with the SiO_2 layer has to be firstly annealed (e.g., at 450 °C for 8 h) to drive off the trapped gases, then polished to reduce the surface roughness (RMS roughness <1 nm), and finally cleaned (e.g., using RCA1 cleaning solution [21]) together with the donor implanted $LiNbO_3$ wafer [2]. Subsequently, the donor and acceptor wafers can be bonded together at room temperature by simply bringing them into contact and pressing them against each other. Then, after thermal treatment at ~200 °C, the ion damaged layer in the implanted wafer will form He bubbles or blisters, resulting in a cleavage layer that defoliates the $LiNbO_3$ thin film from the SiO_2/$LiNbO_3$ substrate. To this step, the initial preparation process for the lithium-niobate-on-insulator (LNOI) thin film has completed (the fabrication sequence is schematically illustrated in Fig. 3.2). However, due to the ion-beam-induced localized lattice damage and the relatively high surface/interface roughness, the overall quality of the as-prepared LNOI thin film is still far from the basic requirements for the subsequent nano-fabrication/structuring of on-chip photonic devices and further practical applications in integrated photonics. In order to further improve the film quality to approach the level of single-crystalline $LiNbO_3$ for commercial production, Hu et al. from Shandong University have developed an optimized LNOI post-processing procedure as following. Firstly, to remove the ion-beam-induced point defects in the implanted thin film layer and to recover its crystalline properties, an optimized post-fabrication thermal annealing is carried out at ~500 °C for 2 h. This also helps to enhance the overall bonding strength and ensures the survival of the whole bonded pairs from further mechanical polishing and chemical etching. Subsequently, an optimized chemical-mechanical polishing (CMP) will be performed to reduce the surface roughness of the thin film. Meanwhile, about

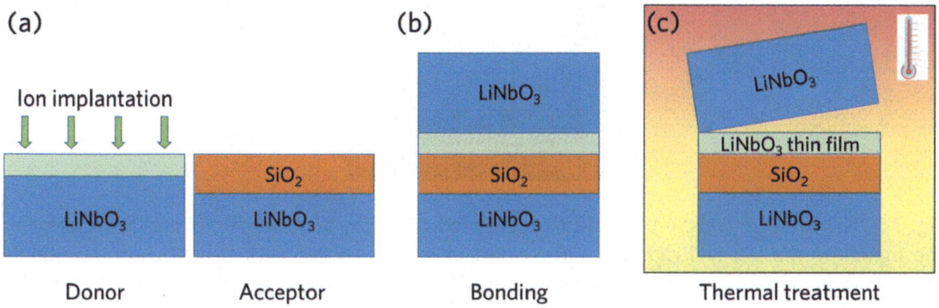

Fig. 3.2 Fabrication sequence of LNOI by combining CIS and wafer bonding: **a** preparation of ion-implanted donor wafer and SiO_2-deposited acceptor wafer; **b** wafer bonding; **c** thermal treatment

100-nm thick $LiNbO_3$ layer will be removed from the surface. Furthermore, prior to the wafer bonding, a surface functionalization procedure together with a further cleaning process to assure perfectly clean and OH-terminated surfaces can be used. The hydrogen bonds at the OH-terminated surface leads to a relatively high bonding strength even at room temperature and a uniform $LiNbO_3/SiO_2$ interface without any residual stress and defects (as shown in Fig. 3.3). Thanks to such an optimized thermally induced exfoliation method, high-quality, large size (3-inch in diameters or even larger [22]), high-quality LNOI thin films (film thickness of 300 nm–1 μm) on $LiNbO_3$ substrate and even on Si substrates have been developed successfully. One of the major advantages of this LNOI technique is the post-annealing treatment at relatively high temperature allows almost full recovery of the damaged lattice structures in $LiNbO_3$ thin film. This can be verified by the measurement of the optical properties of the LNOI film, e.g., electrooptic coefficients (See details in Chap. 6). In addition, the multiple choice of the substrate materials offers new possibility of optoelectronic integration for future on-chip devices. Benefiting from these features, high-quality, large-scale LNOI wafers have been commercially available (NANOLN©, established in 2010).

An alternative approach, firstly proposed as well by Günter's group [20, 23], has similar fabrication sequence to the abovementioned one but utilizes an adhesive polymer benzocyclobutene (BCB) layer instead of a SiO_2 layer. The main intention of using such a polymer adhesive is to substantially relax requirements for the flatness, roughness, and cleanliness of the bonded surfaces. By using this method, LNOI with wafer sizes of several cm^2 can be achieved [2, 20]. In this case, high-temperature thermal treatment, which is able to minimize ion-implantation induced material damage and recover optical properties of implanted layer as well as contribute significantly to a reduction of the surface roughness, is however not possible because polymers are prone to thermal degradation. The highest temperature used

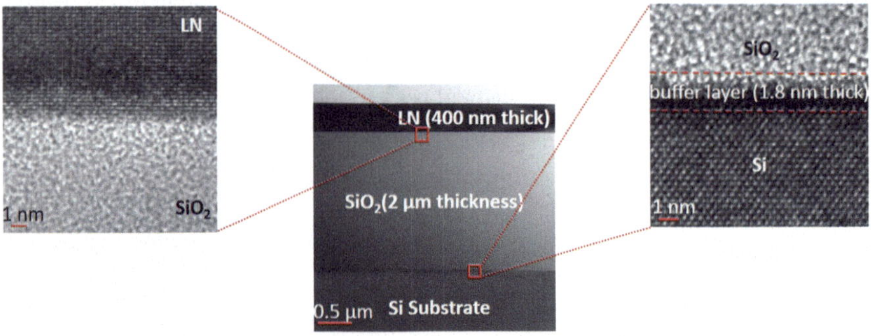

Fig. 3.3 SEM images of the LNOI film, the related LN/SiO_2 and SiO_2/substrate (Si) interfaces. The figures in the left and right show the enlarged images of the marked regions in the middle figure, indicating the high-quality boundaries of the two interfaces. Image courtesy of Prof. H. Hu (Shandong University)

in this method for thermal annealing is usually <300 °C [2, 20]. Nevertheless, good-quality integrated optical devices have been fabricated based on LNOI chips with the $LiNbO_3$/BCB/$LiNbO_3$ combination [2, 20, 23].

A hybrid solution, firstly proposed by Fathpour's group, utilizes SiO_2 layers on Si instead of on $LiNbO_3$ as supporting substrates [24, 25]. The bonded wafer size can be as large as 3-inch in diameters. Because $LiNbO_3$ is notoriously difficult to etch, the configuration of $LiNbO_3$/SiO_2/Si is not only able to significantly simplifies the processing of $LiNbO_3$ photonic devices, but also compatible to the well-developed silicon electronics and photonics. This is the main intention of developing such a heterogeneous photonic platform. However, hetero-structure approach usually suffers from reduced overlap between optical fields and $LiNbO_3$'s nonlinearity and relatively weak bonding strength due to the large thermal expansion mismatch between $LiNbO_3$ and Si. In spite of these challenges, heterogeneous integration of $LiNbO_3$ with SOI technology has attracted significant interests in the past years because of the "ease" of structuring and integration of active photonic elements for EO and nonlinear optical applications [7, 25].

In summary, thermally induced wafer exfoliation method is so far the most successful technique in order to fabricate submicron-thick LNOI thin films with large areas, uniform thicknesses and good quality. It provides the basic material platform for the subsequent micro-/nano-structuring of either an individual optical element or a high-density PIC, enabling compact and complex integrated systems on a single chip [7].

3.3 Micro-/Nano-structuring of LNOI On-Chip Waveguides

The high refractive index contrast (e.g., $n_{o,LiNbO3} \approx 2.21$, $n_{e,LiNbO3} \approx 2.14$ and $n_{SiO2} \approx 1.44$ at 1.55 μm) between the bonding pairs, resembling SOI, enables good confinement and strong guiding of light. Therefore, they are ideally suited for fabrication of LNOI on-chip waveguides and micro-optical devices. Enabled by advanced micro-/nano-structuring methods such as photolithography, chemical and dry etching techniques, high-density LNOI-based PICs consists of on-chip $LiNbO_3$ optical components such as waveguides, micro-cavities, as well as photonic crystal structures are expected [7]. To this end, above all, low-loss waveguides with a strong optical mode confinement are highly required.

Since $LiNbO_3$ is notoriously difficult to etch, proton exchange, a well-established diffusion-based technique for bulk $LiNbO_3$ waveguide fabrication, was well-studied using LNOI shortly after the LNOI chips became commercially available [26–28]. The concept of "exchange" herein means that the diffusant ions (e.g., proton) from certain melt sources (e.g., pyrophosphoric acid) immigrate to the $LiNbO_3$ and substitute relatively mobile ions (e.g., Li^+) in $LiNbO_3$. Subsequently, native ions in

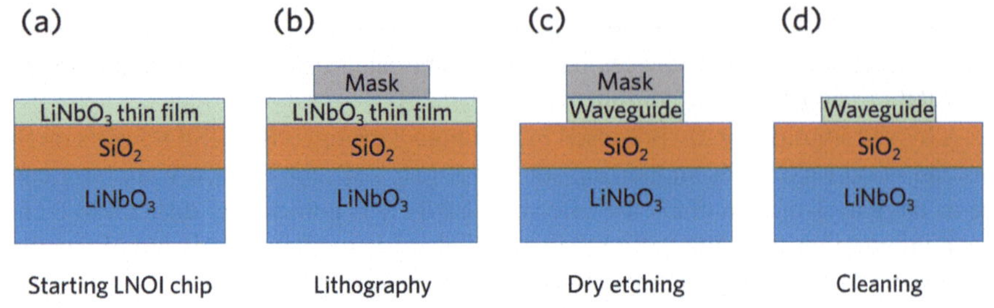

Fig. 3.4 Fabrication sequence of $LiNbO_3$ on-chip waveguides by employing lithography technology (e.g., photolithography or electron-beam lithography) and subsequent dry-etching approach. A starting LNOI chip is shown in (**a**). **b** An etching mask is deposited and patterned on the LNOI chip, followed by a dry-etching process to remove unprotected $LiNbO_3$, forming ridge waveguide structures. **d** A post-cleaning process is necessary in order to remove the residual etching mask

$LiNbO_3$ with smaller size and/or polarizability will be expelled into the melt, completing the "exchange" process. As a result, the desired refractive-index modification in $LiNbO_3$ can be achieved, forming embedded waveguide structures. With an optimized proton exchange and thermal annealing treatment, LNOI waveguide losses down to 0.6 dB/cm for TM modes and 0.2 dB/cm for TE modes with mode sizes of around 3 μm^2 have been experimentally determined [26–28]. In contrast to the dry-etched ridge waveguides (also called photonic wires), proton-exchanged LNOI channel waveguides are cost effective and have smooth boundaries. However, such waveguides suffer from the relatively low refractive index contrast (<0.1 in general) in the in-plane dimension and the polarization-dependent guiding properties, resulting in a waveguide bending radius in a relatively large scale [28].

Dry etching of ridge waveguides, in spite of its generally complicated fabrication sequence, is a more promising technique due to its flexibility and high resolution in terms of removing the unwanted $LiNbO_3$ on either side of the waveguide core [7]. A commonly-used fabrication sequence is schematically illustrated in Fig. 3.4. Briefly, the desired pattern of the resist mask deposited on LNOI surface is firstly defined by lithography technology (e.g., photolithography or electron-beam lithography (EBL)), then inductively coupled plasma reactive ion etching (ICP-RIE) usually with Ar^+ plasma is used to fabricate on-chip ridge waveguides and micro-optical devices [7]. The residual EBL mask is subsequently removed usually by a chemical etching process, e.g., a buffered oxide etching. Finally, a chemical-mechanical polishing and cleaning process is generally necessary with a view to achieving high-quality optical devices with a relatively low surface and side-wall roughness [29]. Specifically, EBL patterning of micro-optical structures usually starts with deposition of a thin-layer titanium or chromium on the LNOI substrate by electron beam evaporation. The titanium/chromium layer serves two purposes: to promote resist adhesion and to provide conduction during subsequent EBL. Following this fabrication sequence, the lowest waveguide loss of the firstly reported on-chip $LiNbO_3$ waveguides by Sohler's group was determined to be around 10 dB/cm (TM mode with an effective mode size of around 0.4 μm^2), being limited primarily by the non-optimized fabrication recipes and the residual side-wall roughness [30]. Before long, with an optimized direct dry-etching recipe proposed by Lončar's group, a straight waveguide (an effective mode size of around 0.52 μm^2) with a propagation loss as low as 3 dB/cm is demonstrated for efficient second harmonic generation (SHG) [31]. In a following work, soon after, waveguides with an unprecedentedly low propagation loss of 2.7 dB/m and a bending loss of 9.3 dB/m are demonstrated, representing by far the lowest level for $LiNbO_3$ on-chip ridge waveguides [32]. The waveguides therein are fabricated with a multi-pass exposure of the photoresist and an improved ICP-RIE process. The abovementioned characteristics of the fabricated waveguides are determined at telecom wavelengths. By applying the identical fabrication procedure, most recently, $LiNbO_3$ on-chip ridge waveguides with an extremely-low propagation loss of 6 dB/m at visible wavelength have been demonstrated [33]. Furthermore, the side-wall roughness can be effectively reduced by either an optimized etching process (resulting in a propagation loss as low as 0.4 dB/cm and an RMS roughness of <2 nm) [34] or a side-wall fine polishing post-process (resulting in an ultralow propagation of 4 dB/m) [29]. Therefore, a proper combination of an optimized ICP-RIE process and a side-wall fine polishing post-process may result in absorption-limited $LiNbO_3$ waveguides. With such high-quality on-chip waveguides, additional optical functionalities and advanced nanostructures, e.g., Bragg gratings [35, 36], optical metasurafaces [37, 38], quantum dots [39], and photodetectors [40], can be easily incorporated on the same device/chip, enabling LNOI-based PICs with high integration density and on-chip multi-functionalities. A shortcoming of dry-etching approach is that the waveguide sidewalls can have a significant angle of >30° with respect to the vertical direction, which must be well considered when designing waveguides and can cause issues when fabricating strongly coupled waveguides [7].

An alternative approach that may have the potential to overcome the sidewall angle issue more straightforwardly is precise diamond blade dicing, which is able to define almost vertical sidewalls in bulk $LiNbO_3$ [41]. On-chip $LiNbO_3$ ridge waveguides fabricated by this method exhibit a propagation loss of 1.2 dB/cm for TE and 2.8 dB/cm for TM mode at 1.55 μm [42]. Most recently, on-chip PPLN ridge waveguides with an estimated propagation loss as low as 0.3 dB/cm by using optical grade dicing has been demonstrated [43]. It is found that better verticality of on-chip waveguide sidewalls can be achieved by defining relatively deep cut ridges, however, they are more prone to cracking and exhibit a more pronounced tilt of their end faces [42]. Another shortcoming of such a purely mechanical method is that it is usually useful for definition of only straight cuts and therefore only straight waveguides.

In summary, considering all the fabrication techniques introduced in this section, a proper combination of different micro-/nano-structuring and processing approaches may provide the optimal solution to absorption-limited $LiNbO_3$ waveguide fabrication and to ultimately reaching LNOI's true potential as well as eventually finding practical applications.

3.4 LNOI On-Chip Whispering-Gallery Resonators

Micro-resonators or whispering-gallery resonators (WGRs), usually with geometries of micro-disks or micro-rings, enable compact and wavelength-sensitive optical devices, finding a variety of applications in photonics such as wavelength filters, lasers, nonlinear frequency converters, and sensors [44, 45]. In such resonators, the light is guided by total internal reflection in a spheroidally-shaped material, allowing for a remarkably high optical confinement and an enhanced intracavity intensity. Chip-integrated WGRs are highly promising based on, for example, LNOI platform since $LiNbO_3$ itself is a multi-functional ferroelectric crystal, opening up exciting opportunities and possibilities for active PICs and lab-on-a-chip platforms [7].

First studies on fabrication of on-chip $LiNbO_3$ micro-resonators utilized the dry etching approach identical to that discussed in Sect. 3.3 [20, 23]. However, the quality factor (Q-factor), which is the general criterion of measuring a WGR's optical quality, of the fabricated micro-ring (with a radius of 100 μm) therein was only 4×10^3 at 1.55 μm due to the un-optimized etching process and high surface/sidewall roughness. Since then, quite a great effort has been made on exploring the Q-factor upper limit of the on-chip WGRs, approaching step by step the desired absorption-limited $LiNbO_3$ micro-resonators via continually optimizing/adjusting micro-/nano-structuring and post-processing techniques [2, 7]. For example, a Q-factor elevated up to 1×10^4 was deduced in a micro-ring (with a radius of 20 μm) fabricated by using laser lithography technique proposed by Günter's group [2]. In a subsequent work, by heating the 20 μm-diameter micro-disk close to its melting temperature after device fabrication, local melting on the disk edge could smoothen the surface through surface tension reshaping, resulting an enhanced Q-factor of 2.6×10^4 [46]. Based on the alternative hetero-structure approach, a Q-factor of a 300 μm-diameter micro-ring was determined to be 7.2×10^4 [24]. The first demonstrations of $LiNbO_3$ on-chip micro-resonators with Q-factors exceeding 10^5 were obtained in 2014 by using femtosecond (fs) laser ablation followed by two-step focused ion beam (FIB) milling proposed by Cheng's group [47] and by using the ICP-RIE approach introduced in Sect. 3.3 [48, 49], reaching 2.5×10^5 [47], 4.84×10^5 [48] and 1.02×10^5 [49], respectively. Since then, these techniques have been gradually developed into the dominating fabrication approaches nowadays for high-quality $LiNbO_3$ on-chip micro-resonators [7]. And these works represent the starting point of their practical applications in photonics. So far, by employing optimized fabrication recipes and post-processing method such as CMP (as shown in Fig. 3.5, SEM views of an on-chip micro-ring WGR with a Q-factor of 3.0×10^6 and its sidewall before and after polishing), optical Q-factors exceeding 10^6 and up to 10^7 at both the telecom and the visible wavelengths have been achieved in $LiNbO_3$ on-chip micro-resonators fabricated by the abovementioned dry-etching and fs-laser ablation approaches [29, 32, 33, 50, 51], demonstrating a variety of interesting on-chip photonic applications such as nonlinear optical frequency conversion, EO modulation, and frequency comb generation [7].

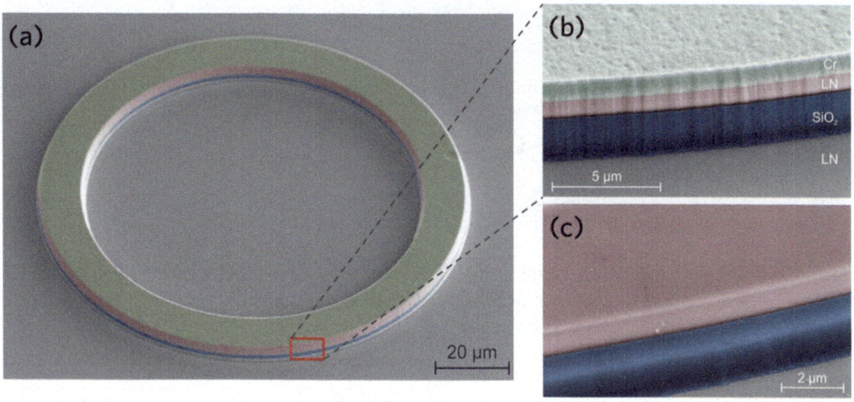

Fig. 3.5 **a** Colored SEM image of a $LiNbO_3$ on-chip micro-ring WGR fabricated by RIE. **b** A zoomed SEM view of WGR sidewall before polishing. **c** Sidewall after polishing and removal of Cr and Ti layer. Grey: $LiNbO_3$ substrate; Blue: SiO_2 layer; Red: $LiNbO_3$ thin film; Yellow: Cr and Ti layer. Reproduced with permission from [50], ©2017 Optical Society of America

3.5 LNOI On-Chip Photonic Crystal Based Structures

Photonic crystal (PhC) based structures, such as 1D PhC nanobeam cavities and 2D PhC slabs, are periodic dielectric structures that are designed to define the energy band structure for photons in the same way as the periodic potential in a semiconductor crystal for defining allowed and forbidden electronic energy bands [52]. Serving as miniaturized and efficient filtering as well as nonlinear optical devices, PhC structures are highly promising for further diversifying the functionality of high-density PICs based on, for example, LNOI chips [7]. First studies on $LiNbO_3$ on-chip PhC structures employed FIB milling, firstly proposed by Günter's group [2, 53], for drilling holes on LNOI chips to define the desired dielectric periodicity, which has been proved to be a very suitable method for fast prototyping of PhC structures in LNOI, demonstrating a 2D PhC slab with a photonic bandgap in the telecom wavelength range with an extinction ratio of up to 20 dB [53, 54] and a compact 1D PhC micro-cavity with a Q-factor of 156 patterned on a $LiNbO_3$ ridge waveguide [55]. In parallel, an alternative approach utilized a combination of FIB milling and the so-called ion beam enhanced etching (IBEE) [56] for fabrication of PhC structures, for example PhC L3 cavities, in self-suspended $LiNbO_3$ membranes [57, 58]. Similar to the CIS technique, the IBEE relies on as well the lattice damages induced by ion-beam bombardment, which results in highly-damaged areas where the stability against acid etching is strongly degraded, enabling micro- or submicrometric scale patterning [3]. As a result, the $LiNbO_3$ PhC L3 cavities fabricated by this approach feature a high refractive index contrast and a Q-factor of 500–600 in the telecom wavelength range, allowing for effective SHG performance [58, 59]. Recently, the optical Q-factors were elevated up by two orders of magnitude in a 1D $LiNbO_3$ on-chip PhC nanobeam cavity, demonstrated by Lin's group, fabricated

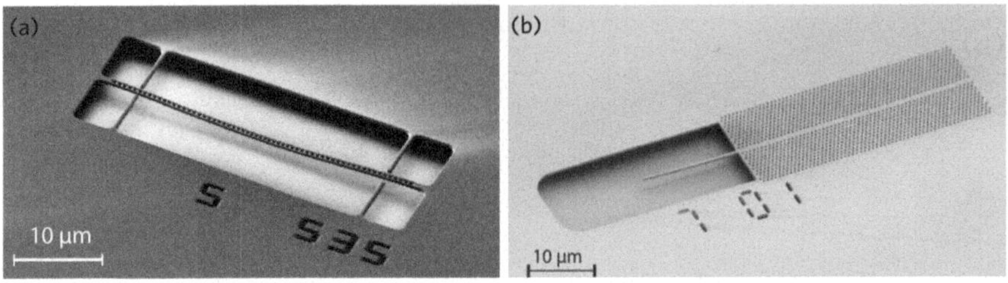

Fig. 3.6 SEM images of $LiNbO_3$ on-chip. **a** 1D PhC nanobeam cavity and **b** 2D PhC slab fabricated by the "EBL+ dry-etching" approach. Reproduced with permission from [60, 63], ©2017 Optical Society of America and ©2019 WILEY-VCH

by employing the advanced "EBL + dry etching" approach, giving a Q-factor as high as 10^5 (an SEM view is shown in Fig. 3.6a) [60]. Together with a tight optical mode confinement, such a device exhibits intriguing nonlinear optical phenomena as well as a strong optomechanical coupling [60–62]. By using the identical fabrication approach, a high-Q (a Q-factor of 3.51×10^5) 2D $LiNbO_3$ on-chip PhC slab has been demonstrated recently as well (an SEM view is shown in Fig. 3.6b) [63].

3.6 LNOI On-Chip Optical Couplers/Interfaces

A major challenge for practical applications of integrated $LiNbO_3$ photonics is the lack of an efficient optical coupler/interface between the standard optical fibers and micrometer-scale $LiNbO_3$ devices. For many applications, including EO modulators, nonlinear optical frequency converters and frequency comb generators, the system performance is directly associated with the total optical power coupled into the on-chip active devices. Due to the large mode mismatch between LNOI on-chip devices and standard optical fibers, the widely-used end-fire coupling method usually require additional optical designs/structures, such as lensed optical fibers [31] and tapered optical fibers [49], for achieving effective coupling. However, the fiber-to-chip loss is still >10 dB/facet in most cases for typical LN photonic integrated circuits [7].

Grating couplers provide a versatile method for optical in-/out-coupling between optical fibers and on-chip waveguides. The coupling can be done anywhere on the surface of the optical circuit, enabling wafer scale optical testing with excellent alignment tolerance. Although these advantages usually come at the cost of an increased insertion loss and somewhat limited spectral range. The first LNOI grating couplers in on-chip waveguides were fabricated by dry etching, demonstrating relatively high experimentally determined coupling losses of $\approx$12 dB [64] and $\approx$10 dB [65] with either an upper SiO_2 cladding layer or with a buried SiO_2 layer, respectively. In both cases, there are still some optical power leakage into the substrate. Therefore, adding a bottom reflector layer (such as a metal bottom reflector) in the LNOI for minimizing the power leakage is an effective approach to further improve the coupling efficiency, which has been realized by depositing an additional Cr/Au/Cr metal layer prior to the SiO_2 deposition on the acceptor LN wafer. Based on such a $LiNbO_3$ film/SiO_2 layer/metal layer/$LiNbO_3$ substrate combination and a proper design, the grating couplers fabricated by FIB technique reduce the coupling loss from 9.1 dB to 6.9 dB [66]. Most recently, a monolithic on-chip $LiNbO_3$ bilayer inversely tapered mode size converter was designed and fabricated by EBL and dry etching, demonstrating an extremely low coupling loss of 1.7 dB/facet [67]. The mode size converter consists of a two-layer taper that gradually decreases the heights of both ridge and slab portions of the on-chip $LiNbO_3$ waveguides. The optical coupling therein is based on the gradual variation of the waveguide cross section to a certain point, such that a better size overlap and index match between modes of the input optical fiber and on-chip waveguides are achieved. Such a high coupling efficiency will enable optical packaging of LNOI devices and opens exciting opportunities for practical applications in $LiNbO_3$-based integrated photonics.

References

1. M. Alexe, U. Gösele, *Wafer Bonding: applications and Technology* (Springer, Berlin/Heidelberg, 2004)
2. G. Poberaj, H. Hu, W. Sohler, P. Günter, Lithium niobate on insulator (LNOI) for microphotonic devices. Laser Photonics Rev. **6**, 488 (2012)
3. F. Chen, Micro-and submicrometric waveguiding structures in optical crystals produced by ion beams for photonic applications. Laser Photonics Rev. **6**, 622 (2012)
4. C. Grivas, Optically pumped planar waveguide lasers, Part I: fundamentals and fabrication techniques. Prog. Quant. Electron. **35**, 159 (2011)
5. C. Grivas, Optically pumped planar waveguide lasers: Part II: gain media, laser systems, and applications. Prog. Quant. Electron. **45**, 3 (2016)
6. Y. Jia, F. Chen, Compact solid-state waveguide lasers operating in the pulsed regime: a review. Chin. Opt. Lett. **17**, 012302 (2019)
7. A. Boes, B. Corcoran, L. Chang, J. Bowers, A. Mitchell, Status and potential of lithium niobate on insulator (LNOI) for photonic integrated circuits. Laser Photonics Rev. **12**, 1700256 (2018)
8. M. Bruel, Silicon on insulator material technology. Electron. Lett. **31**(14), 1201–1202 (1995)
9. M. Levy, R.M. Osgood Jr., R. Liu, L.E. Cross, G.S. Cargill III, A. Kumar, H. Bakhru, Fabrication of single-crystal lithium niobate films by crystal ion slicing. Appl. Phys. Lett. **73**, 2293 (1998)
10. J. Lindhard, M. Scharff, H.E. Schiøtt, Range concepts and heavy ion ranges. Mat. Fys. Medd. K. Dan. Vidensk. Selsk. **33**, 1 (1963)
11. J. Ziegler, Computer code SRIM. www.srim.org
12. A.M. Radojevic, M. Levy, R.M. Osgood Jr., A. Kumar, H. Bakhru, C. Tian, C. Evans, Large etch-selectivity enhancement in the epitaxial liftoff of single-crystal $LiNbO_3$ films. Appl. Phys. Lett. **74**, 3197 (1999)
13. R.M. Rotha, D. Djukic, Y.S. Lee, R.M. Osgood Jr., S. Bakhru, B. Laulicht, K. Dunn, H. Bakhru, L. Wu, M. Huang, Compositional and structural changes in $LiNbO_3$ following deep He^+ ion implantation for film exfoliation. Appl. Phys. Lett. **89**, 112906 (2006)
14. A. Ofan, O. Gaathon, L. Vanamurthy, S. Bakhru, H. Bakhru, K. Evans-Lutterodt, R.M. Osgood Jr., Origin of highly spatially selective etching in deeply implanted complex oxides. Appl. Phys. Lett. **93**, 181906 (2008)
15. T.A. Ramadan, M. Levy, R.M. Osgood Jr., Electro-optic modulation in crystal-ion-sliced *z*-cut $LiNbO_3$ thin films. Appl. Phys. Lett. **76**, 1407 (2000)

16. H.-C. Huang, J.I. Dadap, G. Malladi, I. Kymissis, H. Bakhru, R.M. Osgood, Helium-ion-induced radiation damage in $LiNbO_3$ thin-film electro-optic modulators. Opt. Express **22**, 19653 (2014)
17. F. Schrempel, T. Gischkat, H. Hartung, T. Höche, E.-B. Kley, A. Tünnermann, W. Wesch, Ultrathin membranes in *x*-cut lithium niobate. Opt. Lett. **34**, 1426 (2009)
18. A.M. Radojevic, M. Levy, R.M. Osgood, D.H. Jundt, A. Kumar, H. Bakhru, Second-order optical nonlinearity of 10 μm-thick periodically poled $LiNbO_3$ films. Opt. Lett. **25**, 1034 (2000)
19. P. Rabiei, P. Günter, Optical and electro-optical properties of submicrometer lithium niobate slab waveguides prepared by crystal ion slicing and wafer bonding. Appl. Phys. Lett. **85**, 4603 (2004)
20. G. Poberaj, M. Koechlin, F. Sulser, A. Guarino, J. Hajfler, P. Günter, Ion-sliced lithium niobate thin films for active photonic devices. Opt. Mater. **31**, 1054 (2009)
21. Q.Y. Tong, U. Gosele, *Semiconductor Wafer Bonding* (Wiley, New York, 1999)
22. H. Hu, L. Gui, R. Ricken, W. Sohler, Towards nonlinear photonic wires in lithium niobate. Proc. of SPIE **7604**, 76040R (2010)
23. A. Guarino, G. Poberaj, D. Rezzonico, R. Degl'Innocenti, and P. Günter, Electro-optically tunable microring resonators in lithium niobate, Nat. Photonics **1**, 407 (2007)
24. P. Rabiei, J. Ma, S. Khan, J. Chiles, S. Fathpour, Heterogeneous lithium niobate photonics on silicon substrates. Opt. Express **21**, 25573 (2013)
25. A. Rao, S. Fathpour, Heterogeneous thin-film lithium niobate integrated photonics for electrooptics and nonlinear optics. IEEE J. Sel. Top. Quantum Electron. **24**, 1 (2018)
26. L. Cai, S. Li, H. Han, H. Hu, Waveguides in single-crystal lithium niobate thin film by proton exchange. Opt. Express **23**, 1240 (2015)
27. L. Cai, Y. Wang, H. Hui, Low-loss waveguides in a single-crystal lithium niobate thin film. Opt. Lett. **40**, 3013 (2015)
28. L. Cai, R. Kong, Y. Wang, H. Hu, Channel waveguides and *y*-junctions in *x*-cut single-crystal lithium niobate thin film. Opt. Express **23**, 29211 (2015)
29. R. Wolf, I. Breunig, H. Zappe, K. Buse, Scattering-loss reduction of ridge waveguides by sidewall polishing. Opt. Express **26**, 19815 (2018)
30. H. Hu, R. Ricken, W. Sohler, Lithium niobate photonic wires. Opt. Express **17**, 24261 (2009)
31. C. Wang, X. Xiong, N. Andrade, V. Venkataraman, X.-F. Ren, G.-C. Guo, M. Lončar, Second harmonic generation in nano-structured thin-film lithium niobate waveguides. Opt. Express **25**, 6963 (2017)
32. M. Zhang, C. Wang, R. Cheng, A. Shams-Ansari, M. Lončar, Monolithic ultra-high-*Q* lithium niobate microring resonator. Optica **4**, 1536 (2017)
33. B. Desiatov, A. Shams-Ansari, M. Zhang, C. Wang, M. Lončar, Ultra-low-loss integrated visible photonics using thin-film lithium niobate. Optica **6**, 380 (2019)
34. I. Krasnokutska, J.-L.J. Tambasco, X. Li, A. Peruzzo, Ultra-low loss photonic circuits in lithium niobate on insulator. Opt. Express **26**, 897 (2018)
35. M.A. Baghban, J. Schollhammer, C. Errando-Herranz, K.B. Gylfason, K. Gallo, Bragg gratings in thin-film $LiNbO_3$ waveguides. Opt. Express **25**, 32323 (2017)
36. M.R. Escalé, D. Pohl, A. Sergeyev, R. Grange, Extreme electro-optic tuning of Bragg mirrors integrated in lithium niobate nanowaveguides. Opt. Lett. **43**, 1515 (2018)
37. Z. Li, M.-H. Kim, C. Wang, Z. Han, S. Shrestha, A.C. Overvig, M. Lu, A. Stein, A.M. Agarwal, M. Lončar, N. Yu, Controlling propagation and coupling of waveguide modes using phase-gradient metasurfaces. Nat. Nanotechnol. **12**, 6750683 (2017)
38. C. Wang, Z. Li, M.-H. Kim, X. Xiong, X.-F. Ren, G.-C. Guo, N. Yu, M. Lončar, Metasurface-assisted phase-matching-free second harmonic generation in lithium niobate waveguides. Nat. Commun. **8**, 2098 (2017)
39. S. Aghaeimeibodi, B. Desiatov, J.-H. Kim, C.-M. Lee, M.A. Buyukkaya, A. Karasahin, C.J.K. Richardson, R.P. Leavitt, M. Lončar, E. Waks, Integration of quantum dots with lithium niobate photonics. Appl. Phys. Lett. **113**, 221102 (2018)

40. B. Desiatov, M. Lončar, Silicon photodetector for integrated lithium niobate photonics. Appl. Phys. Lett. **115**, 121108 (2019)
41. N. Courjal, B. Guichardaz, G. Ulliac, J.-Y. Rauch, B. Sadani, H.-H. Lu, M.-P. Bernal, High aspect ratio lithium niobate ridge waveguides fabricated by optical grade dicing. J. Phys. D Appl. Phys. **44**, 305101 (2011)
42. M.F. Volk, S. Suntsov, C.E. Rüter, D. Kip, Low loss ridge waveguides in lithium niobate thin films by optical grade diamond blade dicing. Opt. Express **23**, 1386 (2016)
43. T. Ding, Y. Zheng, X. Chen, On-chip solc-type polarization control and wavelength filtering utilizing periodically poled lithium niobate on insulator ridge waveguide. J. Lightwave Technol. **37**, 1296 (2019)
44. A.B. Matsko, V.S. Ilchenko, Optical resonators with whispering gallery modes I: basics. IEEE J. Sel. Top. Quantum Electron. **12**, 3 (2006)
45. V.S. Ilchenko, A.B. Matsko, Optical resonators with whispering-gallery modes-part II: applications. IEEE J. Sel. Top. Quantum Electron. **12**, 15 (2016)
46. T.-J. Wang, J.-Y. He, C.-A. Lee, H. Niu, High-quality $LiNbO_3$ microdisk resonators by undercut etching and surface tension reshaping. Opt. Express **20**, 28119 (2012)
47. J. Lin, Y. Xu, Z. Fang, J. Song, N. Wang, L. Qiao, W. Fang, Y. Cheng, Second harmonic generation in a high-Q lithium niobate microresonator fabricated by femtosecond laser micromachining. arXiv:1405.6473 (2014)
48. R. Wang and S. A. Bhave, Free-standing high quality factor thin-film lithium niobate microphotonic disk resonators. arXiv:1409.6351 (2014)
49. C. Wang, M.J. Burek, Z. Lin, H.A. Atikian, V. Venkataraman, I. Huang, P. Stark, M. Lončar, Integrated high quality factor lithium niobate microdisk resonators. Opt. Express **22**, 30924 (2014)
50. R. Wolf, I. Breunig, H. Zappe, K. Buse, Cascaded second-order optical nonlinearities in on-chip micro rings. Opt. Express **25**, 29927 (2017)
51. R. Wu, J. Zhang, N. Yao, W. Fang, L. Qiao, Z. Chai, J. Lin, Y. Cheng, Lithium niobate micro-disk resonators of quality factors above 10^7. Opt. Lett. **43**, 4116 (2018)
52. J.D. Joannopoulos, S.G. Johnson, J.N. Winn, R.D. Meade, *Photonic Crystals: molding the Flow of Light* (Princeton University, New York, 2008)
53. F. Sulser, G. Poberaj, M. Koechlin, P. Günter, Photonic crystal structures in ion-sliced lithium niobate thin films. Opt. Express **17**, 20291 (2009)
54. L. Cai, H. Han, S. Zhang, H. Hu, K. Wang, Photonic crystal slab fabricated on the platform of lithium niobate-on-insulator. Opt. Lett. **39**, 2094 (2014)
55. L. Cai, S. Zhang, H. Hu, A compact photonic crystal micro-cavity on a single-mode lithium niobate photonic wire. J. Opt. **18**, 035801 (2016)
56. F. Schrempel, T. Gischkat, H. Hartung, E.-B. Kley, W. Wesch, Ion beam enhanced etching of $LiNbO_3$. Nucl. Instrum. Methods Phys. Res. Sect. B **250**, 164 (2006)
57. H. Hartung, E.-B. Kley, T. Gischkat, F. Schrempel, W. Wesch, A. Tünnermann, Ultra thin high index contrast photonic crystal slabs in lithium niobate. Opt. Mater. **33**, 19 (2010)
58. S. Diziain, R. Geiss, M. Zilk, F. Schrempel, E.-B. Kley, A. Tünnermann, T. Pertsch, Second harmonic generation in free-standing lithium niobate photonic crystal L3 cavity. Appl. Phys. Lett. **103**, 051117 (2013)
59. S. Diziain, R. Geiss, M. Steinert, C. Schmidt, W.-K. Chang, S. Fasold, D. Füßel, Y.-H. Chen, T. Pertsch, Self-suspended micro-resonators patterned in Z-cut lithium niobate membranes. Opt. Mater. Express **5**, 2081 (2015)
60. H. Liang, R. Luo, Y. He, H. Jiang, Q. Lin, High-quality lithium niobate photonic crystal nanocavities. Optica **4**, 1251 (2017)
61. H. Jiang, H. Liang, R. Luo, X. Chen, Y. Chen, Q. Lin, Nonlinear frequency conversion in one dimensional lithium niobate photonic crystal nanocavities. Appl. Phys. Lett. **113**, 021104 (2018)
62. M. Li, H. Liang, R. Luo, Y. He, J. Ling, Q. Lin, Photon-level tuning of photonic nanocavities. Optica **6**, 860 (2019)

63. M. Li, H. Liang, R. Luo, Y. He, Q. Lin, High-*Q* 2D Lithium Niobate photonic crystal slab nanoresonators. Laser Photonics Rev. **13**, 1800228 (2019)
64. M. Mahmoud, S. Ghosh, G. Piazza, Lithium niobate on insulator (LNOI) grating couplers. in *Proceedings of IEEE Conference on Lasers and Electro-Optics* (IEEE, 2015), pp. 1–2
65. M.S. Nisar, X.J. Zhao, A. Pan, S. Yuan, J.S. Xia, Grating coupler for an on-chip lithium niobate ridge waveguide. IEEE Photonics J. **9**, 6600208 (2017)
66. Z. Chen, R. Peng, Y. Wang, H. Zhu, H. Hu, Grating coupler on lithium niobate thin film waveguide with a metal bottom reflector. Opt. Mater. Express **7**, 4010 (2017)
67. L. He, M. Zhang, A. Shams-Ansari, R. Zhu, C. Wang, L. Marko, Low-loss fiber-to-chip interface for lithium niobate photonic integrated circuits. Opt. Lett. **44**, 2314 (2019)

Chapter 4
Nanoparticles Synthesized by Ion Implantation

Abstract In this chapter, the formation processes and properties of nanoparticles (NPs) by ion implantation, which are embedded in transparent insulators such as SiO_2, are described. In Sect. 4.1, the processes of the NP formation from implanted ions are demonstrated: the neutralization of the implanted ion charges, nucleation and growth of NPs, the nonmetal-metal transition, etc. Section 4.2 describes the compound NPs formed by multiple implantations, which includes the compound semiconductor NPs formation and implantation doping to the NPs. Section 4.3 is devoted to oxide NPs formed by a combination of metal-ion implantation and following thermal oxidation with attention of the applications for the vacuum fluorescent display and the single photon emitters. In Sect. 4.4, the amorphous NPs and embedded liquid NPs are presented. In the amorphous NPs, much shorter scattering lengths play a role. Solid-liquid transition of embedded NPs is highlighted. Section 4.5 reviews NPs with two or more resonance peaks. Using the ab-initio band calculation combined with Mie theory, the origins of the UV and IR peaks of Zn NPs are discussed. In Sect. 4.6, the magneto-optical effect of Ni NPs in SiO_2 and the modified Curie transition are described after introducing the super-paramagnetism of Ni NPs. Section 4.7 summarizes some selected topics, including the additional peak appearance at 600 nm in Ag NPs in SiO_2 due to the inter-particle interaction. The high flux implantation effect, the co-irradiation effect of laser, cavity NPs, sandwiched NPs, and nano-planets are described.

4.1 Formation of Elemental Nanoparticles

In this section, NPs consisting of only one element, i.e., elemental NPs, are described. One of the oldest reports on this topic was given by Arnold [1] in 1975 concerning Au NPs formed in lithia-almina-silica glass by implantation with 285 keV Au^+ ions to 1×10^{16} cm^{-2}. An optical absorption peak was observed at ~550 nm, i.e., the localized surface plasmon resonance (LSPR or SPR) of Au NPs (even in as-implanted state), and also with an increment of the peak intensity after thermal annealing. Since then a huge number of studies have been reported, particularly stimulated by the large optical nonlinearity of metal NPs.

F. Chen et al., *Ion Irradiation of Dielectrics for Photonic Applications*,
Springer Series in Optical Sciences 231, https://doi.org/10.1007/978-981-15-4607-5_4

4.1.1 Charge States of Implants in Solids

The formation processes of NPs by ion implantation is schematically depicted in Fig. 4.1. The implantation introduces accelerated ions to a target material. If the implanted ions would conserve their charge states, i.e., exist as ions, the aggregation of the implants could not be proceeded in solids because of Coulomb repulsion between the ions. It seems that the charge states of the implants are accommodated to the local environment in the solids. Since the ionization is frequently induced in the solid during the implantation, the implanted ions have many chances to exchange the charge states even in insulators. Also defects introduced by the implantation contribute (transient) electric conduction for charge equilibration. It does not mean that the accommodated states are always neutral. For example, Er ions are often implanted in non-metals with the charge state of Er^{+} or Er^{2+} [2] for infrared (IR) photoluminescence (PL) applications. The observation of the well-known IR luminescence at 1.5 μm witnesses that the charge state changed to Er^{3+} state. The luminescence emission

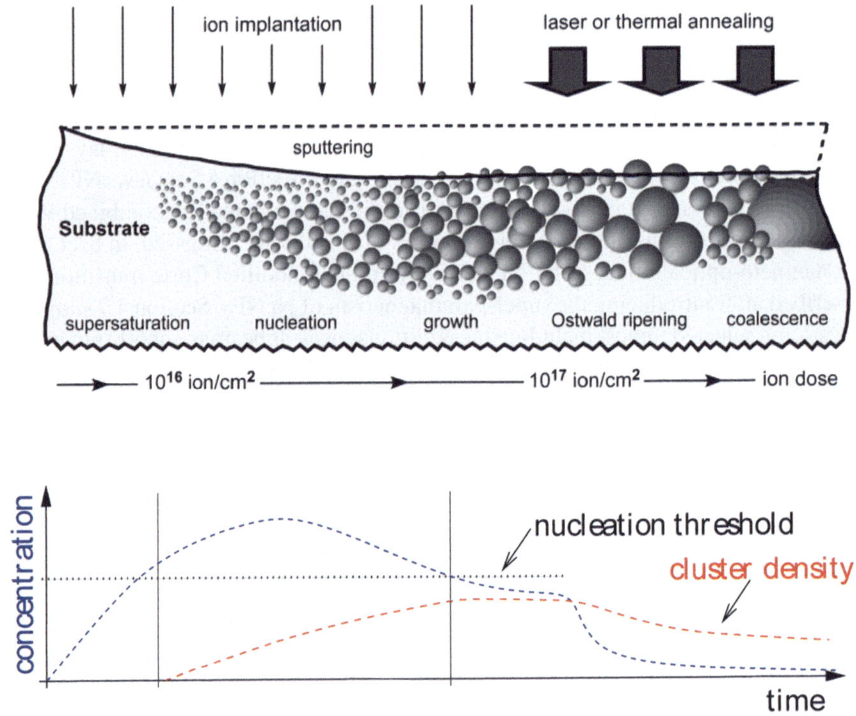

Fig. 4.1 General scheme of ion beam synthesis of NPs in a solid. The evolution of the system during ion implantation and subsequent annealing is sketched. In the lower part, the evolution of the monomer concentration *c* and of the NP number density *n* is shown. A kink of the monomer concentration in the right side is due to the termination of the implantation. Reproduced with permission from [4], ©2005 De Gruyter Poland Ltd

wavelengths from rare-earth (RE) elements strongly depend on the charge state. Only the Er ions in the 3^+ state are able to emit the 1.5 μm luminescence. However, the charge state of Er implants also changes to the neutral at high fluence because Er implants form NPs of Er [2]. Another evidence is obtained from Mossbauer spectroscopy of Fe ions implanted to Al_2O_3. Portion of the charge states Fe^0, Fe^{2+}, and Fe^{3+} changed with the fluence [3].

4.1.2 Nucleation, Growth, and Nonmetal-Metal Transition of Metallic Nanoparticles

Even though the charge state is accommodated, the aggregation of monomers is not driven when the monomer concentration is below the solubility limit. Monomers dispersed in the matrix are more stable than the aggregates of the monomers blow the solubility limit. When the monomer concentration exceeds the solubility limit, a supersaturated state is formed but it does not drive the aggregation of monomers by itself. Always it is not easy for monomers to form dimers, trimers, etc., because the aggregation is energetically unfavorable for monomers. According to the elementary nucleation theory for crystal growth, there exists a critical number of constitute atoms n_c against the aggregation of the monomers. While the aggregates smaller than n_c are energetically unstable and tend to be destroyed even if formed by fluctuation, those larger than n_c are stable and continue to grow ("Nucleation regime"). The monomers need to overcome the critical number by thermal fluctuation.

While the NP formation under ion implantation is a process far from equilibrium, some results from the elementary nucleation theory are described here: First, the free energy change ΔF induced by the formation of an aggregate which consists of N atoms is considered:

$$\Delta F = -(\mu_{\text{agg}} - \mu_{\text{ss}})N + \gamma A \tag{4.1}$$

where μ_{agg} and μ_{ss} denote the chemical potentials of the aggregate and of solid solution, respectively. The second term denotes the energy loss due to the formation of boundary between the two phases, where γ the interface energy and A the area of the boundary. With assuming an aggregate of radius R in a spherical shape, the Eq. (4.1) is described as,

$$\Delta F = -\frac{4\pi}{3v_0}(\mu_{\text{agg}} - \mu_{\text{ss}})R^3 + 4\pi\gamma R^2 \tag{4.2}$$

where v_o denotes the atomic volume. As long as $\mu_{\text{agg}} < \mu_{\text{ss}}$ (i.e., below the supersaturated state), the free energy ΔF is a monotonically increasing function of R. Any sizes of aggregates are not stable except the monomers. When $\mu_{\text{agg}} > \mu_{\text{ss}}$ (i.e., the supersaturated state), the free energy ΔF increases with increasing R up to R_c, i.e., the critical size n_c, and turns to decrease, where

$$R_C = \frac{2\gamma v_0}{\mu_{agg} - \mu_{ss}} \Delta F(R = R_C) = \frac{16\pi}{3} \frac{\gamma^3 v_0^2}{(\mu_{agg} - \mu_{ss})^2} \tag{4.3}$$

These equations show that the aggregates smaller than n_c are energetically unstable because ΔF decreases with decreasing R in the region $R < R_C$. The aggregates tend to be destroyed even if formed by fluctuation. The aggregates larger than n_c are stable and continue to grow because ΔF decreases with increasing R in the region $R > R_C$. The monomers need to overcome the critical number by thermal fluctuation. With increasing the difference between μ_{agg} and μ_{ss}, both R_C and $\Delta F(R = R_C)$ decrease as shown in the Eq. (4.3), which is favorable for the aggregation. Figure 4.1 (lower) depicts the time (fluence) dependence of the monomer concentration c and the NP concentration n. While the monomer concentration increases almost linearly in the beginning of the implantation, the NP density is null because of the critical number for the nucleation. When the monomer density overcomes the nucleation threshold, the NP formation begins.

Once the nucleation begins, the monomer density shows saturation and finally turns to decrease, while the monomers are introduced at the same rate as before by the implantation. The monomers are absorbed by NPs, because it is energetically favorable. When the monomer concentration decreases below the nucleation threshold, new NPs are no longer formed ("Growth regime"). Furthermore, when the monomer concentration decreases below the equilibrium concentration of certain NPs $c(R)$, the NPs are dissolved. The equilibrium concentration $c(R)$ is given [5],

$$c(R) = c_0 \exp\left(\frac{R_{cap}}{R}\right), \; R_{cap} = \frac{2\sigma\Omega}{k_B T} \tag{4.4}$$

where c_0 the equilibrium concentration at the flat surface ($R \to \infty$), R_{cap} the capillary length, σ the surface tension, and Ω the atomic volume. Since smaller NPs have higher $c(R)$ as indicated by Eq. (4.4), smaller NPs are dissolved and larger NPs are further enlarged. This is called "Ostwald ripening regime". See Ref. [6] for more details.

These behaviors have been experimentally confirmed by in-situ optical spectroscopies under implantations with 60 keV Cu^- at various fluxes [7]. The monomer concentration was monitored by the ion-induced luminescence (IIL) from Cu^+ monomer ions in SiO_2. Figure 4.2 shows that the monomer concentration steeply increases up to 0.03×10^{17} ions/cm^2, turns to decrease, and come to nearly constant exceeding 0.3×10^{17} ions/cm^2. The steady state concentration of the monomers seems to depend on the ion flux. These behaviors are similar with the theoretical monomer concentration shown in Fig. 4.1.

The NP concentration was monitored by optical absorption at the SPR peak wavelength. The SPR absorption exhibited a threshold at 3×10^{16} ions/cm^2. While the absorption was almost null below the threshold, it increased almost linearly with the

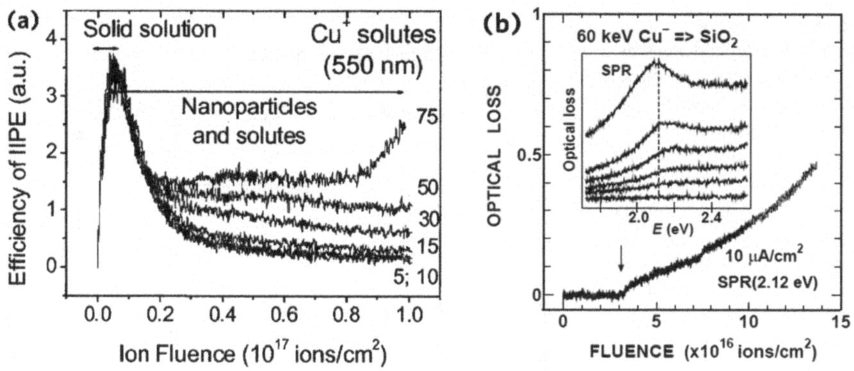

Fig. 4.2 Fluence dependences of **a** ion-induced luminescence from Cu^+ ions in SiO_2 implanted with 60 keV Cu^- ions at various ion fluxes ($\mu A/cm^2$), i.e., of Cu monomer concentration, and **b** the absorption intensity at the SPR peak, i.e., of NP concentration. Reproduced with permission from [7], ©2008 American Institute of Physics

fluence above the threshold. This is qualitatively similar with the theoretical dependence shown in Fig. 4.1. However, Fig. 4.1 predicted that the threshold of the NP density should locate at the fluence where the IIL show the peak. This can be ascribed to the fact that the SPR is observed only in the NPs having itinerant electrons. Small aggregates of metal atoms, e.g., dimmer, trimer, etc., are no longer metals because the density of states around the Fermi energy is not continuous. With increasing the size, the NPs experience the nonmetal-metal transition. While the IIL shows that the aggregation begins at 3×10^{15} ions/cm^2, the SPR shows that the metallic NPs are formed at 3×10^{16} ions/cm^2. According to a simple criterion [8], the nonmetal-metal transition is induced at the NP including 50, 39, and 34 atoms of Fe, Co, and Ni, respectively.

Amekura et al. studied the annealing temperature dependences of the number density of Ni atoms in SiO_2, which were previously introduced by 60 keV Ni ion implantation, by three different methods: Rutherford back scattering spectrometry (RBS) detected total number of Ni atoms in the sample. The absorption intensity of SPR detected Ni atoms forming Ni metallic NPs. The saturated magnetization detected by a superconducting quantum interference device (SQUID) magnetometer represented Ni number density forming ferromagnetic Ni NPs. The results are shown in Fig. 4.3, and are interpreted under an assumption that higher annealing temperature corresponds to larger mean Ni NP diameter. In addition, relatively wide size distributions are confirmed by TEM observation. Even before the annealing, approximately 80% of Ni atoms form metallic Ni NPs and the ratio increases gradually with increasing the annealing temperature. Total Ni concentration detected by RBS slightly decreases with annealing, which could be ascribed to evaporation loss. Contrary Ni atoms forming ferromagnetic Ni NPs were 60% but steeply increases to 75% at 800 °C annealing. Supported by the different temperature dependences, Amekura et al. claimed that the

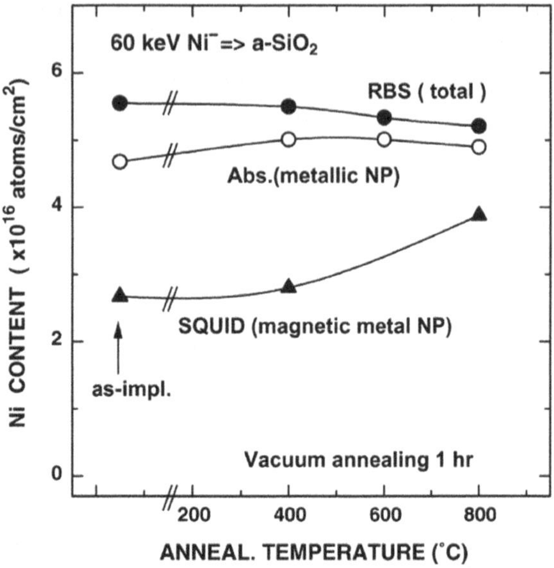

Fig. 4.3 Annealing temperature dependences of Ni content detected by three different methods: RBS (total Ni atom number density), SPR absorption (Ni atoms forming metallic NPs), and SQUID (Ni atoms forming magnetic NPs). Reproduced with permission from [9], ©2004 Elsevier

nonmetal-metal transition and the nonmagnetic-magnetic transition of Ni NPs are induced at different NP sizes [9].

Figure 4.4 displays the fluence dependence of the normalized optical absorption (extinction) spectra of silica implanted with 160 keV Cu ions [10]. The samples implanted to 3.0×10^{16} ions/cm^2 or higher (curves (a)–(c)) exhibit the SPR of Cu NPs at 2.2 eV, indicating the formation of metallic Cu NPs in the samples. It should

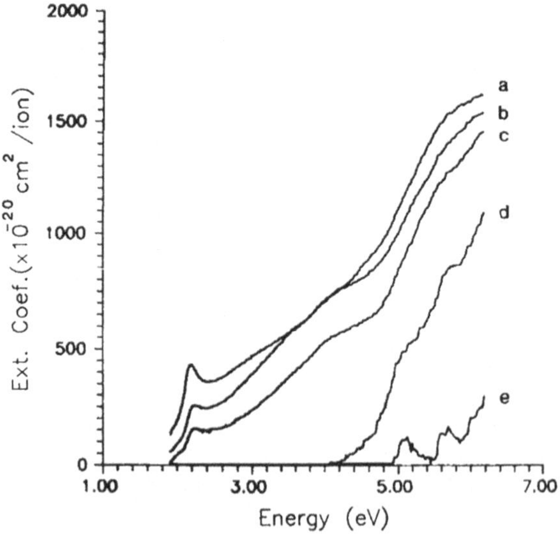

Fig. 4.4 Extinction coefficient per implanted ion in silica with 160 keV Cu ions at a substrate temperature of 296 K. Nominal doses are **a** 12.0, **b** 6.0, **c** 3.0, **d** 1.0, and **e** 0.5 $\times 10^{16}$ ions/cm^2. Reproduced with permission from [10], ©1991 Elsevier

be noted again that SPR is an evidence of the metallic nature of NPs. The samples implanted to 1.0×10^{16} ions/cm^2 or lower (curves (d), (e)) do not exhibit the SPR peak but the absorption at 5 eV and higher. Missing of the SPR at the low fluences is consistent with Fig. 4.2. While peaks at 4.8, 5.1, and 5.7 eV are resolved, the 4.8 and 5.1 eV peaks can be ascribed to the B_2 and E' defect centers of silica. The 5.1 eV peak could be ascribed to Cu^+ ions in silica, probably forming ligand centers. The threshold fluence Φ_{th} for the NP formation is located between 1.0 and 3.0×10^{16} ions/cm^2.

4.1.3 Nucleation Threshold and Control

Figure 4.5 displays the temperature dependence of the nucleation threshold concentration c_e of Au atoms in silica [6, 11]. An upper limit of c_e at room temperature (RT) was evaluated as 2.8×10^{21} atoms/cm^3. Since the atomic density of silica glass is 6.62×10^{22} atoms/cm^3, c_e corresponds to 4.2 at. % in silica at RT. When Au atoms are homogenously introduced in a layer of x nm thick, the threshold fluence Φ_{th} for the NP formation is estimated as (for Au in SiO_2)

$$\Phi_{th} = xc_e/1 \times 10^7 = 2.8\, x\ (\text{nm}) \times 10^{14}\ \left(\text{ions/cm}^2\right) \tag{4.5}$$

where x denotes the thickness of homogenously implanted layer in the unit of nm. In the case of Fig. 4.4, the thickness of the implanted layer is estimated as ~100 nm from RBS data. The Eq. (4.5) gives the threshold fluence Φth of 2.8×10^{16} ions/cm^2, which is a good agreement with the experiments ($1.0 \times 10^{16} < \Phi_{th} < 3.0 \times 10^{16}$ ions/cm^2), while the Eq. (4.5) is only true for Au.

Nie et al. formed Au NPs in a Nd:YAG crystal as saturable absorbers for Q-switching lasing by Au^+ ion implantation at 180 keV to ~2×10^{16} ions/cm^2, i.e., below the threshold [12]. Consequently, the SPR peak was not observed in as-implanted state, while the SPR appeared after thermal annealing.

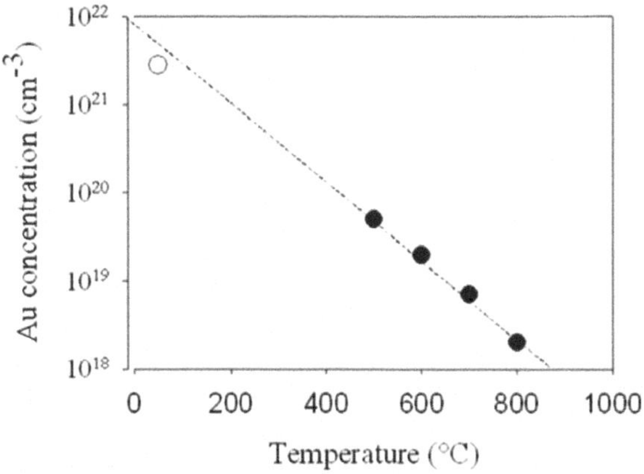

Fig. 4.5 Temperature dependence of the threshold concentration of gold nucleation in silica. Closed and open circles are from Ramaswamy et al. [11] and Rizza et al. [6]. Reproduced with permission from [6], ©2007 American Physical Society

Extrapolation of the temperature dependence shown in Fig. 4.5 indicates that the threshold fluence Φ_{th} for the NP formation becomes higher than 1×10^{17} ions/cm^2, if the irradiation temperature is reduced to 100 K. As shown in Fig. 4.6, a silica sample was implanted at liquid nitrogen temperature (LNT) with multi-energies of 0.7, 0.9, 1.2, and 1.5 MeV to fluences of 2.1, 4.2, 5.2, 7.2×10^{16} ions/cm^2, respectively, in order to obtain a nearly-flat concentration profiles of ~3.6 atomic % [13]. Although the sample was implanted to a high fluence of ~2×10^{17} ions/cm^2 in total, it looks transparent in the visible region as if a virgin silica and does not show the SPR peak, indicating Cu NPs are not formed inside but stay as Cu atoms. The extended X-ray absorption fine structure (EXAFS) of this sample was evaluated. The results manifested that almost no Cu–Cu bonds are detected but mostly Cu–O bonds, confirming Cu NPs are not formed [13]. However, thermal annealing at 650 °C for 1 h drastically changed the color from transparent to strong brown. Cu NPs were formed by the annealing. Therefore, the formation of NPs during the implantation to high fluence was completely controlled by the irradiation temperature and thermal annealing.

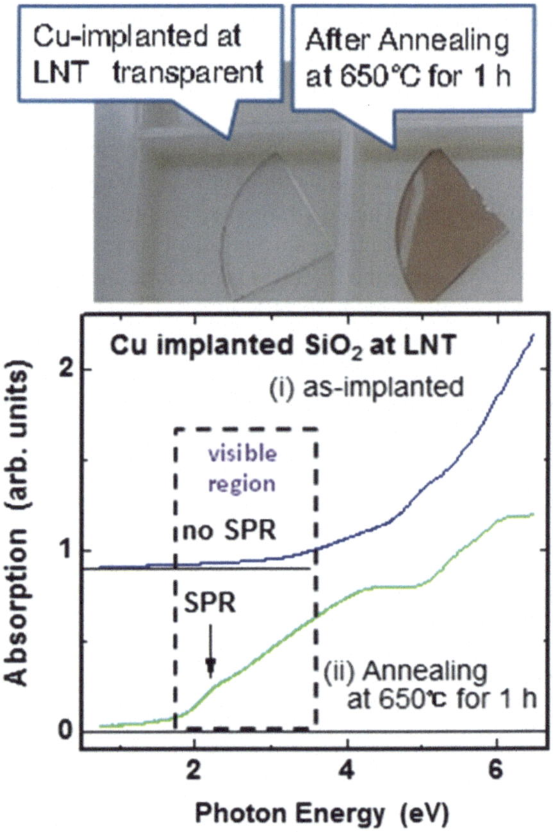

Fig. 4.6 Photoimages and optical absorption spectra of silica implanted with Cu ions of multi-energies at *liquid nitrogen temperature (LNT)*, in as-implanted state (transparent) and after thermal annealing at 650 °C for 1 h (brown color). The implanted energies and fluences are shown in text. The total fluence was 1.87×10^{17} ions/cm^2. Reproduced with permission from [13], ©2011 American Institute of Physics

In the most cases, NPs formed by ion implantation exhibit large size distributions, except some cases, e.g., shown in Sect. 4.7(b), which is ascribed to widely distributed "growth stages" of NPs. Some NPs are "nucleated" almost in the beginning of the irradiation period, which experience long period for the "growth" of NPs. Other NPs are nucleated almost in the end of the irradiation period, which experience only short time for the "growth". Ramaswamy et al. proposed to isolate the nucleation and the growth of NPs. They implanted a low dose of Au ions to SiO_2, ranging from 1×10^{14} to 5×10^{14} ions/cm^2. Successive flash annealing at 1100 °C for 5 min in an Ar-4%-H_2 atmosphere completed the nucleation of NPs. After then the samples were implanted again at 500–700 °C up to 1×10^{15} ions/cm^2 in order to induce the growth of NPs only, but not the nucleation nor the Ostwald ripening. Utilizing this methodology, they succeeded in improving the homogeneity of the size distribution [11].

4.1.4 Very Small Non-metallic Metal Nanoparticles as Luminescence Sensitizers

Very small metal NPs, which do not exhibit SPR, can be described better as molecules. The Padova group found that very-small non-metallic Au or AuAg NPs (molecules of Au (and Ag) atoms) were excellent sensitizers of the 1.5 μm luminescence from Er^{3+} ions. First, they implanted SiO_2 by Er^+ ions with three different energies of 50, 100, and 190 keV to a total fluence of 7×10^{14} Er^+/cm^2 in order to form a box-like profile. After annealing in N_2 gas to convert Er ions to luminescent, the samples are implanted with Au (and Ag) ions at three energies of 60, 110, and 190 keV at various fluences. Then the samples were annealed at various temperatures to control the sizes of metal aggregates. The average nearest neighbor Au-Au distance $d_{\mathrm{Au-Au}}$ determined by X-ray diffraction is in a relationship with the number n of atoms in Au NP.

$$d_{Au-Au}(n) = d_{Au-Au}(\infty)\left(1 - \frac{A}{n^B}\right) \tag{4.6}$$

where $A = 0.128$ and $B = 0.37$ [14].

As shown in Fig. 4.7a, the intensity of the 1.5 μm luminescence from Er^{3+} ions increases with introduction of AuAg NPs compared with the samples implanted with Er ions only (Er800). The introduction of NPs not only enhanced the intensity of the 1.5 μm luminescence but also changed the excitation spectra as shown in Fig. 4.7c. Without the NPs, discrete lines were observed, indicating that the excitation was directly induced via transitions between discrete 4f levels in Er ions. With introduction of NPs, the excitation spectra changed to continuous ones. This observation indicates that the excitation is once absorbed by NPs and then transferred to Er ions. These processes could be called "the energy transfer from NPs to Er ions" or "the sensitization of Er ions by NPs".

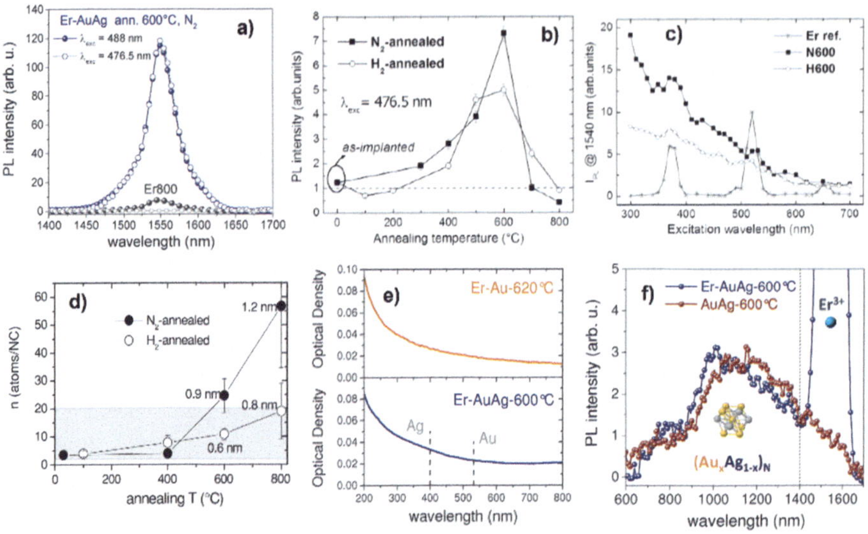

Fig. 4.7 **a** Photoluminescence spectra of Er ions in SiO_2 with/without AuAg small NPs, under resonant (488 nm) and non-resonant (476.5 nm) excitation. **b** Annealing temperature dependence of the 1.5 μm PL intensity of Er ions at under excitation at 476.5 nm. **c** PL excitation spectra of Er ions in SiO_2 with (closed and open circles) and without (crosses) small Au NPs, monitored at 1540 nm. Closed and open circles are from annealed samples at 600 °C in nitrogen and diluted hydrogen gas, respectively. **d** Annealing temperature dependence of the number of atoms *n* consisting one Au NP determined from XRD results using the relation (4.6). **e** Absorption spectra of SiO_2 with Er ions and small Au NPs (upper) and small AuAg NPs (lower). Broken lines indicate the wavelengths of SPRs of Au NPs and Ag NPs. **f** PL spectra from small AuAg NPs with/without Er ions. Reproduced with permission from [15], ©2015 Royal Society of Chemistry and [14], ©2011 American Physical Society and [14], ©2011 American Physical Society and [14], ©2011 American Physical Society and [15], ©2015 Royal Society of Chemistry and [15], ©2015 Royal Society of Chemistry

With changing the annealing temperature, the sensitization efficiency is controlled through the size changes of the NPs as shown in Fig. 4.7b. The maximum efficiency was given at the annealing temperature of 600 °C, i.e., at the optimum sizes of 10–20 atoms per NP inferred from Fig. 4.7d. These very small NPs are no longer metallic NPs. In fact, the SPR was not observed in these samples as evidenced in Fig. 4.7e. Rather, these very small AuAg NPs emit luminescence as they were molecules with a very broad width between 600 and 1700 nm, as shown in Fig. 4.7f. Similar but not identical spectra were observed from small Au NPs. However, samples implanted with Xe ions under the same conditions do not show the broad PL band, excluding a defect origin of this band [15]. These observations again indicate that the small NPs contributing the sensitization are not metallic but like molecules emitting luminescence.

4.1.5 Miscellaneous

A frequently happened misunderstanding is that metal NPs could be formed in any materials if the metal ions would be implanted to high fluence. The prerequisites for metal NP formation is low reactivity between the implanted metal ions and the matrix and non-existence of stable compounds between them. For example, high fluence Co ion implantation to Si substrate (and thermal annealing) does not result in the formation of Co NPs in Si but the formation of continuous $CoSi_2$ layer as shown in a cross-sectional TEM image (Fig. 4.8) [16, 17]. This example indicates another exception: the shapes of the products are not always in NPs but could be a continuous layer [17].

It has been discussed whether metal (M) NPs or oxide (MO) phase is formed when M ions are implanted to SiO_2. Hosono attributed this problem to Gibbs energy change ΔG of Si–O bond formation and that of M–O bond formation. Since it has been known that implanted SiO_2 showed atomic network at high fictive temperature ~3000 K, Hosono adopted the value of ΔG at 3000 K [18]. Later Hosono and coworkers generalized this idea using the electronegativity [91, 92]. Contrary, Cattaruzza pointed out that the high temperature continued only for 10–20 ps, and the local temperature returned within a few hundred ps. He divided the radiation effects into two categories, i.e., high-energy ballistic effects and low-energy chemically guided effects. He applied ΔG at the substrate temperature [19].

Amekura et al. have implanted silica with 60 keV Zn^+ ions to a fluence of 1.0×10^{17} ions/cm^2, and evaluated the chemical states of the implanted Zn ions embedded in silica by X-ray photoelectron spectroscopy (XPS) with sputtering depth profiling. While the major component was metallic Zn NPs, a minor portion of Zn_2SiO_4 phase was detected [20]. If the ion implantation into SiO_2 would purely follow the thermodynamics, only metallic Zn phase could be formed but not ZnO phase. This is because the formation Gibbs energy change of SiO_2 is much higher, i.e., more stable, than that of ZnO. However, the experiments indicate that the ZnO phase is formed

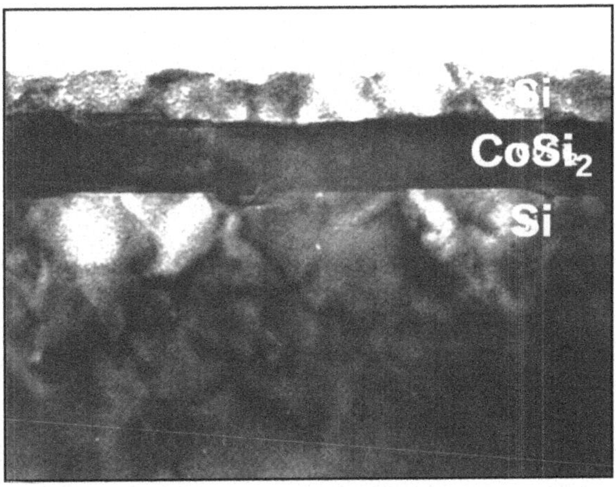

Fig. 4.8 XTEM images of a *continuous* $CoSi_2$ layer buried in Si, formed by 40 keV Co implantation to 5.75×10^{17} ions/cm^2 at 400 °C and annealing at 900 °C for 5 min. Reproduced with permission from [16], ©1991 Elsevier

due to the non-equilibrium nature of the ion implantation. The formed ZnO phase soon reacts with SiO_2 matrix and is transformed to the Zn_2SiO_4 phase.

4.2 Semiconductor and Compound Nanoparticles

4.2.1 Elemental Semiconductor Nanoparticles

Since Canham observed visible luminescence from porous silicon fabricated by electrochemical route in 1990 [21], huge attention was paid to the visible luminescence from silicon nanostructures in connection with the quantum size confinement. The first observation of visible photoluminescence from Si NPs fabricated by ion implantation was reported by Shimizu-Iwayama et al. in 1993 [22]. They implanted Suprasil fused silica with 1 MeV Si^+ ions to fluences of 1–4 × 10^{17} ions/cm^2 and observed a broad luminescence band around 2.0 eV. Since the bandgap energy of bulk Si is ~1.1 eV, the visible luminescence was tentatively ascribed to the quantum size effect.

After then a number of papers were published concerning the Si NPs fabricated by ion implantation. However, many of them reported that the behaviors are different from what expected from the simple quantum size effect: Even increasing the NP sizes by increasing annealing duration, the luminescence peak shows almost no red shift [23]. This behavior can be explained by "the three-region model" reported by Kanemitsu et al. [24], where the luminescence wavelength is determined by the interface region between Si NPs and surrounding SiO_2. However, the fluence dependence cannot be explained by the three-region model: A clear red shift of the luminescence peak was observed with increasing the fluence, where an increase of the NP sizes was confirmed by TEM. To consistently explain the observations, Shimizu-Iwayama et al. assumed that the interface states change their energies with inter-particle interaction, i.e., "the reactive nanocluster model [23]." On the contrary, Brongersma et al. reported a simple size dependence of luminescence from Si NPs using the size reduction via oxidation [25].

An observation on optical gain in silicon NPs was reported in 2000 [26], and was accepted with enthusiasm. However, a problem in the evaluation method was pointed out later [27]. According to a review paper published in 2011 [28], the optical gain in silicon NPs is still an open question.

4.2.2 Compound Semiconductor Nanoparticles

In order to synthesize compound semiconductor NPs, e.g., GaAs NPs, in a matrix, sequential implantations of Ga and As ions (and thermal annealing) are usually applied [29]. An exception is NPs of oxide semiconductor such as ZnO. The NPs can be synthesized by thermal oxidation of pre-formed metal Zn NPs [30, 31], which will be described in 4.3. NPs of various III–V and II–VI compound semiconductors were synthesized in Oak Ridge National Laboratory (ORNL) in late 1990s [32]. Table 4.1 summarizes semiconductor NPs embedded in substrates, which were synthesized at ORNL via ion beam synthesis by 1998 [32].

Microscopic optical properties of II–VI semiconductor NPs synthesized by sequential implantation up to four sequences (elements) as shown in Table 4.2 were evaluated by low temperature scanning near-field optical microscopy (SNOM). CdS NPs were formed in α-Al_2O_3 crystals with sequential implantation of two elements 450 keV Cd^+ (4.3×10^{16} ions/cm^2) and 164 keV S^+ (the same fluence). Luminescence of highly excited single CdS NPs were studied by SNOM [33]. Furthermore, the optical properties of Mn impurity in CdS NPs were studied, which were formed in α–Al_2O_3 crystals by sequential implantation of three elements, i.e., 430 keV Cd^+ (4.5×10^{16} ions/cm^2), 164 keV S^+ (4.5×10^{16} ions/cm^2), and 250 keV Mn^+ (2×10^{14} ions/cm^2) [34].

Donor-acceptor (D-A) pair luminescence from single ZnS NPs doped with Cu and Al ions were evaluated. Impurities of Al and Cu ions act as electron donors and acceptors in ZnS NPs. The electron-hole recombination processes between Al donors and Cu acceptors were observed as typical D-A pair luminescence as shown in Fig. 4.9. Sequential implantations of four elements, 420 keV Zn^+ (1.0×10^{17} ions/cm^2), 225 keV S^+ (1.0×10^{17} ions/cm^2), 400 keV Cu^+ (0.2–1.0×10^{16} ions/cm^2), and 170 keV Al^+ (0.2–1.0×10^{16} ions/cm^2) [37], were applied.

Table 4.1 Semiconductor NPs synthesized in substrates at ORNL by ion implantations. Y: the NP was formed. N: the desired NP was not.

Nanocrystal	Substrate		
	SiO_2-(glass)	Al_2O_3	Si
Si	Y	Y	
Ge	Y	Y	
SiGe	Y	Y	
GaAs	Y	Y	Y
GaP	Y	Y	Y
GaN	N	Y	N
InAs	Y		Y
InP	Y		Y
CdS	Y	Y	Y
CdSe	Y	Y	Y
$CdSe_{0.5}S_{0.5}$	Y		
CdTe			Y
$CdSe_{0.5}Te_{0.5}$			Y
ZnS	Y	Y	Y
ZnSe	Y		
PbS	Y	Y	N

Reproduced with permission from [32], ©1998 Elsevier

Table 4.2 Impurity-doped compound-semiconductor nanoparticles fabricated with multiple sequential implantations. (Sequence = 2, 3, and 4).

Substrate	Nanoparticles (NPs)		Dopant(s) in NPs	Implant. Sequences	Refs.
α-Al_2O_3	CdS	Cd^+ 450 keV 4.3×10^{16} cm^{-2}	non	2	[33]
		S^+ 164 keV 4.3×10^{16} cm^{-2}			
α-Al_2O_3	CdS	Cd^+ 450 keV 4.5×10^{16} cm^{-2}	Mn^+ 250 keV 2×10^{14} cm^{-2}	3	[34]
		S^+ 164 keV 4.5×10^{16} cm^{-2}			
α-Al_2O_3	ZnS	Zn^+ 420 keV 1.0×10^{17} cm^{-2}	Cu^+ 400 keV 0.2–1.0×10^{16} cm^{-2}	4	[37]
		S^+ 225 keV 1.0×10^{17} cm^{-2}	Al^+ 170 keV 0.2–1.0×10^{16} cm^{-2}		

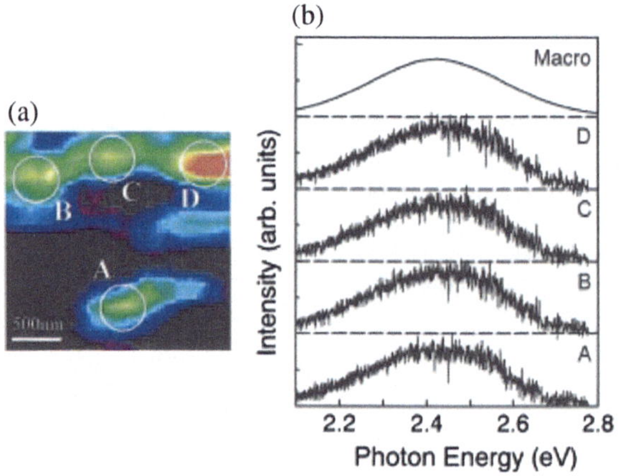

Fig. 4.9 **a** Two-dimensional SNOM-PL mapping of Cu,Al-codoped ZnS NPs in Al_2O_3 at RT, fabricated by sequential implantations of four elements of Zn, S, Cu, and Al. **b** micro PL spectra at each site and macro one. Reproduced with permission from [37], ©2004 American Institute of Physics

In spite of the success of the formation of various compound NPs, the fundamental processes of compound NP formation by sequential implantation have not been well understood. Here the binary compound NP formation is discussed, which do not limit only to semiconductor NPs but general binary material NPs [38].

Let's suppose that ions of element A are firstly implanted to an inert matrix such as SiO_2. Since NP formation requires high concentration of material introduction, the implanted element A may form NPs of pure element A in the matrix. Then the element B is implanted. Since the element B is implanted as energetic ions, parts of the pure A NPs are destroyed by the ballistic intermixing (BI) effect. If the BI effect and the chemical reactivity between A and B are both weak, the implanted element B may form the pure B NPs at any sites without spatial correlation with the pre-formed A NPs. In this case, the compound NPs AB are not formed but the pure A and pure B NPs are independently observed at different sites. The products of the sequential implantation are mixture of the pure A and pure B NPs.

If the reactivity between A and B slightly increases, the B atoms aggregate around the pre-formed A NPs. Since the A NPs have been already formed and not destroyed ballistically or chemically, the element B aggregates around the pre-formed A NPs and may form the core/shell A/B NPs. The formation of compound AB NPs may require destruction of the pre-formed A NPs to some extent. Therefore, the physics of the compound NP formation by sequential implantation is important but has not been clarified yet. Mattei categorized the consequence of sequential implantations as (i) separated A and B NPs, (ii) alloy NPs, and (iii) core/shell NPs [38].

Hosono observed NP formation in silica depending on the implantation sequence after sequential implantation of 160 keV Cu (6×10^{16} ions/cm^2) and 40 keV F (1×10^{17} ions/cm^2) [39]. When the silica was implanted with F first and then Cu, the SPR intensity of Cu NPs was almost the same as the Cu single implantation as shown in Fig. 4.10. Contrary, the reverse sequence, i.e., Cu first and then F, the SPR disappeared. TEM observation showed that the products of the latter sequence were Cu NPs with Cu_2O shells, while those of the former sequence were Cu NPs. Because of the higher electronegativity of F atom than O atom, the implanted F ions break Si-O bonds and form Si-F bonds. The liberated O atoms or O_2 molecules partly oxidize the Cu NPs, which results in the formation of Cu_2O shells.

The sequential implantations with metal (M) ions followed by F ions present a new route to form M/M_xO_y core-shell NPs in SiO_2. In fact, Ren et al. have succeeded in the formation of Zn/ZnO core/shell NPs in SiO_2 by Zn and F sequential implantation [40]. The high-resolution TEM image is shown in Fig. 4.11.

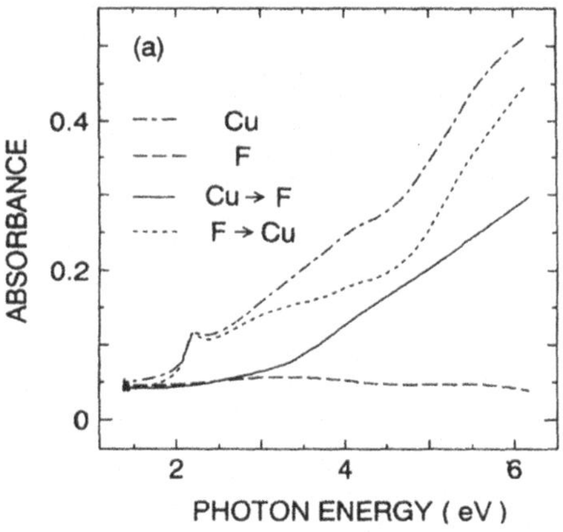

Fig. 4.10 Implantation sequence dependence of optical absorption spectra of SiO_2 with 160 keV Cu and 40 keV F ions. In the sequence of Cu → F, the SPR absorption at 2.2 eV disappears. Reproduced with permission from [39], ©1995 American Physics Society

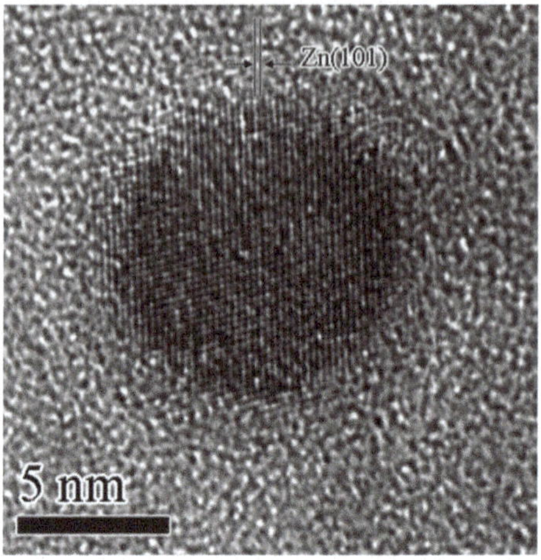

Fig. 4.11 High resolution TEM image of Zn/ZnO core/shell NPs formed by sequential implantation of Zn and F ions. Reproduced with permission from [40], ©2006 Institute of Physics

Computer simulation study of sequential implantation using the kinetic lattice Monte Carlo method was reported by Strobel et al. [5, 41]. Figure 4.12 shows some examples. While the implantation depth profiles of both the ions are adjusted to

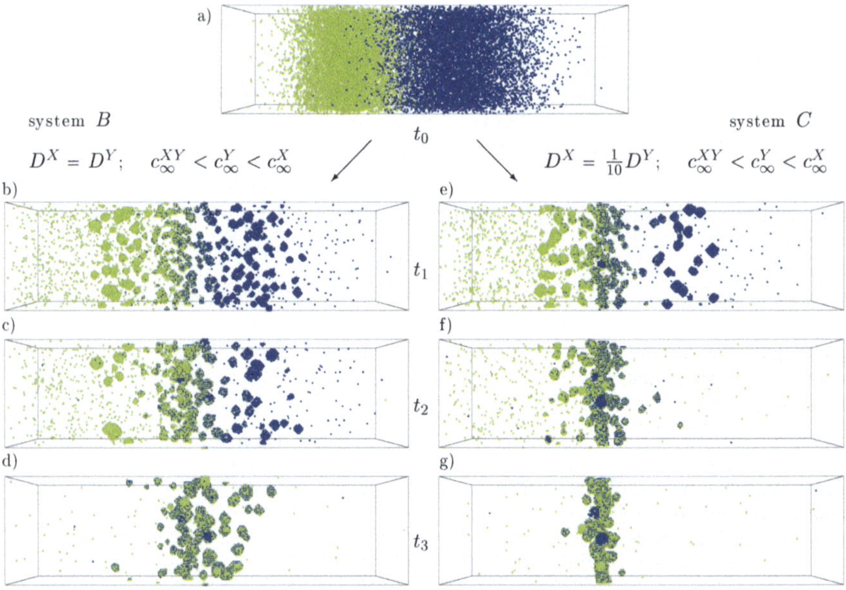

Fig. 4.12 Kinetic lattice Monte Carlo simulations of sequential implantation, starting from "shifted" implantation profiles of elements X and Y. Left: $D^X = D^Y$. Right: $D^X = 0.1D^Y$. Reproduced with permission from [5], ©2001 Materials Research Society

overlap for most of experiments, shifted depth profiles were adopted in this calculation [41]. The results show that compound NPs are formed even under the "shifted" implantation profiles in certain conditions.

Liu and coworkers studied sequential implantation effects on metal NPs, in order to tailor the size distribution [42], to enhance the SPR [43], to form brass (Cu–Zn alloy) NPs [44], and to poly-crystallized Ag NPs [45], and etc.

4.3 Oxide Nanoparticles

As described, compound NPs are synthesized by sequential implantations. However, the products could depend on the implantation sequences; i.e., the products of A-ion implantation followed by B-ion implantation could be different from those of B-ion implantation followed by A-ion implantation. Furthermore, the compound formation processes could have non-thermal equilibrium nature, which have not been clarified yet.

Instead, Amekura et al. studied oxide NP formation via thermal oxidation of implanted metal NPs [31, 94], which was called ion implantation and thermal oxidation (IITO) method. While this process is limited to oxide NPs, the processes are mainly governed by thermodynamics, which could be easier to understand than the sequential implantations. See Fig. 4.13. First, metal NPs (e.g., Zn NPs) are formed with ion implantation (at ~100 keV or less). After then, the implanted samples are thermally annealed in oxygen atmosphere. While the metal NPs are formed inside of the matrix, the oxygen migrates inside of the matrix and transforms the metal

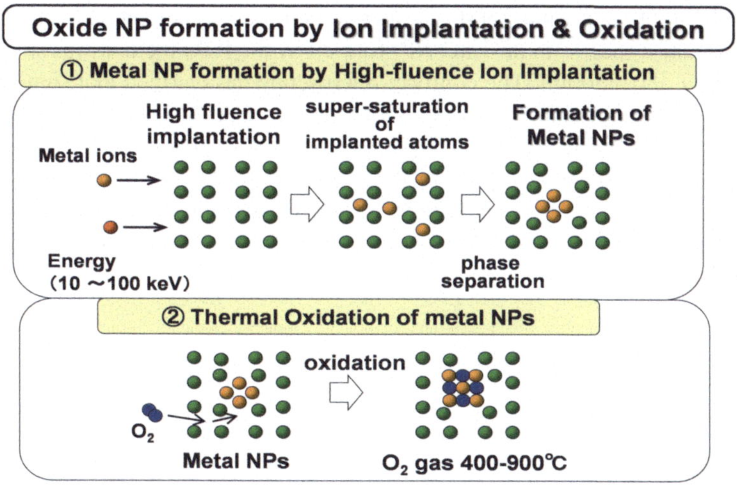

Fig. 4.13 Schematically depicted processes of oxide NP formation by ion implantation and thermal oxidation (IITO) method

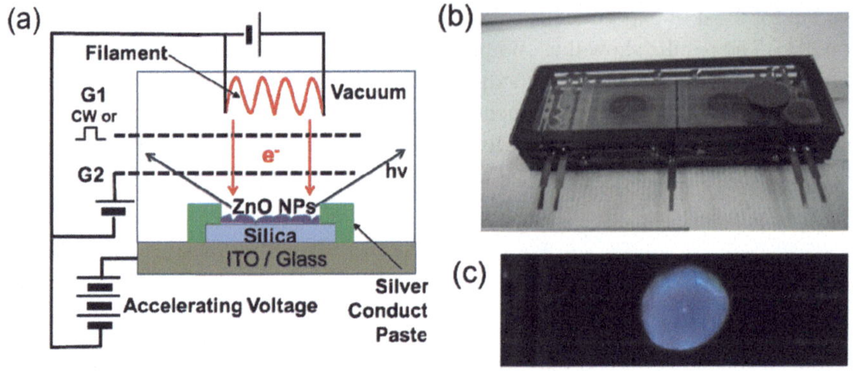

Fig. 4.14 **a** Schematic cross section of fluorescent display utilizing ZnO NPs. **b** Device image, **c** purple-blue luminescence from the devices. Reproduced with permission from [47], ©2011 American Institute of Physics

NPs to oxide NPs. The low implantation energy (~100 keV or less) is important for the oxidation of shallowly embedded NPs. See the review paper concerning this methodology for details [31].

Here the recent three applications of ZnO NPs formed by the IITO method are shown: (i) vacuum fluorescent display utilizing ZnO NPs [47], (ii) single-photon emission from defects in ZnO NPs [48], and (iii) ZnO NP aggregates uptaken by human epidermal keratinocyte (HaCaT) cells as stable luminescent bio-markers [48].

ZnO NPs formed by the IITO method were installed in a vacuum fluorescent display as shown in Fig. 4.14, and were irradiated with cathode rays of 600–3000 V to excite luminescence. Depending on the oxidation temperature, luminescence of purple-blue (650 °C), white (800 °C), and green (900 °C) is emitted [47]. Defect-related luminescence from the ZnO NPs formed by the IITO method was studied in the range of 580–850 nm. The confocal microscopy has clarified that some of the centers show anti-bunching behaviors due to the single-photon emitters, i.e., the 2nd order correlation function of emitted photons becomes zero with $\tau \to 0$, as shown in Fig. 4.15. ZnO NPs were formed by the IITO method and removed from the silica substrate. The aggregates of the removed ZnO NPs were uptaken by HaCaT cells. The aggregates exhibited stable luminescence without decay of the intensity for long time [48].

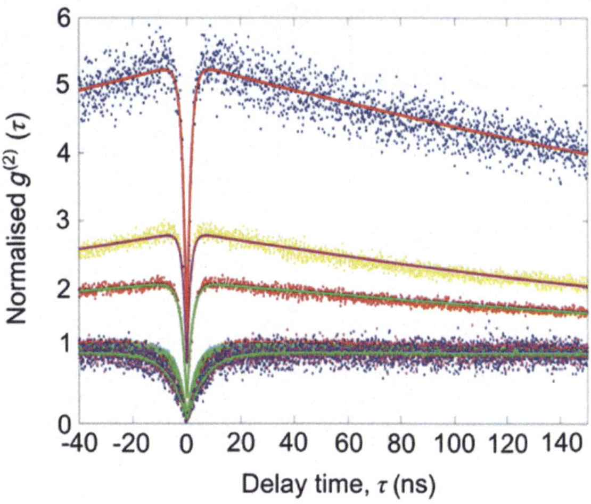

Fig. 4.15 Second-order autocorrelation function of luminescence from ZnO NP defect under four different pump powers. Reproduced with permission from [48], ©2017 The authors

4.4 Amorphous and Liquid Nanoparticles

4.4.1 Amorphous Nanoparticles

Because the metal bondings are direction-insensitive, metals generally have structure flexibility against randomness introduced in the crystalline structures [49]. This is in contrast with tetrahedrally bonded semiconductors, e.g., Si, Ge, III-Vs. The direction-sensitive bondings are weak against structural disorder, and the semiconductors are easily amorphized. Ion irradiation induces amorphization of semiconductors, which is frequently induced below 1 displacement per atom (dpa). This fact indicates that the structural transition is induced only by the accumulated frustration of the direction-sensitive bondings, without all the atoms in the semiconductor being displaced. Contrary, amorphization of *bulk* metals which consist of only one element has never been successfully induced to date. While amorphous metal alloys in bulk forms have been synthesized, the alloys consist of multi-elements, and mostly include non-metallic element(s).

In cases of nano-structures, non-negligible parts of atoms exist on the surface, which is a favorable condition for amorphization. Furthermore, ion irradiation induces extremely rapid thermal quenching with the cooling rate of ~10^{16} K/s for 0.1 keV/u ions and ~10^{15} K/s for 1 MeV/u ions, which is favorable for amorphization to exclude thermal recrystallization at high temperatures.

It is known that Ge NPs are amorphized under ion irradiation at much lower fluence than the bulk counterpart. The degree of the amorphization was monitored as the Debye-Waller (DW) factor which was determined from the extended X-ray absorption fine structures (EXAFS) analysis using synchrotron radiation (Fig. 4.16). Higher DW factor corresponds higher randomness. The DW factor in unirradiated state is shown by broken lines for NPs and bulk. With the first shot of 2×10^{13}

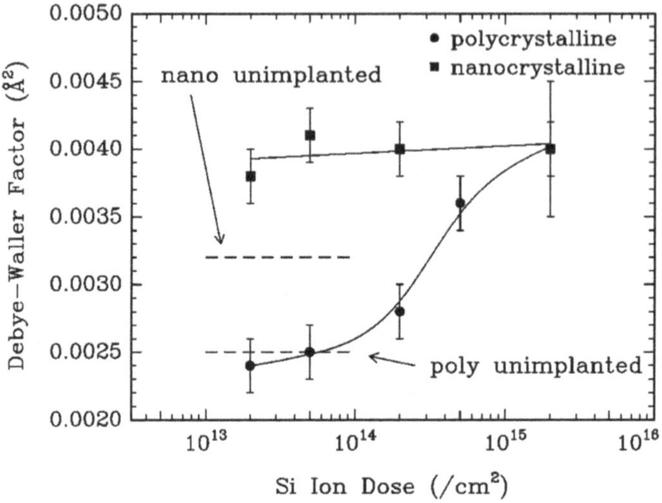

Fig. 4.16 Fluence dependence of Debye-Waller factors of Ge nanocrystals (NPs) and polycrystals (bulk) irradiated with 5 MeV Si ions. Reproduced with permission from [50], ©2005 American Physics Society

ions/cm^2 of 5 MeV Si ions, the Ge NPs, denoted as nanocrystalline, transformed to the amorphous state. Contrary, the bulk, denoted as polycrystalline, required the fluence of 2×10^{15} ions/cm^2 for total amorphization. The Ge NPs have more than 100 times higher susceptibility for the amorphization [50].

Johannessen et al. irradiated Cu NPs in SiO_2 with 5 MeV Sn ions at liquid nitrogen temperature, and evaluated the pair-correlation function between atoms using EXAFS. They observed that 2nd and higher order peaks except the first one diminished, indicating the loss of the long-range orders but maintenance of the short-range order, i.e., attainment of an amorphous structure [51, 52]. Amekura et al. studied the optical absorption of the amorphous Cu NPs with comparing the crystalline Cu NPs [13], as shown in Fig. 4.17. While both the crystalline and amorphous Cu NPs show the SPR at 2.2 and 2.1 eV, respectively, another structure at 4 eV is observed only in crystalline NPs. The missing of the 4 eV structure is reproduced by Maxwell-Garnett theory with assuming much shorter electron mean-free-path (EMFP) than the NP diameter, indicating the scattering of electrons not only at the NP surfaces but also inside of NPs due to loss of the lattice periodicity in amorphous structure.

4.4.2 *Molten Nanoparticles*

Another transition similar to the crystalline-amorphous transition is the solid-liquid transition in crystals at high temperatures. Also this transition is presumed to be involved in the inelastic thermal spike effects discussed in Chap. 5. Changes on the

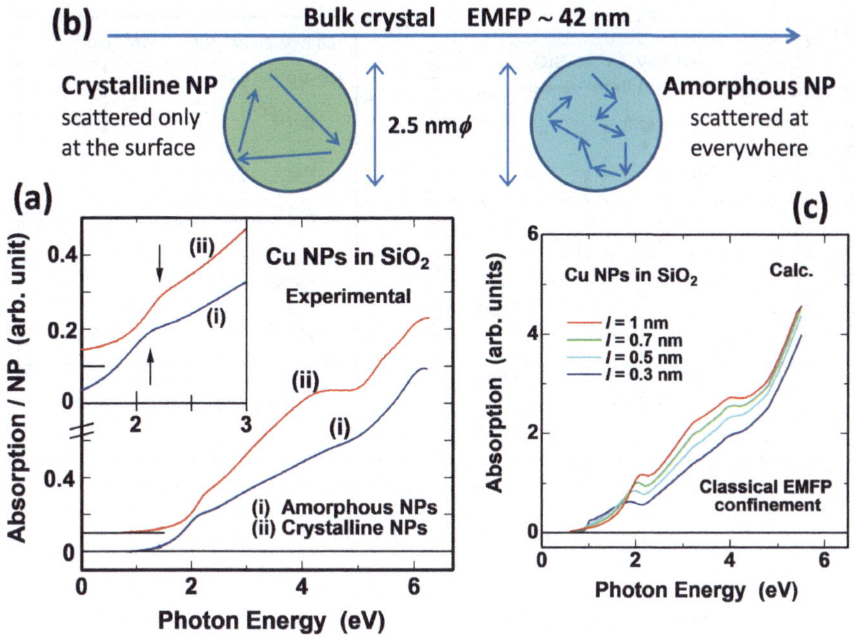

Fig. 4.17 **a** Experimental absorption spectra of (i) amorphous and (ii) crystalline Cu NPs. (c) Calculated spectra with various electron mean-free path (EMFP). **b** schematic depicted relationships between the NP sizes and EFMP in amorphous and crystalline NPs. Reproduced with permission from [13], ©2011 American Institute of Physics

SPR with the solid-liquid transitions have been already reported in Au [53] and Cu [54] NPs in silica. However, the temperature was raised up to 900 °C and 900 K for the former and latter, respectively. Coarsening of the NPs due to thermally enhanced diffusion are not negligible at such high temperatures. Furthermore, the melting points were not determined on the same samples but were referred from literature values.

Amekura et al. selected NPs of Zn for the experiments, which has low *bulk* melting point (m.p.) of 420 °C [35, 55]. The m.p. of the *NPs* in SiO_2 was determined by high temperature (HT) X-ray diffraction (XRD) on the same sample. To reduce the thermal diffusion during the measurements, synchrotron radiation was used to shorten the measurement time. After the high temperature measurements, negligible size change of NPs were confirmed by small angle X-ray scattering (SAXS). The temperature dependence of 101_{Zn} XRD peak intensity is exhibited in Fig. 4.18a [55]. A clear XRD hysteresis loop with a width of ~100 °C was observed, which was ascribed to the melting-resolidification transition of Zn NPs in SiO_2. This was contrast with results of Zn thin film, which showed much narrower hysteresis. In the film, once an inhomogeneous nucleation begins at any sites, the transition propagates all over the film because all the parts of the film are connected. Contrary, NPs dispersed in SiO_2 are not connected with each other. Even if the inhomogeneous nucleation begins in a

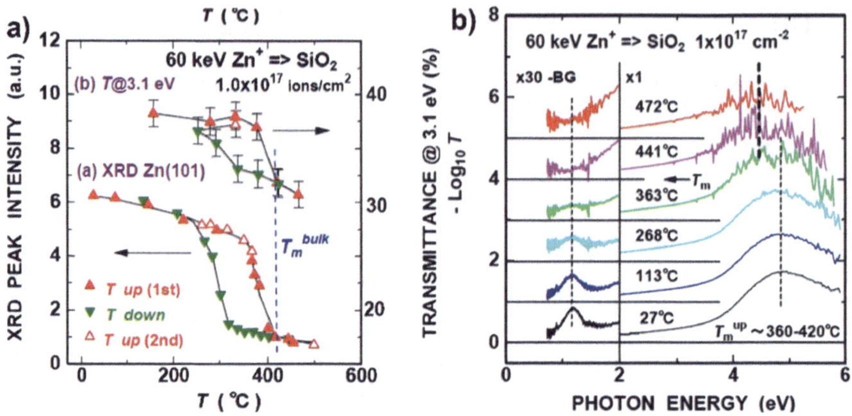

Fig. 4.18 **a** Temperature dependence of the peak intensity of Zn (101) diffraction peak from Zn NPs in SiO_2 (**a**) and the hysteresis of the optical transmittance at 3.1 eV (**b**). **b** The optical density spectra, $-\mathrm{Log}_{10}T$, from near-infrared to ultraviolet region, at various temperatures of Zn NPs in SiO_2. Reproduced with permission from [55], ©2010 American Institute of Physics

NP, it does not propagate to other NPs. This could be an origin of the wide hysteresis loop. The hysteresis loop closes around the bulk m.p. T_m. The NPs have a certain size distribution. The m.p. of larger NPs are close to the bulk one.

As shown in Fig. 4.18b, Zn NPs show two resonant peaks at 1.2 and 4.8 eV at RT. The former is ascribed to free electron origin, i.e., the SPR in the narrow definition. The latter is a resonantly enhanced band-to-band transitions, which are sometimes called "electronic resonance" [36]. (The details are discussed in the next section.) With increasing the temperature, the 1.2 eV peak disappeared around the m.p. This is the similar situation with the amorphization: With the melting, the periodicity of the lattice is lost, then the EMFP steeply reduced. Consequently, huge broadening of the peak makes itself invisible. Contrary, the 4.8 eV band survived up to the bulk m.p., but showed a red shift of ~0.3 eV exceeding the m.p. As shown in Fig. 4.18a, the peak shift (the optical transmittance at 3.1 eV) exhibits almost the same hysteresis loop as the XRD, indicating that the red shift of the 4.8 eV peak ascribes to the melting of Zn NPs.

4.5 Metal Nanoparticles with Two or More Resonances

As shown in Fig. 4.18b, Zn NPs show two absorption peaks at 1.2 and 4.8 eV at room temperature [56]. From the thermal stability, the resistivity to oxidation, and the compatibility with experimentally determined dielectric function, both the peaks are ascribed to intrinsic, i.e., not due to impurity effect [56]. However, still a question remains, i.e., which peak is the SPR of Zn NPs. To answer this question, the first principle (*ab-initio*) band calculation code WIEN2k [57] was applied [58],

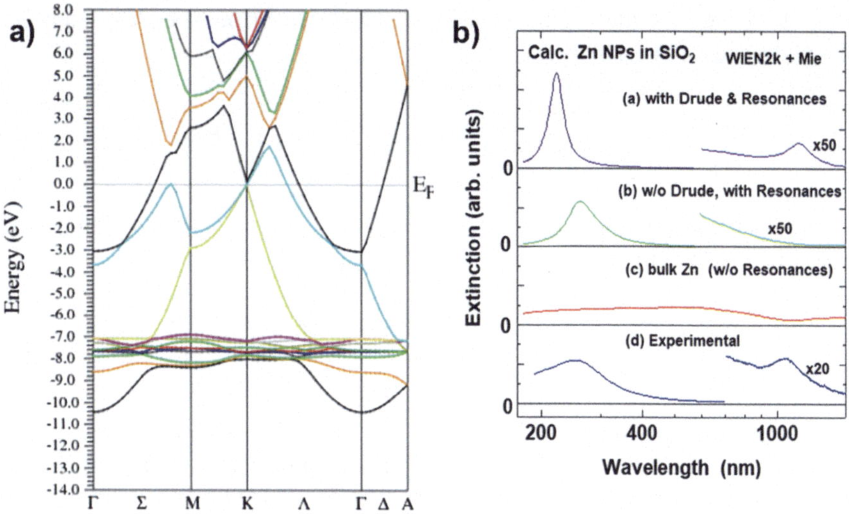

Fig. 4.19 (**left**) Energy band diagram of bulk Zn calculated by WIEN2K code. Inner-core bands of 3 s, 3p, and lower are out of the scale. (**right**) **a** calculated optical absorption spectra using the dielectric function derived from the WIEN2K. **b** The Drude term was excluded. The bulk formula was used (**c**). The experimental result for reference (**d**). Reproduced with permission from [58], ©2017 Institute of Physics

which was based on the density functional theory (DFT) with the local density approximation using the full potential linearized augmented plane wave method. This code calculated the dispersion relations of the electronic bands of bulk Zn, as shown in Fig. 4.19 (left). The dielectric function was calculated from the band diagram using the Lindhard formula with the random phase approximation [59], which is given as a sum of two parts, i.e., the inter-band and the intra-band components. The inter-band term is given as,

$$\varepsilon_{ij}^{\text{intra}}(\omega) = \sum\nolimits_{n,n'} \frac{\hbar^2 e^2}{\pi m^2 \omega^2} \int_k p_{i;n',n,k} p_{j;n',n,k} \left(f_0(\varepsilon_{n,k}) - f_0\left(\varepsilon_{n',k}\right)\right) \delta\left(\varepsilon_{n',k} - \varepsilon_{n,k} - \omega\right) \tag{4.7}$$

where $p_{i;n',n,k}$, f_0, and $\varepsilon_{n,\mathrm{k}}$ denote the momentum matrix element between states of bands n and n' with momentum **k**, the Fermi distribution function, and the single particle energy in the band n and **k**, respectively. The intra-band component is given with assuming the Drude form to avoid the singularity problem,

$$\varepsilon_{ij}^{\text{intra}}(\omega) = 1 - \frac{\omega_{p;i,j}^2}{\omega^2 + i\omega\Gamma} \tag{4.8}$$

where Γ denotes the lifetime broadening, which is a free parameter. The (squared) plasma frequency is calculated as,

$$\omega_{p;i,j}^2 = \frac{\hbar^2 e^2}{\pi m^2} \sum_n \int_k p_{i;n',n,k} p_{j;n,n,k} \delta(\varepsilon_{n,k} - \varepsilon_F) \tag{4.9}$$

The optical absorption spectra of Zn NPs in SiO_2 were calculated with the 1st order Mie theory [60],

$$\gamma(\omega) = \frac{18\pi f \left(\varepsilon^{SiO_2}\right)^{3/2}}{\lambda_0} \frac{\varepsilon_2^{Zn}}{\left(\varepsilon_1^{Zn} + 2\varepsilon^{SiO_2}\right)^2 + \left(\varepsilon_2^{Zn}\right)^2} \tag{4.10}$$

The calculated spectrum is shown in Fig. 4.19 right (a). Compared with the experimental result (d), the UV peak (~220 nm) and the NIR peak (~1100 nm) shift to shorter and longer wavelengths, respectively, which could indicate that the ab-initio calculations are not always the best. It is not theoretically guaranteed that the DFT calculations always predict the excited states precisely.

One of the advantages of the WIEN2K calculations is that the inter- and intra-band contributions can be separated. Figure 4.19b shows the calculated spectrum without the intra-band contribution. The NIR peak disappeared, indicating that the NIR peak is ascribed to the intra-band transitions, i.e., the free electron resonance, "the SPR in the narrow meaning."

Since the UV peak survived without the intra-band transitions, the peak comes from the inter-band transitions. However, the Eq. (4.10) is not free from the resonance effect. To clarify it, the absorption spectrum of the interband transitions was calculated from the relation,

$$\alpha = \frac{4\pi k}{\lambda} \tag{4.11}$$

where k is the imaginary part of the refractive index obtained from the calculated dielectric function. The result was a flat spectrum as shown in Fig. 4.18c, indicating that the UV peak is the inter-band transitions enhanced by the resonance effect. Gartz and Quinten observed two absorption peaks from Y NPs, and pointed out the possibility which we have proven using the WIEN2K code. They call the higher energy resonance as "electronic resonance" [36]. Two absorption peaks have also been observed in Ni NPs in SiO_2 [61], which could be ascribed to the SPR and the electronic resonance.

Two or more resonances of metal NPs could be relatively common phenomena: The Drude dielectric function is described and approximated in the clean limit:

$$\varepsilon_1(\omega) = 1 - \frac{\omega_p^2}{\omega^2 + i\omega\Gamma_0} \approx 1 - \frac{\omega_p^2}{\omega^2} \tag{4.12}$$

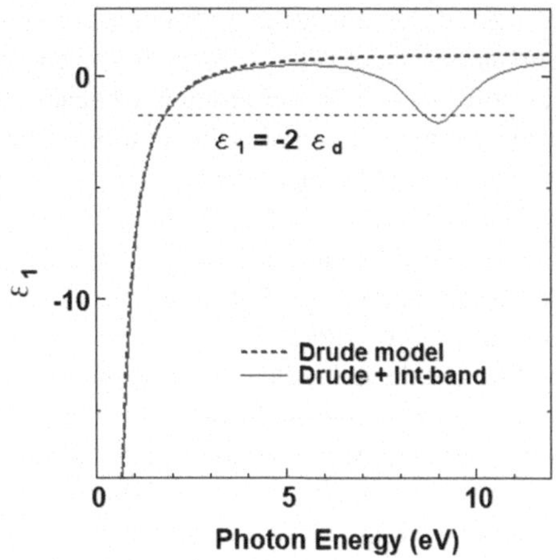

Fig. 4.20 Broken and solid curves denote the Drude dielectric function and its modification by the inter-band transition. A horizontal dotted line denotes the simplified SPR condition, $\varepsilon_1(\omega) = -2\varepsilon_d$

As indicated by the most right-hand side of the equation, the dielectric function $\varepsilon_1(\omega)$ is a monotonically increasing function of frequency ω, as shown by a blue broken curve in Fig. 4.20. The resonance is induced when the relation $\varepsilon_1(\omega) = -2\varepsilon_d$ is hold, where ε_d is the dielectric function of the matrix. Since ε_d does not strongly depend on the photon energy, it gives a nearly horizontal line as shown in Fig. 4.20. The intersection between them provides the resonance. (While this approximated criterion for the SPR is accepted in some cases, it could be too simplified and make confusions in other cases [58, 61]) Since the Drude (real) dielectric function (4.12) is a monotonically increasing function of the photon energy ω, the intersection happens only once. However, the real dielectric function can bend and sometimes have peaks. The 2nd (and possibly 3rd) intersections are possible in the optical frequencies as shown in Fig. 4.20. However, the 3rd SPRs have been rarely observed, which is probably due to enhancement in the imaginary part $\varepsilon_2(\omega)$. The enhanced imaginary part largely broadens the peaks to invisible.

4.6 Magnetic Nanoparticles and Magneto-Optical Effects

4.6.1 Superparamagnetic Nanoparticles

Only three bulk element metals, Fe, Co, and Ni, are ferromagnetic at room temperature. (Gd is ferromagnetic up to 16 °C). Bulk ferromagnetic materials consist of multi magnetic domains. While the formation of the domain walls requires energy loss, the existence of the magnetic domains having different orientation results in reduction of the total static magnetic energy. The static magnetic energy and the domain-wall formation energy compete with each other, which are proportional to d^3 and d^2, respectively, where d is the size of the domains. With decreasing the size *d*, a mono-domain structure becomes more stable than the multi-domain structures. This is observed as an increase in the coercive force. Reversal of magnetization is easier for multi-domain structures, because this is attained by the shift of the domain walls. Contrary, it requires the rotation of the magnetic moment in the mono-domain, which demands higher energy.

With further decrease of the size of NP, the coercive force again decreases due to the thermal fluctuation of the magnetization of the mono-domains, i.e., super-paramagnetism. For simplicity, a mono-domain NP with uniaxial magnetic anisotropy is considered.

Without the field, the magnetization points to the direction of the easy magnetization. In order to induce the magnetic inversion, the magnetic anisotropy energy $E_A = KV$ is required, where K and V denote the magnetic anisotropy constant and the NP volume, respectively. With decreasing the size, E_A decreases. When E_A is comparable to thermal energy kT, the magnetization inversion is induced by thermal fluctuation, as if they were paramagnetic. While each atom consisting of the NP exchanges the ferromagnetic interaction with other Ni atoms in the NPs, the NPs themselves behave like paramagnetic. This phenomenon is called super-paramagnetism.

Since a mono-domain NP consists of 10^2–10^5 atoms, the magnetic moment per NP is huge. This is a big difference between the paramagnetism and the super-paramagnetism. According to quantum mechanics, the expectation value of the square of the magnetic moment $<\psi| L^2|\psi>$ is $L(L + 1)$, but not L^2. However, in the super-paramagnets, $<\psi| L^2|\psi>$ is well approximated by L^2, because $L \gg 1$. Consequently, the super-paramagnets can be described by the classical theory.

Figure 4.21 shows the field dependence of magnetization of Ni NPs in silica monitored by SQUID magnetometer, which were fabricated with 60 keV Ni ion implantation without post-implantation annealing [62]. In Fig. 4.21a, the magnetization of the implanted and of un-implanted samples were shown. Always strong

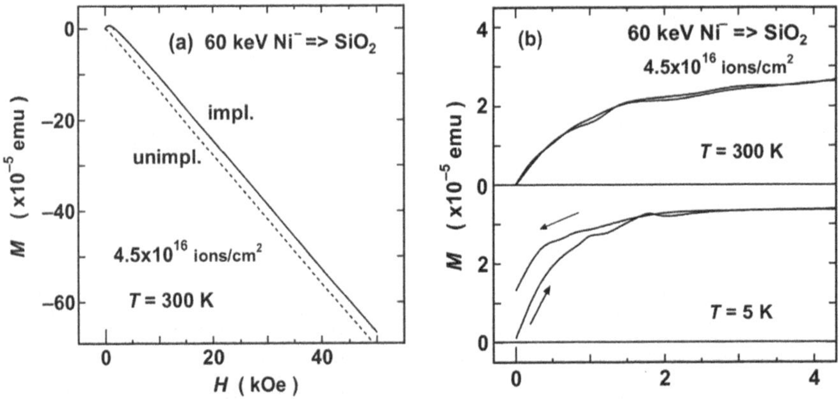

Fig. 4.21 Field dependence of magnetization of Ni NPs in silica, which were formed by 60 keV Ni⁻ implantation. **a** Implanted and unimplanted. **b** Measured at 5 and 300 K. The magnetization of silica substrate was subtracted. Reproduced with permission from [62], ©2004 Elsevier

diamagnetic signal from the silica substrate was observed, in addition to weak but non-negligible signal from Ni NPs. The Ni NPs exist only in the surface layer of ~100 nm thick but the silica sample was 0.5 mm thick. Even though the large volume ratio, the signal from Ni NPs was observed. This fact indicates much higher magnetization per atom of the ferromagnetic NPs compared to that of the diamagnetic silica substrate.

Magnetization curves only from the Ni NPs at 5 and 300 K are shown in Fig. 4.21b after subtraction of the silica signal. While any magnetic hysteresis was not observed, i.e., super-paramagnetic, within the experimental errors at 300 K, a clear ferromagnetic hysteresis was observed at 5 K. The field dependence of Ni NPs was compared at RT with that of ferromagnetic Ni continuous film of ~120 nm thick as shown in Fig. 4.22 [63]. While the Ni film exhibits a square-shaped hysteresis, Ni NPs exhibit an S-shape curve without hysteresis within the experimental error, which is expected from super-paramagnetism.

Temperature dependences of the magnetization with field cooling (FC) of 300 Oe and zero field cooling (ZFC) are shown in Fig. 4.23. A clear deviation was observed between them below the blocking temperature of ~50 K. Below the blocking temperature, the flip of the magnetic moment in the NPs cannot follow the applied field, because of the reduction of thermal fluctuation.

Fig. 4.22 Hysteresis loops of a Ni film on SiO_2 and Ni NPs in SiO_2 detected as Kerr ellipticity. Reproduced with permission from [63], ©2004 MRS-Japan

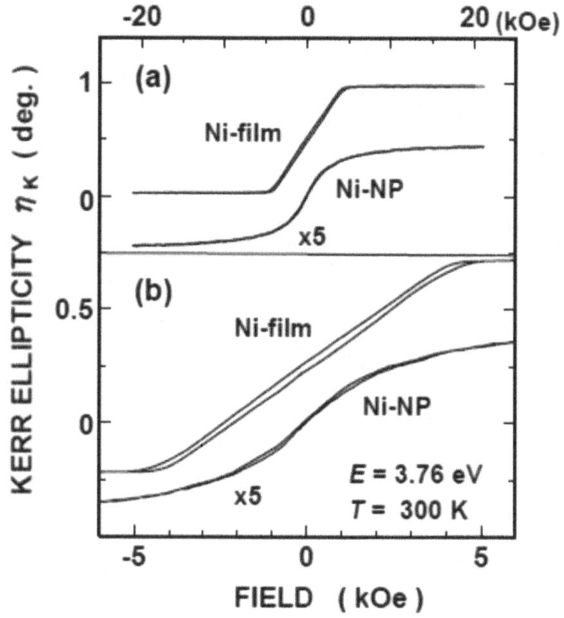

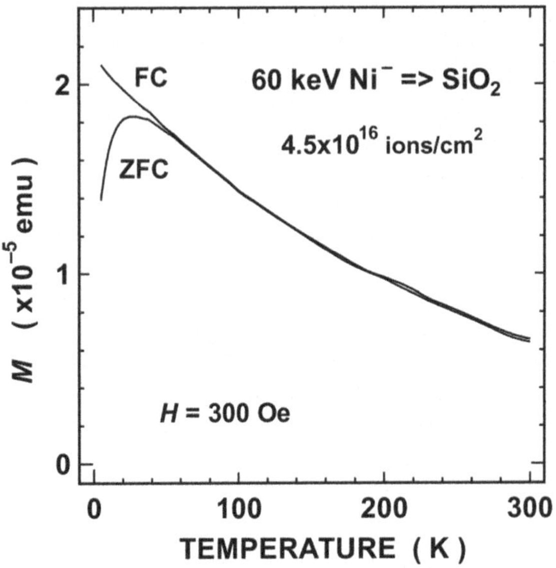

Fig. 4.23 Temperature dependence of magnetization of Ni NPs in silica at 300 Oe. FC: field cooling, and ZFC: zero-field cooling. Reproduced with permission from [62], ©2004 Elsevier

4.6.2 Magneto-Optical Kerr Effect

When magnetic field is applied to a magnetically isotropic medium, along, say z-direction, (i) off-diagonal components ε_{xy} appear and (ii) the zz component is no longer the same as xx and yy components.

$$\begin{pmatrix} \varepsilon_{xx} & 0 & 0 \\ 0 & \varepsilon_{xx} & 0 \\ 0 & 0 & \varepsilon_{xx} \end{pmatrix} \xrightarrow[H\neq 0]{} \begin{pmatrix} \varepsilon_{xx} & \varepsilon_{xy} & 0 \\ -\varepsilon_{xy} & \varepsilon_{xx} & 0 \\ 0 & 0 & \varepsilon_{zz} \end{pmatrix} \tag{4.13}$$

The former (i) induces the Faraday effect and the Kerr effect, while the latter (ii) induces the Cotton-Mouton effect [64]. The Faraday effect and the (magneto-optical) Kerr effect are induced in the transmission and the reflection configuration, respectively. When linearly polarized light is incident, the polarization state of the transmitted/reflected light is modified: the polarization plane is rotated and simultaneously the linearly polarized light changes to elliptically polarized light. Here the Kerr effect of Ni NPs in SiO_2 is described. With measuring the rotation angle and the ellipticity of reflected light, the magnetization of NPs is monitored as shown in Fig. 4.24. Strong diamagnetic signal was observed in the Kerr rotation θ_K for the NP sample. Magnetization loops from Ni NPs were obtained with subtracting the diamagnetic signal. The diamagnetic signal was also visible in θ_K for the film sample. Contrary, the diamagnetic signal in the Kerr ellipticity η_K was negligible in both the NPs and the films. Figure 4.25 exhibits the Kerr ellipticity loops at various photon energies. The loop changes not only the magnitude but also the sign. However, every

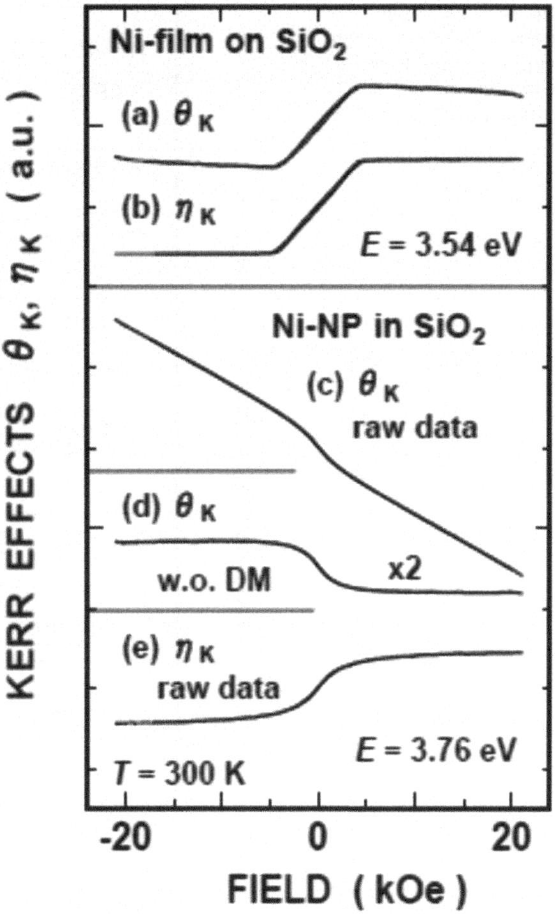

Fig. 4.24 Magnetization loops of Ni film on SiO_2 and Ni NPs in SiO_2, detected via Kerr rotation θ_K and ellipticity η_K. Diamagnetic contribution from SiO_2 is strong in θ_K. Reproduced with permission from [63], ©2004 MRS-Japan

loop shows saturation at high field. The Kerr rotation and ellipticity at the high field limit are extrapolated and plotted as circles in Fig. 4.26.

Regarding to optical absorption of non-magnetic metal NPs in glass, Ruppin [66] has succeeded in describing the spectrum using the Maxwell-Garnett (MG) [67] effective medium theory with the mean-free-path confinement (MFPC). He divided the experimentally determined dielectric function into two parts, i.e., the free electron contribution described by the Drude model and the residual part due to the bound electrons. He applied the MFPC to the Drude component and reproduced the experimental spectra well. Lissberger and Saunders (LS) [68] extended the MG theory to include the off-diagonal parts to calculate Kerr spectra of Co NPs in insulators. However, they approximated the dielectric tensor by the Drude type only, and neglected the bound electron contribution. Their theory failed since the bound electron component cannot be neglected in actual magnetic metals.

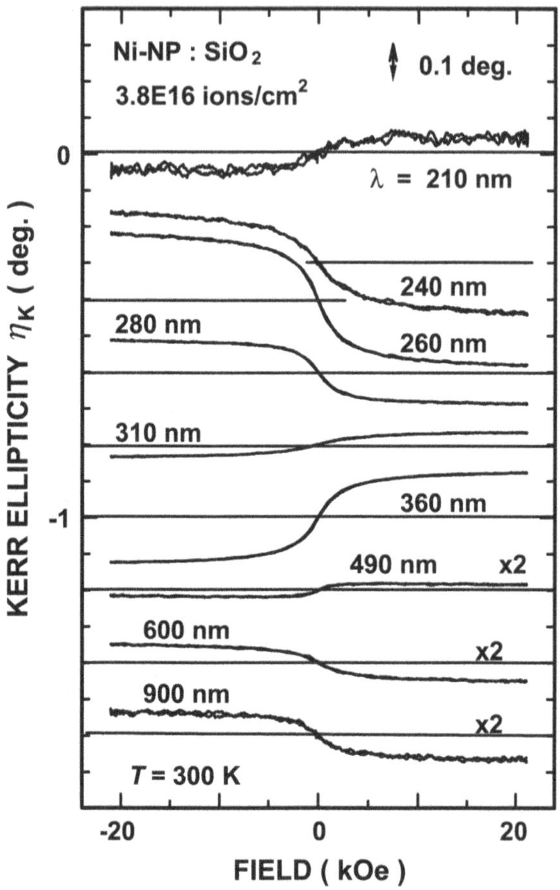

Fig. 4.25 Kerr ellipticity loops of Ni NPs at various photon energies. Reproduced with permission from [63], ©2004 MRS-Japan

Amekura et al. [65] improved the LS theory with including the bound electron component. The *xx* and *xy* components of the dielectric function of metal NPs are, respectively, given as,

$$\begin{aligned} \varepsilon_{m,xx}(d) &= \varepsilon_{f,xx}(d) + \varepsilon_{b,xx} \\ \varepsilon_{m,xy}(d) &= \varepsilon_{f,xy}(d) + \varepsilon_{b,xy} \end{aligned} \tag{4.14}$$

where the 1st and 2nd terms in the right-hand side of each equation denote the free and bound electron contributions, respectively. The off-diagonal free-electron component is given as,

$$\varepsilon_{f,xy}(d) = -i\frac{\omega\omega_c\omega_p^2}{\left\{\omega^2 + i[\omega/\tau(d)]\right\}^2 - \omega^2\omega_c^2} \tag{4.15}$$

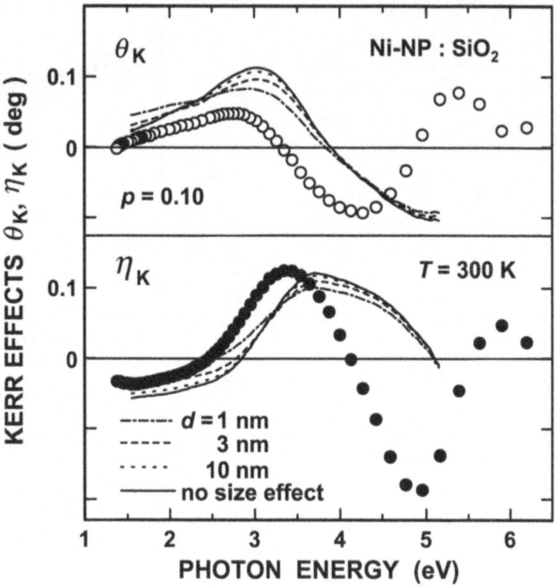

Fig. 4.26 Spectra of Kerr rotation and ellipticity of Ni NPs dispersed in SiO_2. Circles are experimental data and curves are calculations assuming different diameters. Reproduced with permission from [65], ©2004 Elsevier

where ω_c and ω_p denote cyclotron frequency and plasma frequency, respectively. The size-dependent relaxation time $\tau(d)$ is given as,

$$\tau^{-1}(d) = \tau_0^{-1} + \frac{2v_F}{d} \tag{4.16}$$

where v_F denotes the Fermi velocity of the meal. The dielectric tensor of the effective media is given as an extension of MG theory,

$$\begin{aligned} \varepsilon_{\mathrm{eff},xx} &= \varepsilon_{d,xx}\frac{\varepsilon_{m,xx}(1+2p)+2\varepsilon_{d,xx}(1-p)}{\varepsilon_{m,xx}(1-p)+2\varepsilon_{d,xx}(2+p)} \\ \varepsilon_{\mathrm{eff},xy} &= p\varepsilon_{m,xy}\left\{\frac{2\varepsilon_{d,xx}}{\varepsilon_{m,xx}(1-p)+2\varepsilon_{d,xx}(2+p)}\right\}^2 \end{aligned} \tag{4.17}$$

where p denotes volume fraction of the NPs. From the calculated dielectric tensor of the effective media, the Kerr rotation and ellipticity spectra of Ni NPs in SiO_2 are derived, and shown in Fig. 4.26 with the experimental results.

$$\begin{aligned} \theta_{\mathrm{eff},K} &= \frac{n\left(1-n_{\mathrm{eff}}^2+3k_{\mathrm{eff}}^2\right)\varepsilon'_{\mathrm{eff},xy}+k\left(1-3n_{\mathrm{eff}}^2+k_{\mathrm{eff}}^2\right)\varepsilon''_{\mathrm{eff},xy}}{\left(n_{\mathrm{eff}}^2+k_{\mathrm{eff}}^2\right)\left\{\left(1-n_{\mathrm{eff}}^2-k_{\mathrm{eff}}^2\right)+4k_{\mathrm{eff}}^2\right\}} \\ \eta_{\mathrm{eff},K} &= \frac{-k\left(1-3n_{\mathrm{eff}}^2+k_{\mathrm{eff}}^2\right)\varepsilon'_{\mathrm{eff},xy}+n\left(1-n_{\mathrm{eff}}^2+3k_{\mathrm{eff}}^2\right)\varepsilon''_{\mathrm{eff},xy}}{\left(n_{\mathrm{eff}}^2+k_{\mathrm{eff}}^2\right)\left\{\left(1-n_{\mathrm{eff}}^2-k_{\mathrm{eff}}^2\right)+4k_{\mathrm{eff}}^2\right\}} \end{aligned} \tag{4.18}$$

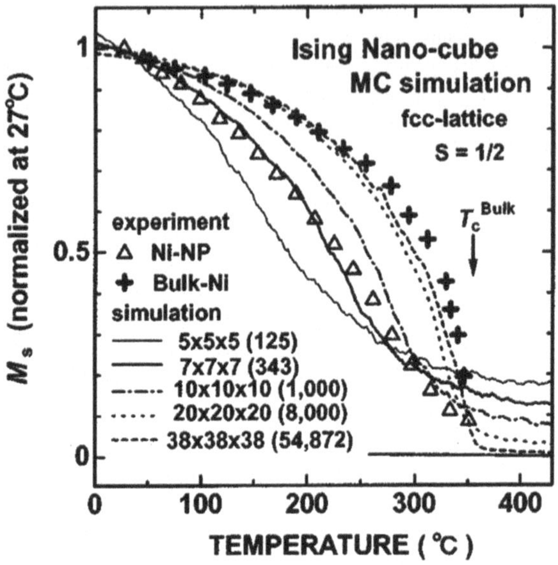

Fig. 4.27 Temperature dependence of magnetization of Ni NPs (triangles) and of bulk Ni (crosses). Curves are simulation results from Ising nano-cubes including spins of d^3, where $d = 5, 7, 10, 20,$ and 38. Reproduced with permission from [69], ©2005 American Physics Society

See Ref. [65] for details of the calculation.

4.6.3 Curie Transition Modified by the Finite-Size Effect in Magnetic Nanoparticles

Monitoring the Kerr rotation and ellipticity, temperature dependences of magnetization of Ni NPs and Ni film up to the Curie points were determined as shown in Fig. 4.27 [69]. The dependence of Ni film was almost the same as literature data of bulk Ni. The Ni NPs show different temperature dependence from the bulk, whose dependence is reproduced by the Ising nano-cubes ($s = 1/2$). A Ni atom has a spin of 1/2. The Curie transition of Ni NPs is highly modified as a phase transition of finite-size system [69].

4.7 Some Phenomena to Be Noted

4.7.1 Inter-particle Interaction Between Nanoparticles and Appearance of a New Band

Liu et al. implanted a single crystal of SiO_2 with 200 keV Ag^+ ions to various fluences between 5×10^{15} and 9×10^{16} ions/cm^2. The ion flux was about 20 μA/cm^2. They observed an appearance of a new peak around 600 nm in wavelength with increasing the fluence, in addition to the SPR peak of Ag NPs around 400 nm as shown in Fig. 4.28a–e [70, 71]. While they used crystalline SiO_2 for implantation substrates, the amorphization of the substrates are sure because of the high fluences. In fact, the 600 nm absorption band has been observed also in amorphous SiO_2 implanted with high fluence Ag ions as shown in Fig. 4.29 [72].

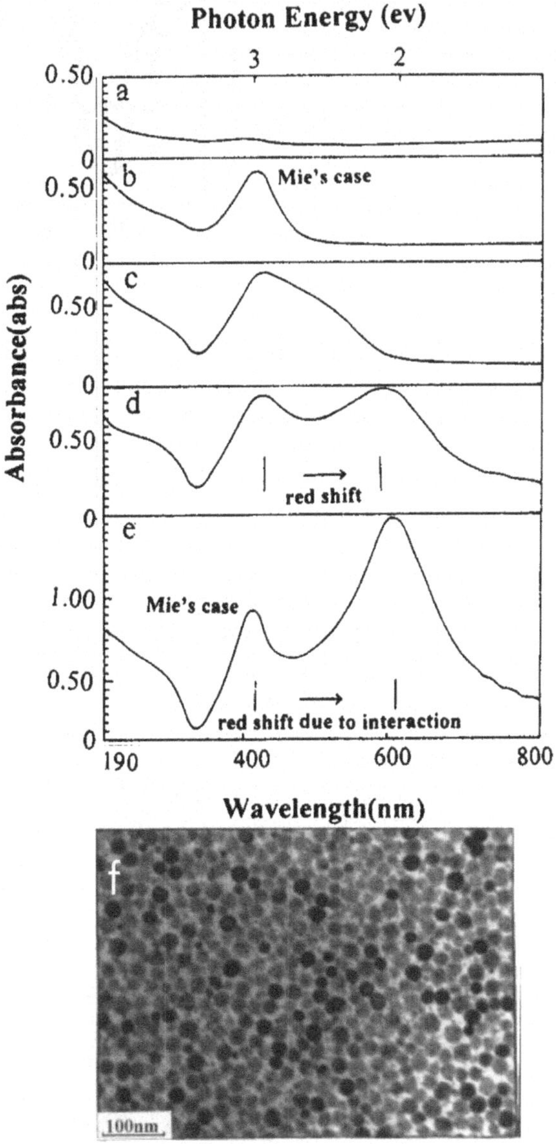

Fig. 4.28 Optical absorption spectra of SiO_2 implanted with 200 keV Ag ions to the fluences of **a** 0.5, **b** 2.3, **c** 4.5, **d** 5.6, and **e** 6.7 × 10^{16} ions/cm^2. **f** A planar TEM image of Ag NPs in SiO_2 at the fluence of 6.7 × 10^{16} ions/cm^2. Reproduced with permission from [70], ©1998 American Institute of Physics

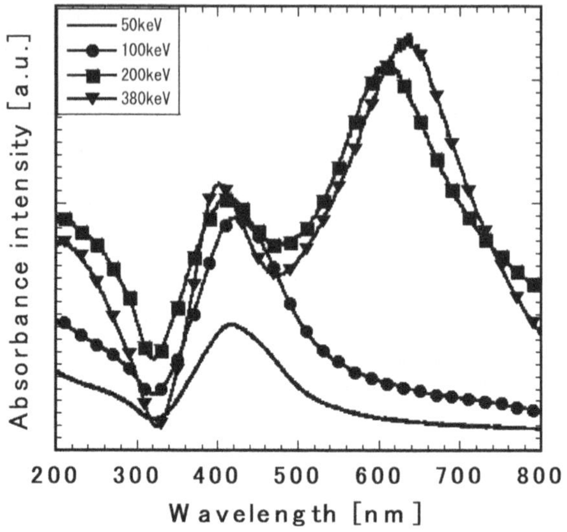

Fig. 4.29 Optical absorption spectra of SiO_2 implanted with Ag ions of four energies, 50, 100, 200, and 380 keV, to a fixed fluence of 7×10^{16} ions/cm^2. Reprint from Yamada et al. [72]

The origin of the 600 nm was ascribed to the multipole interaction induced by the extremely high density of NPs as shown in planar TEM image of Fig. 4.28f, i.e., the inter-particle interaction. While the reason has not been clarified, high dense but very thin NP layer along the depth, i.e., say 2-dimensional NP layer, is formed around the ion-range depth, which was confirmed by cross-sectional TEM [72]. Yamada et al. implanted amorphous SiO_2 with Ag ions of four different energies, 50, 100, 200, and 380 keV up to 7×10^{16} ions/cm^2 as shown in Fig. 4.29 [72]. The 600 nm band and the 2D-NP layer were confirmed at 200 and 380 keV but not for the lower energies. Particularly XTEM images of the 50 keV implantation indicated relatively high sputtering loss. One of the scenarios is that the 2D-NP layer is formed only when the ions are implanted deep enough to avoid the sputtering. However, this scenario explains why the 2D-NP layers are not formed under 50 and 100 keV implantation, but does not explain why the 2D-NP layers are formed under the high energy implantations.

The 600 nm band was semi-quantitatively reproduced by the model developed by Garcia et al. [73], who introduced the effect of the field from adjacent NPs to the Maxwell-Garnett (MG) effective medium theory [67]. The local field acting on the NP, E_{loc}, is described as,

$$\begin{aligned} E_{\text{loc}} &= E_{\text{ext}} + \frac{P}{3\varepsilon_m} + \frac{1}{4\pi\varepsilon_m}\sum_j\left[\frac{3x_{ij}^2 P_{jx}}{r_{ij}^5} - \frac{P_{jx}}{r_{ij}^3}\right] \\ &= E_{\text{ext}} + \frac{P}{3\varepsilon_m} + \frac{K}{4\pi\varepsilon_m}P \end{aligned} \tag{4.19}$$

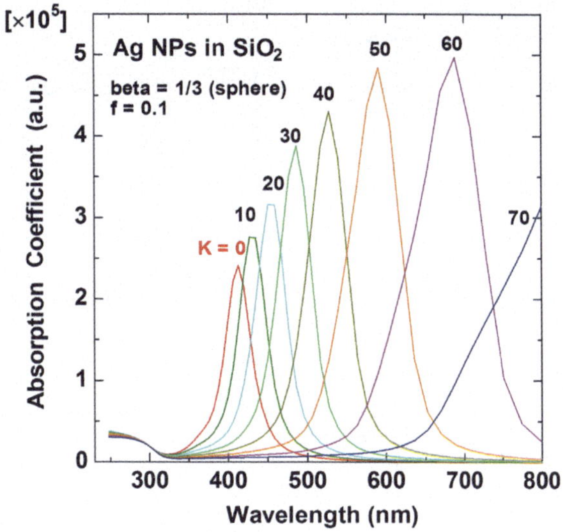

Fig. 4.30 Calculated absorption spectra of Ag NPs in SiO_2 for various particle interaction *K*. Reprint from Yamada et al. [72]

where the first, second, and third term in the right-hand side of the equation denote the external field, the Lorenz field, and the contribution from the adjacent NPs, respectively. See [73] for details. Hereafter the third term is parameterized as *K*. Spectra for various *K* values are calculated by the Garcias' theory and shown in Fig. 4.30. At $K = 0$, the peak matches with the value of the isolated Ag NPs. With increasing *K*, the peak shifts to longer wavelength. At $K \sim 50$, the observed 600 nm band was reproduced. An important finding is that only one peak is observed at any given value of *K*. However, as shown in Fig. 4.28a–e, experimentally two peaks are observed, indicating that two different environments exist for NPs, i.e., the 2D-NP layer with high inter-particle interaction and isolated NPs with almost no inter-particle interaction.

4.7.2 High Flux Implantation Effects

Kishimoto et al. implanted SiO_2 with 60 keV Cu^- ions at various fluxes between 1 and 100 $\mu A/cm^2$, with maintaining the total fluence to 3×10^{16} ions/cm^2, and found that depth distribution of Cu NPs changed with the flux, while fixing the implantation energy to 60 keV [74]. Figure 4.31 shows the depth profiles of Cu atomic density determined from XTEM images. These observations indicate that certain portion of implanted Cu atoms are still in isolated state, and have not been absorbed in the NPs at the flux of 1 $\mu A/cm^2$. With increasing the flux to 3 $\mu A/cm^2$, the aggregation ratio increases almost to 100%, probably due to enhanced diffusion with increasing the flux. Further increase of the flux to 10 $\mu A/cm^2$ shifts the depth profile to the surface side. At the flux of 45 $\mu A/cm^2$, the depth profile moved further to the surface side,

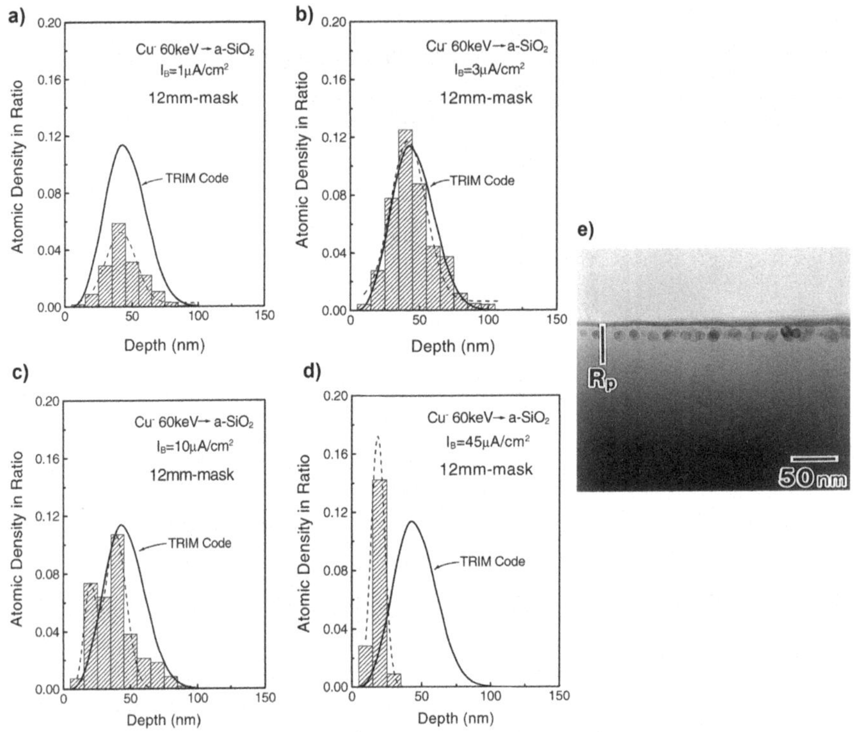

Fig. 4.31 **a–d** Depth profiles of Cu atomic density in SiO_2 implanted with 60 keV Cu ions to 3×10^{16} ions/cm^2 at four different ion fluxes, 1, 3, 10, and 45 μA/cm^2. Solid curves are from TRIM code. **e** Cross-sectional TEM image of a sample with the flux of 45 μA/cm^2. Reproduced with permission from [74], ©2000 Elsevier

and Cu NPs exist within a very thin layer, i.e., the formation of 2-dimensional NP layer. The corresponding XTEM image is shown in Fig. 4.31e [74]. While it is not shown in the figure, a sample implanted with the highest flux of 100 μA/cm^2 to 3×10^{16} ions/cm^2 showed no longer any Cu atoms inside, which was confirmed by RBS [75].

Figure 4.32 shows an XTEM images of SiO_2 implanted with 60 keV Ni^- ions to 3.8×10^{16} ions/cm^2 at high flux of 56 μA/cm^2, instead of 60 keV Cu ions. While the ion energy, the flux, the fluence, and the ion mass are almost the same between 60 keV Cu and 60 keV Ni implantations, the depth profile of Ni NPs does not show any shift toward the sample surface, in spite of high flux of 56 μA/cm^2 comparable to the Cu ions. It seems that the shallowing phenomenon under high flux implantation depends on the NP species, or chemistry between NP species and the matrix. The mean size of the NPs determined from Figs. 4.31 and 4.32 are ~10 nm (Cu) and ~3 nm (Ni), indicating higher diffusion constant of Cu atoms in SiO_2 under high flux implantation than Ni atoms.

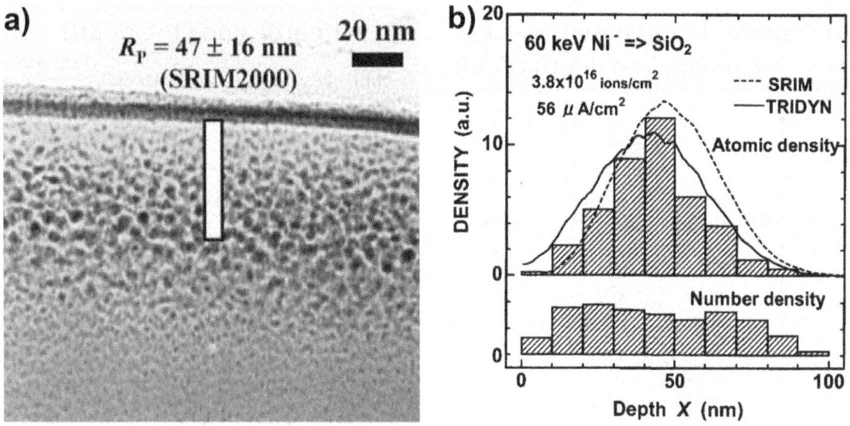

Fig. 4.32 **a** XTEM images of SiO_2 implanted with 60 keV Ni ions to 3.8 × 10^{16} ions/cm^2 at the flux of 56 μA/cm^2. **b** Depth profiles of Ni atomic density determined from Fig. 4.32a. Reproduced with permission from [62], ©2004 Elsevier

This kind of the 2D-NP layers have been also observed by other groups. For example, Ren et al. observed the 2D layer of Ag NPs with 200 keV Ag ions to 5 × 10^{16} ions/cm^2 at 2.5 μA/cm^2 [76]. The previously mentioned Ag NP layer with showing the 600 nm band is also in this case. Liu et al. implanted with 200 keV Ag ions up to 9 × 10^{16} ions/cm^2 at the flux of ~20 μA/cm^2 [72]. It seems that the 2D layer is formed under high ion energies and/or high fluxes.

While the flux of tens μA/cm^2 is practically high, it is still in the static regime where the next ion is implanted after the previously implanted ion comes to rest. The interaction between moving two implanted ions is negligible. Consequently, the most possible mechanism could be heating. However, the strong flux dependence was observed in Cu NPs but not in Ni NPs. (See Figs. 4.31 and 4.32). Chemistry (or diffusion) could play a role. To evaluate the possibility of heating, implanted samples were thermally annealed in a furnace between 873 and 1273 K [77]. However, directional motion of Cu NPs toward the sample surface was not observed except at 1273 K. At the highest annealing temperature of 1273 K, all the Cu NPs disappeared.

From the thermal annealing experiments, homogenous temperature rise in the sample is not the driving force of the profile shallowing. However, as shown in Fig. 4.33 (right), the energy depositions from the electronic and nuclear stopping are not homogenous along the depth. Rather, the peaks of both the energy depositions locate shallower than the peak of the implants. Therefore, the shallower half of the implants are excited higher than the deeper half of the implants. A directional diffusion could be induced.

In addition, NPs prefer to aggregate because of energetically favorable. As shown in Fig. 4.33 (left), the ions were implanted following the depth profile predicted by SRIM code. After then, because of enhanced diffusion due to higher energy or higher flux, diffusion of implants is enhanced to coarsen the NPs. However, the coarsening

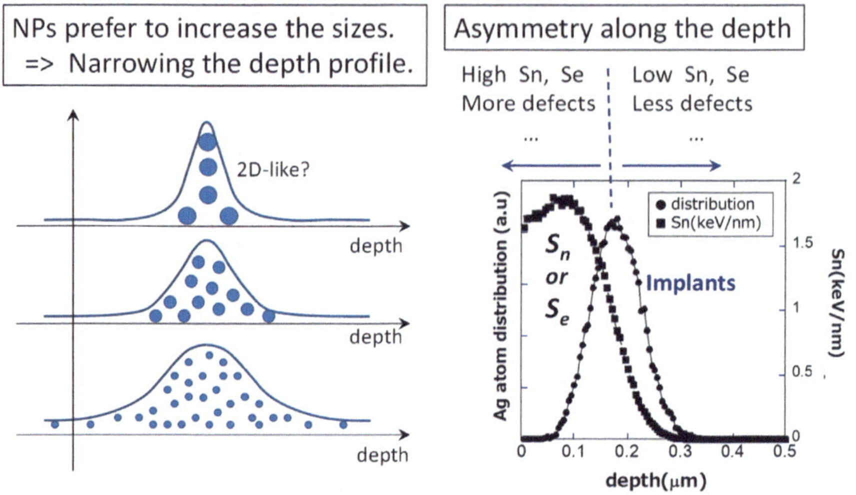

Fig. 4.33 Possible two factors inducing the nearly 2-dimensional NP distribution and shallowing of the depth profiles

of the NPs could result in thinning of the NP layer as schematically depicted in Fig. 4.33 (left). Finally, the 2-D NP layer could form.

4.7.3 Laser Co-irradiation Effects on Nanoparticle Formation

Kishimoto and coworkers studied simultaneous irradiation effects of pulsed YAG laser and 3 MeV Cu^{2+} ions to SiO_2 on the formation of Cu NPs. They carried out three different modes of irradiations. Ion: ion only irradiation for reference, Seq: sequential irradiation, i.e., laser irradiation after terminating the ion irradiation. Co: co-irradiation, i.e., a sample was simultaneously irradiated with the ion beam and the laser at the same time. The fluence of the ion irradiation was set to 3×10^{16} ions/cm^2, where Cu NPs were not observed by TEM and no SPR, when the sample is irradiated with the ion beam only. (Please don't confuse with the 60 keV Cu implantation in the previous sub-section. The fluence of 3×10^{16} ions/cm^2 is enough for Cu NP formation for the low energy 60 keV, but too low for NP formation under 3 MeV implantation. The ion range, i.e., the implanted Cu concentration is much different.) The second- (2.3 eV) and third- (3.5 eV) order harmonics from YAG laser was used. The laser pulse was nano-second width and the repetition was 10 Hz. The intensity was changed between 0.05–0.2 J/cm^2 per pulse.

Figure 4.34a shows XTEM images of the three different irradiation modes using the second harmonics laser line of 2.3 eV. While the ion-only irradiation does not form any Cu NPs, small NPs were observed after the sequential irradiations. After the

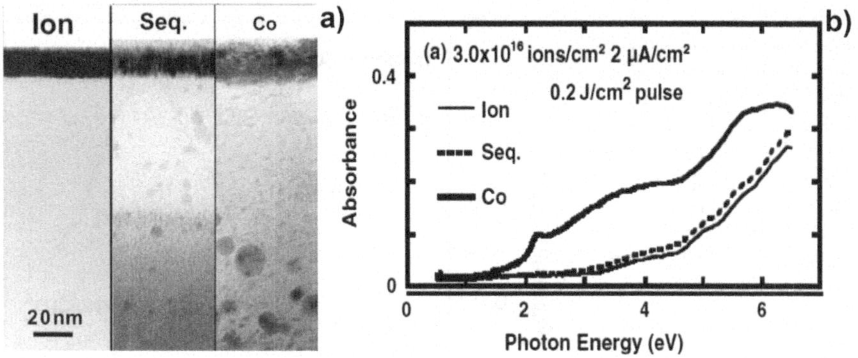

Fig. 4.34 **a** XTEM images of SiO_2 implanted with 3 MeV Cu^{2+} ions to 3×10^{16} ions/cm^2, in three different irradiation modes described in text. **b** Optical absorption spectra of SiO_2 implanted with 3 MeV Cu^{2+} ions to 3×10^{16} ions/cm^2, in three different irradiation modes described in text. Reproduced with permission from [78], ©2003 Elsevier and [79], ©2004 Elsevier

co-irradiation of laser and ions, many large Cu NPs were observed. The difference between the sequential irradiation and the co-irradiation was more clearly shown in the absorption spectra shown in Fig. 4.34b. While the co-irradiation forms the clear SPR at 2.2 eV, either of the sequential irradiation or ion-only irradiation only exhibits almost the same spectrum without SPR. The spectra without SPR is similar to the Cu-implanted sample at liquid nitrogen temperature as shown in Fig. 4.6, where the Cu ions exist but not form the Cu NPs.

Since the duration of the laser pulses was less than 10 ns with the repetition rate of 10 Hz, the duration of the co-irradiation of the ion beam and laser was very limited compared to whole the irradiation period. Even though the co-irradiation enhances the formation and growth of the NPs.

The same experiments were carried out using the third-order harmonics of the YAG laser (3.5 eV). However, the enhancement of NP formation was very weak.

An arising question was what absorbed the 2.3 eV laser light. Before the irradiation, SiO_2 is completely transparent around this energy region. Even when Cu atoms are implanted, they do not absorb photons of this energy region unless they form metallic NPs as shown in Fig. 4.4. Transient optical absorption spectroscopy was carried out at an irradiation chamber of the accelerator facility during the 3 MeV Cu^{2+} ion irradiation. The results are shown in Fig. 4.35: Before the irradiation, there is no absorption in SiO_2. Surprisingly, a broad absorption peak centered at 2.4 eV appears only when the sample is irradiated. It should be noted that the ion fluence was below 1×10^{16} ions/cm^2, i.e., the formation of Cu NPs with SPR is unlikely. After the interruption of the irradiation, the peak at 2.4 eV disappears with leaving a weak tail to higher energy.

Consequently, a transient state which absorbs the laser line of 2.3 eV appears only during the ion irradiation. The absorbed energy is probably transferred to the formation of Cu NPs. Therefore, Cu NPs are efficiently formed only under co-irradiations of the 2.3 eV laser and the Cu ion beam.

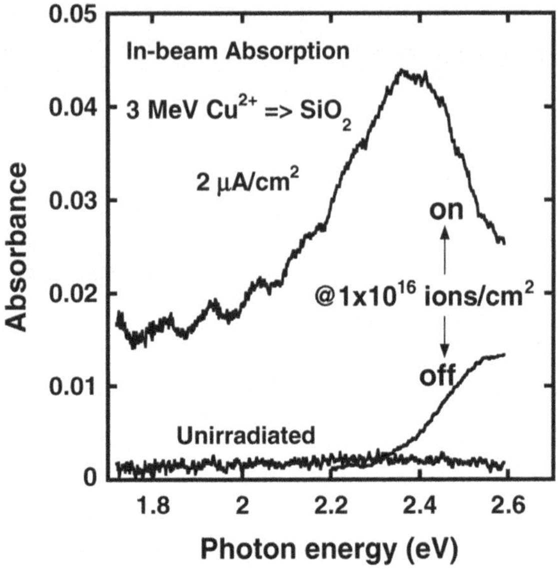

Fig. 4.35 Transient absorption spectra of SiO_2 during the irradiation with 3 MeV Cu^{2+} ions to a low fluence below 1×10^{16} ions/cm^2, where NP formation is not expected. Only a peak appears at 2.4 eV during the irradiation. Reproduced with permission from [80], ©2006 Elsevier

4.7.4 Single Electron Tunneling to Nanoparticles Embedded in Insulator

Metal NPs shallowly embedded in SiO_2 form metal-insulator-metal junctions between a surface electrode and a substrate electrode. When one additional electron stays in a NP, the charging energy of a junction is given as

$$E_C = \frac{e^2}{2C} \tag{4.20}$$

where e and C denote the elemental charge and the electric capacity of the junction, respectively. When the charging energy is higher than the thermal energy, i.e., with small capacitance and/or at low temperature, electron tunneling from the NP to the electrode is no longer allowed. With applying the voltage across the junction to $V_{\rm th} = \pm e/2C$, the current flow suddenly begins. This phenomenon is called "Coulomb blockade" [82, 83, 93].

Amekura et al. formed four different thicknesses of SiO_2 layers of 50, 60, 70, and 100 nm on highly conductive Si substrate by thermal oxidation. Then, they implanted the samples with 60 keV Cu^- ions to 3×10^{16} ions/cm^2 at the flux of ~ 30 μA/cm^2, to form nearly 2D Cu NP layer in the oxide layer (as shown in Fig. 4.31e). The current-voltage (I-V) characteristics were evaluated at room temperature (RT) and down to ~30 K. All the samples were conductive at RT, where the charging energy of the junction $E_C \ll kT$; i.e., the conditions for the Coulomb blockade is not satisfied. The samples of 70 and 100 nm thick turned to non-conductive at low temperatures. Contrary, the sample of 50 nm thick was still conductive even at 30 K. The Coulomb

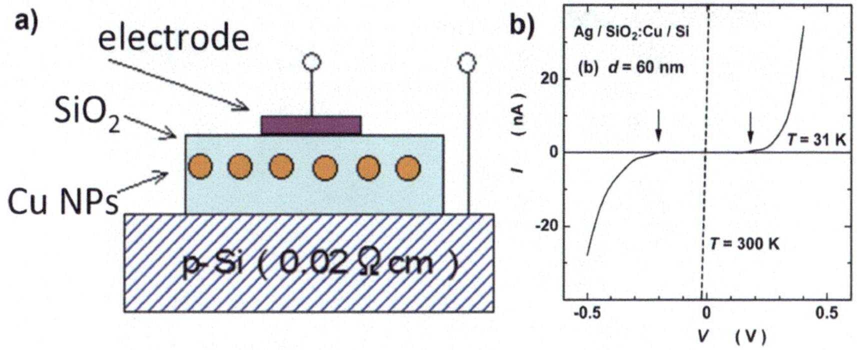

Fig. 4.36 **b** Current-voltage (*I*-*V*) characteristics of a structure of electrode/ insulator/ NP/ insulator/ metallic-Si as depicted (**a**). At 30 K, the Coulomb blockade is observed. Reprint from Amekura et al. [81]

blockade was observed in the sample of 60 nm thick at 30 K, as indicated by arrows in Fig. 4.36 [81]. At any voltages between the arrows, the current was nearly zero within the experimental error. Nakajima et al. formed shallow junctions via NPs of Sb [82] and of Sn [83] fabricated by low energy implantation of 10 keV, and observed the single electron tunneling behaviors [82, 83, 93].

4.7.5 Cavity Nanoparticles, Sandwiched Nanoparticles, and Nanoplanets

Ren and coworkers implanted silica samples with 200 keV Ag ions maintaining the beam current less than 2.5 μA/cm^2, which indicated that this phenomenon is not the high flux effect. Figure 4.37 exhibits planar TEM images at three different fluences [84]. While conventional spherical Ag NPs were observed at 5×10^{16} ions/cm^2, many small NPs were observed, which surrounded large mother NPs at 1×10^{17} ions/cm^2. This kind of the structure is called "core/satellite NPs" or "nanoplanets", which were studied by Padova group [85] and by Rizza et al. [6, 86] When two NPs are very close to each other, highly enhanced field is induced, which is called "hot spots". Since many satellite NPs locate very close to the mother, highly enhanced field is expected at many places [87] (Fig. 4.38).

In addition to the satellite NPs, cores of the mother NPs show weaker color compared with the periphery of the NPs, indicating that the cores are a different state rather than the conventional Ag NPs. First, the cores were considered as vacant. After the detailed studies [76], Ren et al. consider that the cores are filled by something but not vacuum, probably by silica. These NPs are called "cavity NPs."

At the highest fluence of 2×10^{17} ions/cm^2, a small Ag NP is formed inside of the cores of the NPs, which is called "sandwiched NPs". The formation process of the cavity NPs and sandwiched NPs are considered as purely ballistic ones. First, Ag

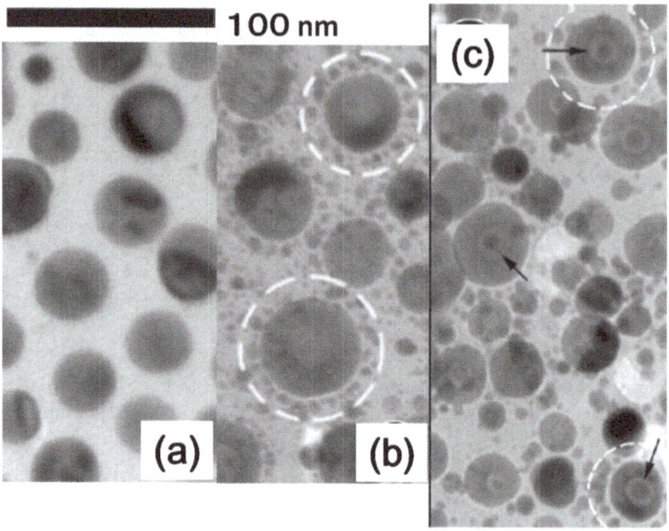

Fig. 4.37 Planar TEM images of SiO_2 implanted with 200 keV Ag ions to **a** 5×10^{16}, **b** 1×10^{17}, and **c** 2×10^{17} ions/cm^2. Cavity NPs and sandwich NPs are observed in (**b**) and (**c**), respectively. Reproduced with permission from [84], ©2006 American Physical Society

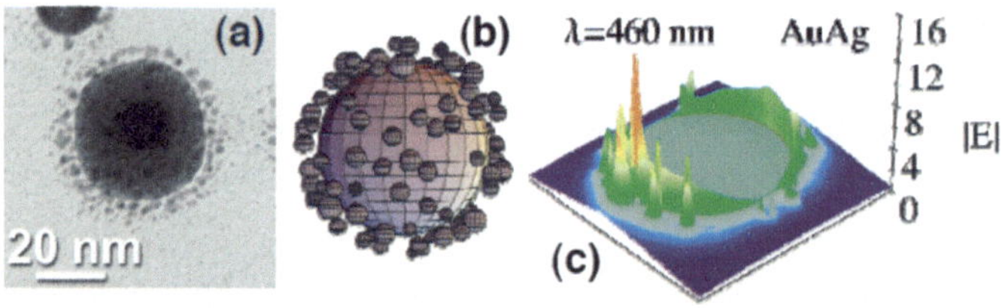

Fig. 4.38 **a** Experimental core/satellite NP formed by ion irradiation to AuAg NPs, **b** its corresponding model, **c** the calculated local field on a section in the equatorial plane at the SPR wavelength. Reproduced with permission from [87], ©2008 Elsevier

NPs are formed as shown in Fig. 4.37a. The Ag NPs were suffered by further ion irradiation, and much defects are introduced. Simultaneously, Si and O atoms are injected into the NPs via ballistic processes. Soon the solubility limit is overcome. The phase separation is induced as the cores and the shells, i.e., cavity NPs. Further irradiation injects Ag ions to the SiO_2 cores of the cavity NPs, which finally results in the aggregation of injected Ag ions in the SiO_2 cores, i.e., the sandwiched NPs.

4.7.6 Catalysis Nanoparticles

One of the other applications of NPs could be catalysis. Ni ions of 100 keV were implanted to thermally oxidized SiO_2 layer of 500 nm thick on Si substrate. Ni NPs were formed and enlarged by post-implantation annealing at 900 °C. After the annealing, the samples were dipped in 1% HF solution to remove the surface SiO_2 layer and to expose the NPs. Then the carbon nanotubes (CNTs) were formed by chemical vapor deposition (CVD) using the implanted substrates. Strong correlation was observed between the sizes of the tube structures and of the NPs [88]. Single-walled CNTs were formed from the NP substrates implanted with 60 keV Fe^+ ions [89, 90].

References

1. G.W. Arnold, Near-surface nucleation and crystallization of an ion-implanted lithia-alumina-silica glass. J. Appl. Phys. **46**, 4466 (1975)
2. A. Polman, Erbium implanted thin film photonic materials. J. Appl. Phys. **82**, 1 (1997)
3. C.J. McHargue, G.C. Farlow, P.S. Sklad, C.W. White, A. Perez, N. Kornilios, G. Marest, Iron ion implantation effects in sapphire. Nucl. Instrum. Methods Phys. Res. B **19/20**, 813 (1987)
4. A.L. Stepanov, R.I. Khaibullin, Fabrication of metal nanoparticles in sapphire by low-energy ion implantation. Rev. Adv. Mater. Sci. **9**, 109 (2005)
5. M. Strobel, K.-H. Heinig, W. Moeller, Understanding ion beam synthesis of nanostructures: Modeling and atomistic simulations. Mat. Res. Soc. Symp. Proc. **647**, O2.3 (2001)
6. G. Rizza, Y. Ramjauny, T. Gacoin, L. Vieille, S. Henry, Chemically synthesized gold nanoparticles embedded in a SiO_2 matrix: a model system to give insights into nucleation and growth under irradiation. Phys. Rev. B **76**, 245414 (2007)
7. O. Plaksin, Y. Takeda, H. Amekura, N. Kishimoto, Electronic excitation and optical responses of metal-nanoparticle composites under heavy-ion implantation. J. Appl. Phys. **99**, 044307 (2006)
8. J. Zhao, X. Chen, G. Wang, Critical size for a metal-nonmetal transition in transition-metal clusters. Phys. Rev. B **50**, 15424 (1994)
9. H. Amekura, H. Kitazawa, N. Kishimoto, Non-magnetic to magnetic and non-metal to metal transitions in nickel nanoparticles in SiO_2 under heat treatment. Nucl. Instrum. Methods Phys. Res. B **219**, 825 (2004)
10. R.H. Magruder III, R.A. Weeks, R.A. Zuhr, G. Whichard, Optical absorption of Cu implanted silica. J. Non-Cryst. Solids **129**, 46 (1991)
11. V. Ramaswamy, T.E. Haynes, C.W. White, W.J. Moberly-Chan, S. Roorda, M.J. Aziz, Synthesis of nearly monodisperse embedded nanoparticles by separating nucleation and growth in ion implantation. Nano Lett. **5**, 373 (2005)
12. W. Nie, Y. Zhang, H. Yu, R. Li, R. He, N. Dong, J. Wang, R. Hubner, R. Bottger, S. Zhou, H. Amekura, F. Chen, Plasmonic nanoparticles embedded in single crystals synthesized by gold ion implantation for enhanced optical nonlinearity and efficient Q-switched lasing. Nanoscale **10**, 4228 (2018)
13. H. Amekura, B. Johannessen, D.J. Sprouster, M.C. Ridgway, Amorphization of Cu nanoparticles: effects on surface plasmon resonance. Appl. Phys. Lett. **99**, 043102 (2011)
14. C. Maurizio, E. Trave, G. Perotto, V. Bello, D. Pasqualini, P. Mazzoldi, G. Battaglin, T. Cesca, C. Scian, G. Mattei, Enhancement of the Er^{3+} luminescence in Er-doped silica by few-atom metal aggregates. Phys. Rev. B **83**, 195430 (2011)

15. T. Cesca, B. Kalinic, N. Michieli, C. Maurizio, A. Trapananti, C. Scian, G. Battaglin, P. Mazzoldi, G. Mattei, Au-Ag nanoalloy molecule-like clusters for enhanced quantum efficiency emission of Er^{3+} ions in silica. Phys. Chem. Chem. Phys. **17**, 28262 (2015)
16. K. Maex, J. Vanhellemont, S. Petersson, A. Lauwers, Formation of ultra-thin buried $CoSi_2$ layers by ion implantation in (100) Si. Appl. Surf. Sci. **53**, 273 (1991)
17. S. Mantl, Ion beam synthesis of epitaxial silicides: fabrication, characterization and applications. Mater. Sci. Rep. **8**, 1 (1992)
18. H. Hosono, Simple criterion on collide formation in SiO_2 glasses by ion implantation. Jpn. J. Appl. Phys. **32**, 3892 (1993)
19. E. Cattaruzza, Quantum-dot composite silicate glasses obtained by ion implantation. Nucl. Instrum. Methods Phys. Res. B **169**, 141 (2000)
20. H. Amekura, M. Yoshitake, O.A. Plaksin, N. Kishimoto, Ch. Buchal, S. Mantl, Implantation-induced nonequilibrium reaction between Zn ions of 60 keV and SiO_2 target. Appl. Phys. Lett. **91**, 063113 (2007)
21. L.T. Canham, Silicon quantum wire array fabrication by electrochemical and chemical dissolution of wafers. Appl. Phys. Lett. **57**, 1046 (1990)
22. T. Shimizu-Iwayama, M. Ohshima, T. Niimi, S. Nakao, K. Saitoh, T. Fujita, N. Ito, Visible photoluminescence related to Si precipitates in Si^+-implanted SiO_2. J. Phys.: Condens. Matter **5**, L375 (1993)
23. T. Shimizu-Iwayama, D.E. Hole, P.D. Townsend, Light emission from ion beam induced silicon nanoclusters in silicon dioxide: role of cluster-cluster interactions via a thin oxide. Nucl. Instrum. Methods Phys. Res. B **148**, 980 (1999)
24. Y. Kanemitsu, T. Ogawa, K. Shiraishi, K. Takeda, Visible photoluminescence from oxidized Si nanometer-sized spheres: Exciton confinement on a spherical shell. Phys. Rev. B **48**, 4883 (1993)
25. M.L. Brongersma, A. Polman, K.S. Min, E. Boer, T. Tambo, H.A. Atwater, Tuning the emission wavelength of Si nanocrystals in SiO_2 by oxidation. Appl. Phys. Lett. **72**, 2577 (1998)
26. L. Pavesi, L.D. Negro, C. Mazzoleni, G. Franzo, F. Priolo, Optical gain in silicon nanocrystals. Nature **408**, 440 (2000)
27. J. Valenta, I. Pelant, J. Linnros, Waveguide effects in the measurement of optical gain in a layer of Si nanocrystals. Appl. Phys. Lett. **81**, 1396 (2002)
28. I. Pelant, Optical gain in silicon nanocrystals: current status and perspectives. Phys. Status Solidi A **208**, 625 (2011)
29. C.W. White, J.D. Budai, J.G. Zhu, S.P. Withrow, R.A. Zuhr, D. Hembree Jr., D.O. Henderson, A. Ueda, Y.S. Tung, R. Mu, R.H. Magruder, GaAs nanocrystals formed by sequential ion implantation. J. Appl. Phys. **79**, 1876 (1996)
30. H. Amekura, N. Umeda, Y. Sakuma, N. Kishimoto, Ch. Buchal, Fabrication of ZnO nanoparticles in SiO_2 by ion implantation combined with thermal oxidation. Appl. Phys. Lett. **87**, 013109 (2005)
31. H. Amekura, N. Kishimoto, Fabrication of oxide nanoparticles by ion implantation and thermal oxidation, in *Lecture Notes in Nanoscale Science and Technology*, vol. 5, ed. by Z. Wang (Springer, New York, 2009), pp. 1–75
32. C.W. White, J.D. Budai, S.P. Withrow, J.G. Zhu, E. Sonder, R.A. Zuhr, A. Meldrum, D. Hembree Jr., D.O. Henderson, S. Prawer, Encapsulated semiconductor nanocrystals formed in insulators by ion beam synthesis. Nucl. Instrum. Methods Phys. Res. B **141**, 228 (1998)
33. Y. Kanemitsu, T.J. Inagaki, M. Ando, K. Matsuda, T. Saiki, C.W. White, Photoluminescence spectrum of highly excited single CdS nanocrystals studied by a scanning near-field optical microscope. Appl. Phys. Lett. **81**, 141 (2002)
34. Y. Kanemitsu, H. Matsubara, C.W. White, Optical properties of Mn-doped CdS nanocrystals fabricated by sequential ion implantation. Appl. Phys. Lett. **81**, 535 (2002)
35. H. Amekura, M. Tanaka, Y. Katsuya, H. Yoshikawa, M. Ohnuma, Y. Matsushita, K. Kobayashi, N. Kishimoto, Melting-solidification transition of Zn nanoparticles embedded in SiO_2: observation by synchrotron X-ray and ultraviolet-visible-near-infrared light. J. Appl. Phys. **108**, 104302 (2010)

36. M. Gartz, M. Quinten, Broadening of resonances in yttrium nanoparticle optical spectra. Appl. Phys. B **73**, 327 (2001)
37. A. Ishizumi, C.W. White, Y. Kanemitsu, Space-resolved photoluminescence of ZnS:Cu, Al nanocrystals fabricated by sequential ion implantation. Appl. Phys. Lett. **84**, 2397 (2004)
38. G. Mattei, Alloy nanoclusters in dielectric matrix. Nucl. Instrum. Methods Phys. Res. B **191**, 323 (2002)
39. H. Hosono, Importance of Implantation sequence in the formation of nanometer size colloid particles embedded in amorphous SiO_2: formation of Composite colloids with Cu core and a Cu_2O shell by coimplantation of Cu and F. Phys. Rev. Lett. **74**, 110 (1995)
40. F. Ren, L.P. Guo, Y. Shi, D.L. Chen, Z.Y. WU, C.Z. Jiang, Formation of Zn–ZnO core-shell nanoclusters by Zn/F sequential ion implantation. J. Phys. D: Appl. Phys. **39**, 488 (2006)
41. M. Strobel, K.-H. Heinig, W. Moeller, Can core/shell nanocrystals be formed by sequential ion implantation? predictions from kinetic lattice Monte Carlo simulations. Nucl. Instrum. Methods Phys. Res. B **148**, 104 (1999)
42. H. Liu, X. Zhang, Y. Shen, L. Zhang, J. Wang, F. Zhu, B. Zhang, C. Liu, Tailoring the size distribution of Ag nanoparticles embedded in SiO_2 by Xe ion postirradiation. Appl. Phys. Express **5**, 105002 (2012)
43. J. Guangyi, W. Jun, Z. Lihong, L. Huixian, X. Rong, L. Changlong, Remarkably enhanced surface plasmon resonance absorption of Cu nanoparticles in SiO_2 by post Zn ion implantation. EPL **101**, 57005 (2013)
44. J. Wang, L. Zhang, X. Zhang, Y. Shen, C. Liu, Synthesis, thermal evolution and optical properties of CuZn alloy nanoparticles in SiO_2 sequentially implanted with dual ions. J. Alloy. Compd. **549**, 231 (2013)
45. G. Jia, H. Liu, X. Mu, H. Dai, C. Liu, Xe ion irradiation-induced polycrystallization of Ag nanoparticles embedded in SiO_2 and related optical absorption property. Opt. Mater. Express **4**, 1303 (2014)
46. H. Amekura, N. Umeda, Y. Takeda, J. Lu, N. Kishimoto, Fabrication of nickel oxide nanoparticles in SiO_2 by metal-ion implantation combined with thermal oxidation. Appl. Phys. Lett. **85**, 1015 (2004)
47. C.H. Aaronson, H. Amekura, Y. Sato, N. Kishimoto, Vacuum fluorescent displays utilizing ZnO nanoparticles. J. Appl. Phys. **109**, 024506 (2011)
48. K. Chung, J. Karle Timothy, A. Khalid, N. Abraham Amanda, R. Shukla, C. Gibson Brant, A. Simpson David, B. Djurišic Aleksandra, H. Amekura, S. Tomljenovic-Hanic, Room-temperature single-photon emission from zinc oxide nanoparticle defects and their in vitro photostable intrinsic fluorescence. Nanophotonics **6**, 269 (2017)
49. W.A. Harrison, *Electronic Structure and the Properties of Solids the Physics of the Chemical Bond* (Dover, New York, 1980)
50. M.C. Ridgway, G.d.M. Azevedo, R.G. Elliman, C.J. Glover, D.J. Llewellyn, R. Miller, W. Wesch, G.J. Foran, J. Hansen, A. Nylandsted-Larsen, Ion-irradiation-induced preferential amorphization of Ge nanocrystals in silica. Phys. Rev. B **71**, 094107 (2005)
51. B. Johannessen, P. Kluth, D.J. Llewellyn, G.J. Foran, D.J. Cookson, M.C. Ridgway, Amorphization of embedded Cu nanocrystals by ion irradiation. Appl. Phys. Lett. **90**, 073119 (2007)
52. B. Johannessen, P. Kluth, D.J. Llewellyn, G.J. Foran, D.J. Cookson, M.C. Ridgway, Ion-irradiation-induced amorphization of Cu nanoparticles embedded in SiO_2. Phys. Rev. B **76**, 184203 (2007)
53. D. Dalacu, L. Martinu, Temperature dependence of the surface plasmon resonance of Au/SiO_2 nanocomposite films. Appl. Phys. Lett. **77**, 4283 (2000)
54. O.A. Yeshchenko, I.M. Dmitruk, A.A. Alexeenko, A.M. Dmytruk, Size-dependent melting of spherical copper nanoparticles embedded in a silica matrix. Phys. Rev. B **75**, 085434 (2007)
55. H. Amekura, M. Tanaka, Y. Katsuya, H. Yoshikawa, H. Shinotsuka, S. Tanuma, M. Ohnuma, Y. Matsushita, K. Kobayashi, Ch. Buchal, S. Mantl, N. Kishimoto, Melting of Zn nanoparticles embedded in SiO_2 at high temperatures: effects on surface plasmon resonances. Appl. Phys. Lett. **96**, 023110 (2010)

56. H. Amekura, N. Umeda, K. Kono, Y. Takeda, N. Kishimoto, Ch. Buchal, S. Mantl, Dual surface plasmon resonances in Zn nanoparticles in SiO_2: an experimental study based on optical absorption and thermal stability. Nanotechnology **18**, 395707 (2007)
57. P. Blaha, K. Schwarz, G.K.H. Madse, D. Kvasnicka, J. Luitz, *WIEN2k, An Augmented Plane Wave + Local Orbitals Program for Calculating Crystal Properties* (Techn. Universität Wien, Austria, 2001)
58. H. Amekura, H. Shinotsuka, H. Yoshikawa, Are the triple surface plasmon resonances in Zn nanoparticles true? Nanotechnology **28**, 495712 (2017)
59. C. Ambrosch-Draxl, J.O. Sofo, Linear optical properties of solids within the full-potential linearized augmented planewave method. Comput. Phys. Commun. **175**, 1 (2006)
60. G. Mie, Beiträge zur Optik trüber Medien, speziell kolloidaler Metallösungen. Ann. Phys. **25**, 377 (1908)
61. H. Amekura, Y. Takeda, N. Kishimoto, Criteria for surface plasmon resonance energy of metal nanoparticles in silica glass. Nucl. Instrum. Methods Phys. Res. B **222**, 96 (2004)
62. H. Amekura, H. Kitazawa, N. Umeda, Y. Takeda, N. Kishimoto, Nickel nanoparticles in silica glass fabricated by 60 keV negative-ion implantation. Nucl. Instrum. Methods Phys. Res. B **222**, 114 (2004)
63. H. Amekura, Y. Takeda, K. Kono, N. Kishimoto, Magneto-optical Kerr effect of nickel nanoparticles in SiO_2 fabricated by negative-ion implantation of 60 keV. Trans. Mat. Res. Soc. Jpn. **29**, 615 (2004)
64. K. Sato, *Light and Magnetism*. 2nd ed. (Asakura Pub., Tokyo, 2001) [Japanese]
65. H. Amekura, Y. Takeda, N. Kishimoto, Magneto-optical Kerr spectra of nickel nanoparticles in silica glass fabricated by negative-ion implantation. Thin Solid Films **464**, 268 (2004)
66. R. Ruppin, Optical absorption of copper colloids in photochromic glasses. J. Appl. Phys. **59**, 1355 (1986)
67. J.C. Maxwell-Garnett, Colors in metal glasses and in metallic films. Phil. Trans. Roy. Soc. A **203**, 385 (1904)
68. P.H. Lissberger, P.W. Saunders, Optical and magneto-optical properties of thin film cermets. Thin Solid Films **34**, 323 (1976)
69. H. Amekura, Y. Fudamoto, Y. Takeda, N. Kishimoto, Curie transition of superparamagnetic nickel nanoparticles in silica glass: a phase transition in a finite size system. Phys. Rev. B **71**, 172404 (2005)
70. Z. Liu, H. Li, X. Feng, S. Ren, H. Wang, Z. Liu, B. Lu, Formation effects and optical absorption of Ag nanocrystals embedded in single crystal SiO_2 by implantation. J. Appl. Phys. **84**, 1913 (1998)
71. Z. Liu, H. Wang, H. Li, X. Wang, Red shift of plasmon resonance frequency due to the interacting Ag nanoparticles embedded in single crystal SiO_2 by implantation. Appl. Phys. Lett. **72**, 1823 (1998)
72. T. Yamada, K. Fukuda, S. Semboshi, F. Hori, Y. Saitoh, H. Amekura, A. Iwase, in preparation
73. M.A. García, J. Llopis, S.E. Paje, A simple model for evaluating the optical absorption spectrum from small Au-colloids in sol-gel films. Chem. Phys. Lett. **315**, 313 (1999)
74. N. Kishimoto, N. Umeda, Y. Takeda, V.T. Gritsyna, T.J. Renk, M.O. Thompson, In-beam growth and rearrangement of nanoparticles in insulators induced by high-current negative copper ions. Vacuum **58**, 60 (2000)
75. T.T. Lay, H. Amekura, Y. Takeda, N. Kishimoto, In-situ spectroscopy of ion-induced photon emission during metal nanoparticle formation in silica glass with high-flux Cu^- implantation. Mat. Res. Soc. Symp. Proc. **569**, 191 (1999)
76. F. Ren, X. Xiao, G. Cai, J. Wang, C. Jiang, Engineering embedded metal nanoparticles with ion beam technology. Appl. Phys. A Mater. Sci. Process. **96**, 317 (2009)
77. N. Umeda, V.V. Bandourko, V.N. Vasilets, N. Kishimoto, Metal precipitation process in polymers induced by ion implantation of 60 keV Cu^-. Nucl. Instrum. Methods. Phys. Res. B **206**, 657 (2003)
78. N. Okubo, N. Umeda, Y. Takeda, N. Kishimoto, Enhancement of metal-nanoparticle precipitation by co-irradiation of high-energy heavy ions and laser in silica glass. Nucl. Instrum. Methods Phys. Res. B **206**, 610 (2003)

79. N. Okubo, N. Kishimoto, Y. Takeda, Dynamics of co-irradiation effects of high-energy ions and photons on nanoparticle precipitation in silica glass. Nucl. Instrum. Methods Phys. Res. B **219**, 830 (2004)
80. N. Kishimoto, O.A. Plaksin, K. Masuo, N. Okubo, N. Umeda, Y. Takeda, Electronic excitation effects on nanoparticle formation in insulators under heavy-ion implantation. Nucl. Instrum. Methods Phys. Res. B **242**, 186 (2006)
81. H. Amekura, N. Umeda, Y. Takeda, N. Kishimoto, H. Nejo, in *Proceedings of the 5th International Symposium on Advanced Physical Fields* (NIRM, Tsukuba, 2000), p. 203
82. A. Nakajima, H. Nakao, H. Ueno, T. Futatsugi, N. Yokoyama, Coulomb blockade in Sb nanocrystals formed in thin, thermally grown SiO_2 layers by low-energy ion implantation. Appl. Phys. Lett. **73**, 1071 (1998)
83. A. Nakajima, T. Futatsugi, H. Nakao, T. Usuki, N. Horiguchi, N. Yokoyama, Microstructure and electrical properties of Sn nanocrystals in thin, thermally grown SiO_2 layers formed via low energy ion implantation. J. Appl. Phys. **84**, 1316 (1998)
84. F. Ren, C. Jiang, C. Liu, J. Wang, T. Oku, Controlling the morphology of Ag nanoclusters by ion implantation to different doses and subsequent annealing. Phys. Rev. Lett. **97**, 165501 (2006)
85. V. Bello, G. De Marchi, C. Maurizio, G. Mattei, P. Mazzoldi, M. Parolin, C. Sada, Ion irradiation for controlling composition and structure of metal alloy nanoclusters in SiO_2. J. Non-Cryst. Solids **345**, 685 (2004)
86. G. Rizza, H. Cheverry, T. Gacoin, A. Lamasson, S. Henry, Ion beam irradiation of embedded nanoparticles: toward an in situ control of size and spatial distribution. J. Appl. Phys. **101**, 014321 (2007)
87. P. Mazzoldi, G. Mattei, Synthesis of metal nanoclusters upon using ion implantation. in *Metal Nanoclusters in Catalysis and Materials Science*, eds. by B. Corain, G. Schmid, N. Toshima (Elsevier, Amsterdam, 2008), p. 269
88. A.R. Adhikari, M.B. Huang, D. Wu, K. Dovidenko, B.Q. Wei, R. Vajtai, P.M. Ajayan, Ion-implantation-prepared catalyst nanoparticles for growth of carbon nanotubes. Appl. Phys. Lett. **86**, 053104 (2005)
89. Y. Choi, J. Sippel-Oakley, A. Ural, Single-walled carbon nanotube growth from ion implanted Fe catalyst. Appl. Phys. Lett. **89**, 153130 (2006)
90. Y. Hoshino, H. Arima, Y. Saito, J. Nakata, Growth of single-walled carbon nanotubes from hot-implantation-formed catalytic Fe nanoparticles assisted by microwave plasma. Appl. Surf. Sci. **258**, 2982 (2012)
91. H. Hosono, N. Matsunami, Structural defects and chemical interaction of implanted ions with substrate structure in amorphous SiO_2. Phys. Rev. B **48**, 13469 (1993)
92. H. Hosono, H. Imagawa, Ion-solid chemistry in implanted amorphous SiO_2. Nucl. Instrum. Methods Phys. Res. B **91**, 510 (1994)
93. A. Nakajima, T. Futatsugi, K. Kosemura, T. Fukano, N. Yokoyama, Si single-electron tunneling transistor with nanoscale floating dot stacked on a Coulomb island by self-aligned process. J. Vac. Sci. Technol., B **17**, 2163 (1999)
94. H. Amekura, Y. Takeda, K. Kono, H. Kitazawa, N. Kishimoto, Modification of metal nanoparticles in SiO_2 by thermal oxidation, Rev. Adv. Mater. Sci. **5**, 178 (2003)

Chapter 5
Shape Elongation of Nanoparticles Induced by Swift Heavy Ion Irradiation

Abstract In this chapter, the shape elongation of nanoparticles (NPs) induced by swift heavy ions (SHIs) and optical applications are described. SHIs are very high energy ions whose energy loss processes are dominated by electronic friction. To understand the elongation mechanism and the optical properties, Sect. 5.1 describes the fundamental phenomena induced by SHI irradiation, including the point defect production, ion track formation, ion-induced compaction, core/shell ion tracks, and ion hammering. In Sect. 5.2, the properties of the shape elongation are demonstrated with important concepts of the minimum width for the elongation, the mass *non*-conservation, the scaling law, etc. In Sect. 5.3, the elongation mechanism, while it is still in debate, is discussed comparing with the synergy model and the thermal pressure model, after introducing the concepts of the inelastic thermal spike, two-temperature molecular dynamics. Section 5.4 reviews the optical applications of the shape elongation, including the linear dichroism, the birefringence, possible application to nanometer-thick polarizer, the second-harmonic generation microscopy, etc. In Sect. 5.5, some comments are presented about the ion tracks, which are important to understand the irradiation effects with SHIs. Finally, the elongation by MeV C_{60} cluster ions is described in Sect. 5.6, which is as efficient as 200 MeV Xe ions, but two orders of the magnitude lower energy.

5.1 Ion Irradiation Effects on Amorphous Silica

5.1.1 Optical Absorption of Point Defects

Various types of point defects are introduced in amorphous silica (SiO_2) under ion irradiation. Some are formed directly and the others are due to consequences of interactions with impurities and/or other defects, etc. The types and concentrations of the radiation-induced defects strongly depend on what impurities/defects already exist in the silica and how dense they are.

There are many methods to synthesize bulk and thin films of amorphous silica, and the types and concentrations of impurities depend on the synthesis methods.

F. Chen et al., *Ion Irradiation of Dielectrics for Photonic Applications*, Springer Series in Optical Sciences 231, https://doi.org/10.1007/978-981-15-4607-5_5

Therefore, the point defects observed after the ion irradiation also depend on the synthesis method.

Once, the synthesis methods were categorized into four types. Since some other methods have been invented after then, these categories could have only a historically meaning: the method type I and II are synthesis by the fusion of naturally-grown quartz (SiO_2) powder, while the type III and IV are direct synthesis from $SiCl_4$ gas. Consequently, the products of the former and the latter are called "fused silica" and "synthetic silica", respectively. However, this terminology is often in confusion. The type I method is due to the fusion of the quartz in vacuum or non-active gas using a furnace or arc-plasma, while the type II is due to the fusion in flames. The type III is a direct deposition via the hydrolysis of $SiCl_4$ gas in oxygen-hydrogen flames, while the type IV is due to the hydrolysis of $SiCl_4$ gas with the radio-frequency induction-coupled plasma of Ar/O_2. While the types I and II contain metal impurities in high concentration, the type III and IV contain less than 1 ppm of metal impurities. The type III contains OH impurities in quite high concentration of 500–1500 ppm, while the type IV contains OH impurities of less than 5 ppm. However, the high concentration of OH impurity improves the resistance to ultra-violet (UV) irradiation and probably to ionization by ion irradiation. The type III silica is used for material for optical apparatuses such as lens. The low OH concentration of the type IV is appropriate for optical fibers. However, the Soot method is recently more applied than the type IV. The synthesis methods which are not included in these classical four categories include the Soot method, the sol-gel, LPD (liquid phase deposition), various chemical vapor deposition (CVD) methods, etc.

Figure 5.1 shows the major optical absorption/luminescence excitation bands and emission bands of defect centers in synthetic silica [1]. See also review papers of

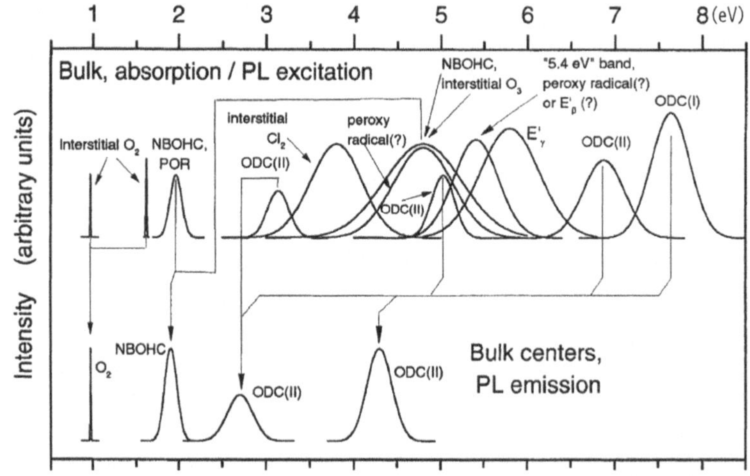

Fig. 5.1 Major optical absorption/luminescence excitation bands (upper) and emission bands (lower) of defect centers in synthetic silica are schematically depicted. The abscissa is photon energy in the unit of eV. Reproduced with permission from [1], ©1998 Elsevier

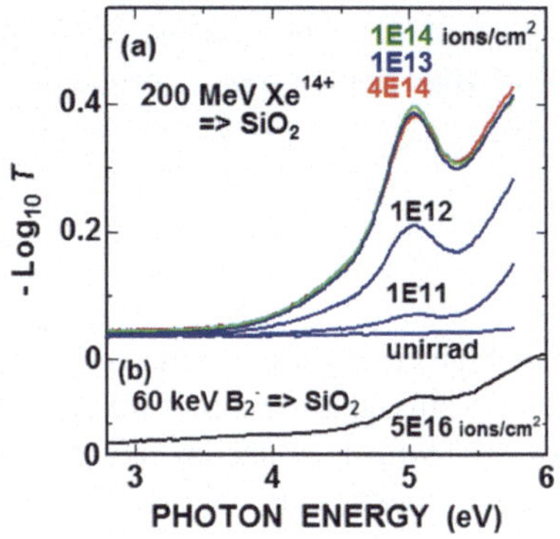

Fig. 5.2 Optical absorption spectra of silica glass irradiated with **a** 200 MeV Xe^{14+} ions and **b** 60 keV B_2^- negative molecular ions. Reproduced with permission from [5, 6], ©2008, 2018 American Institute of Physics. A peak at 5 eV, which could be ascribed to the point defects ODC(II), is observed under both high and low energy ion irradiation. Curves of 1×10^{13}, 1×10^{14}, and 4×10^{14} ions/cm^2 are overlapped with each other, indicating a strong saturation of ODC(II) concentration

defects in SiO_2 [1–4]. As shown in Fig. 5.2, an absorption peak at 5 eV grows with the fluence up to 5×10^{12} ions/cm^2 under 200 MeV Xe^{14+} ions but completely saturates the intensity exceeding 5×10^{12} ions/cm^2. The absorption at 1×10^{13} ions/cm^2 shows the same intensity as that of 4×10^{14} ions/cm^2, i.e., 40 times higher fluence. The fluence Φ dependence of the 5 eV peak absorption $A(\Phi)$ follow the track coverage law,

$$A(\Phi)=A_0\{1-\exp(-\sigma\Phi)\} \tag{5.1}$$

where A_0 and σ denote the saturated absorption and the defect formation cross-section, respectively. A similar peak, which has probably the same origin, was observed from the same species of the silica samples irradiated with low energy 60 keV negative di-boron molecular ions, i.e., B_2^-, as also shown in Fig. 5.2. It should be noted that much different energies of 200 MeV Xe^{14+} ions and 60 keV B_2^- ions generate the same ODC(II) point defects. It should be noted that 60 keV B_2^- is S_e-dominant in SiO_2, since $S_e = 0.28$ keV/nm and $S_n = 0.12$ keV/nm.

Oliver et al. [7] reported a similar peak at 248 nm, i.e., 5 eV, in high-purity amorphous silica irradiated with 4 MeV Si^+ and 3 MeV Au^{2+} ions to 0.3–1.5 × 10^{17} ions/cm^2, which they tentatively ascribed to B_2 centers. (The "B_2" is the name of the centers. Please don't confuse it with the B_2^- molecular ion.) The peak was stable against annealing for 1 h in 10%-H_2/90%-N_2 gas at 600 and 900 °C. Contrary, the peak shown in Fig. 5.2 begins to decay at annealing in a vacuum at 200 °C and mostly disappeared at 400 °C. The remarkable difference in the thermal stability indicates both the defects are different origins, while the peak energy is the same. As shown in Fig. 5.1, there are three candidate defects having peak around 5 eV. From the FWHM of the peak, the peak shown in Fig. 5.2 is ascribed to ODC(II) defects.

In fact, Olivers' samples contain OH concentration of 150 ppm, while our samples contain OH concentration of 1000 ppm or more. Different defect species may form.

5.1.2 *Ion Tracks*

When swift heavy ions, i.e., high energy ions whose energy loss processes are dominated by electronic friction, go through in certain solids, cylindrical modified regions (often amorphized) are formed along the trajectories of the ions, which are called ion tracks. TEM images of the ion tracks in crystalline SiO_2 are shown in Fig. 5.3. Ion tracks are formed in certain materials, e.g., amorphous materials including metallic glasses, (crystalline) oxides and certain semiconductors and insulators. Tracks are difficult to be formed in metals. Even if the tracks can be formed in a material, energy deposition (S_e) higher than the threshold is required. While the S_e dependence of track radius in crystalline SiO_2 is exhibited in Fig. 5.4a, the radius becomes null, i.e., the tracks are not formed, below ~2 keV/nm. Table 5.1 summarizes the track formation thresholds of various materials. For semiconductors, the tracks can be formed but require higher threshold S_e. Both crystalline and amorphous SiO_2 show particularly low track registration threshold. The radii of the tracks are in the range of nm. The length is, of course, less than the ion range. A track terminates when the ion energy decreases with increasing the depth so that the S_e decreases below the threshold. The depth dependence of track radii was experimentally studied by TEM [8] and small angle X-ray scattering (SAXS) [9]. Even in amorphous materials, ion tracks are formed. However, it is difficult to observe the tracks in amorphous materials by TEM. Instead, SAXS and FTIR are used.

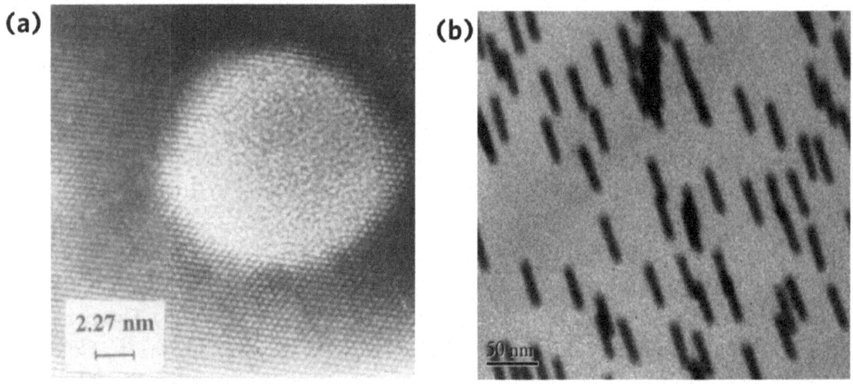

Fig. 5.3 Transmission electron microcopy images of ion tracks in crystalline SiO_2. **a** with 62 MeV Pb ions, in a cross-sectional configuration [10], and **b** with 1 GeV Pb ions, in an inclined configuration [11]. Track length is limited by TEM sample thickness. Reproduced with permission from [10, 11], ©1994 American Physical Society, ©2013 IOP

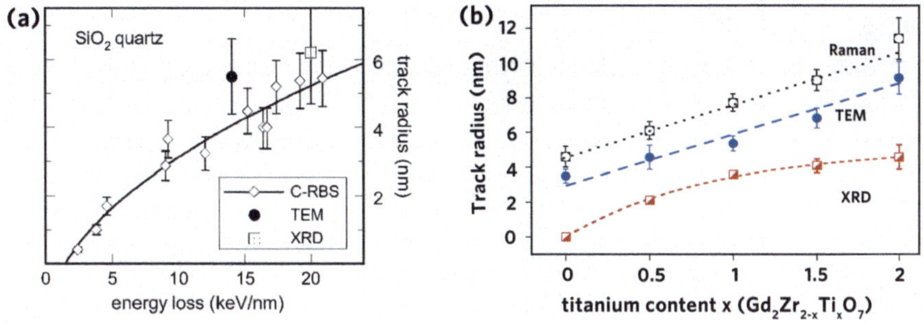

Fig. 5.4 Ion track radii of **a** quartz (SiO_2) crystal in the electronic energy loss dependence, and of **b** pyrochlore $Gd_2Zr_{2-x}Ti_xO_7$ in the Ti content dependence. The radii were determined by c-RBS, TEM, XRD, and Raman scattering. In **a**, a solid curve is a square-root fitting of S_e. Reproduced with permission from [10], ©1996 and [13], ©2009 American Physical Society

Table 5.1 Track formation thresholds of various materials. * denotes values of C_{60} projectiles. # denotes surface data

Threshold range (keV nm^{-1})	Target (threshold, keV nm^{-1})	Authors and published year
0–5	SiO_2 (2)	Meftah et al. (1994)
	a-Ge (3)	Furuno et al. (1996)
	SiO_2 (3.5 #)	Arnoldbik et al. (2003)
	LiF (4)	Trautmann et al. (2003)
	$BaFe_{12}O_{19}$ (4)	Meftah et al. (1994)
	$Pd_{80}Si_{20}$ (amorphous)	Klaumünzer et al. (1994)
	Mica (5)	Toulemonde et al. (1994)
	CaF_2 (5)	Boccanfuso et al. (2002)
5–10	$Y_3Fe_5O_{12}$ (6.5)	Meftah et al. (1994)
	$LiNbO_3$ (7)	Meftah et al. (1994)
	SiO_2 (7 #)	Khalfaoui et al. (2003)
	$MgAl_2O_4$ (7.5)	Zinkle and Skuratov (1998)
	$Y_3Al_5O_{12}$ (7.5)	Meftah et al. (1994)
10–15	InP (14*)	Kamarou et al. (2008)
15–20	Si_3N_4 (15)	Zinkle et al. (2002)
	MgO (15.8#)	Skuratov et al. (2003)
	$Y_3Fe_5O_{12}$ (16#)	Meftah et al. (1994)
	a-Si (17)	Furuno et al. (1996)
	GeS_2 (18)	Vetter et al. (1994)
	U_3Si (19)	Hou and Klaumunzer (2003)
20–25	InP (20)	Gaiduk et al. (2000)
	MgO (20)	Beranger et al. (1996)
	Al_2O_3 (21)	Canut et al. (1995)

(continued)

Table 5.1 (continued)

Threshold range (keV nm^{-1})	Target (threshold, keV nm^{-1})	Authors and published year
25–30	Al_2O_3 (25 #)	Skuratov et al. (2003)
	GaSb (28)	Szenes et al. (2002)
	InSb (28)	Szenes et al. (2002)
	InAs (28)	Szenes et al. (2002)
	UO_2 (29)	Matzke et al. (2000)
30–35	Zr (30)	Dunlop and Lesueur (1993)
	Bi (31, 31^*)	Wang et al. (1996)
	GaAs (31^*)	Kamarou et al. (2008)
	Ge^* (33)	Kamarou et al. (2008)
	GaN (<34)	Kucheyev et al. (2008)
	SiC (>34)	Zinkle et al. (2002)
	AlN (>34)	Zinkle et al. (2002)
	$Si_{0.5}Ge_{0.5}$ (34)	Gaiduk et al. (2002)
35–40	Si (37^*)	Kamarou et al. (2008)
	Co (37)	Dunlop and Lesueur (1993)
	Ge (38)	Colder et al. (2001)
	GaAs (38^*)	Colder et al. (2001)
	Fe (40)	Dumlop et al. (1994)
Above 40	Ge (42)	Komarov (2003)
	Si (46*)	Dunlop et al. (1998)

Reproduced with permission from [12], ©2009 IOP

While the track radii determined from c-RBS, TEM, and XRD were almost the same in Fig. 5.4a, the radii determined from other evaluation methods should not be the same, because other methods detect different physical properties. Figure 5.4b exhibits an example where the track radii of pyrochlores $Gd_2Zr_{2-x}Ti_xO_7$ determined by different methods, i.e., XRD, TEM, and Raman scattering, give different values [13]. The tracks in the pyrochlores (except $x = 0$) consist of amorphous core, disordered defect-fluorite-structured shell, and strained defect-rich pyrochlore halo. While XRD detects only the amorphous cores, the bright-field TEM observes both the amorphous cores and the defect-fluorite structured shells. Further, Raman scattering also detects the strained pyochlore halos. The combination of different methods could provide detailed information of the tracks.

5.1.3 Compaction

Ion irradiation to crystals induce damage regions, which usually result in a decrease in density because of introduction of defects such as Frenkel pairs (vacancies and interstitials). According to the Lorentz-Lorenz's formula, the relative change of the refractive index $\Delta n/n$ is described as the following relation [14],

$$\frac{\Delta n}{n} = \frac{(n^2-1)(n^2+2)}{6n}\left[-\frac{\Delta V}{V} + \frac{\Delta\alpha}{\alpha} + F\right] \tag{5.2}$$

where $\Delta V/V$, $\Delta\alpha/\alpha$, and F denote relative changes of volume, of polarizability, and other factors, respectively. The decrease in density results in a decrease in refractive index. This has been utilized to form optical waveguides as described in [14]. In the cases of silica glass, ion irradiations induce an increase in density, which results in an increase in refractive index. The increase in density is called ion-induced "compaction" or "densification".

The compaction of amorphous silica is induced by not only swift heavy ions but also MeV protons, keV heavy ions, neutrons, and even ~10 keV electrons [2, 4]. An example of fluence dependence of the compaction is exhibited in Fig. 5.5 [2]. The density increases with the fluence but saturates at the increment of ~3%.

The compaction is induced by both the nuclear collisions and the ionization. Devine pointed out that the required fluence for the saturated change corresponds to the energy deposition of 3×10^{23} eV/cm^3 for nuclear displacement and ~2×10^{26} eV/cm^3 for ionization [2]. He concluded that the ionization would be 660 times less efficient for the compaction than the nuclear displacement for low S_e ion irradiation. For high S_e irradiations (swift heavy ion irradiation), the energy deposition for the saturated electronic compaction is comparable to the value of the saturated nuclear compaction of 3×10^{23} eV/cm^3.

The saturation value of 3×10^{23} eV/cm^3 corresponds to the energy deposition ($E_{\text{Si–O}}$) of 3.4 eV per Si–O bond, which is close to the Si–O bonding energy. This fact indicates that a large fraction of all the bonds have been once broken when

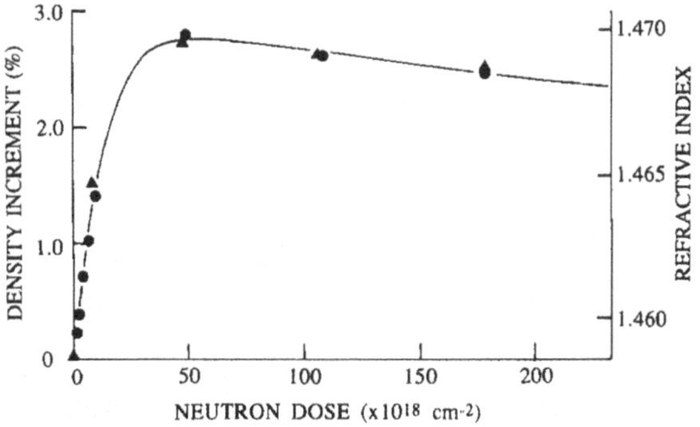

Fig. 5.5 Fluence dependence of relative density change (circles) and refractive index (triangles) of silica glass exposed to fast neutrons. Reproduced with permission from [2], ©1993 Elsevier

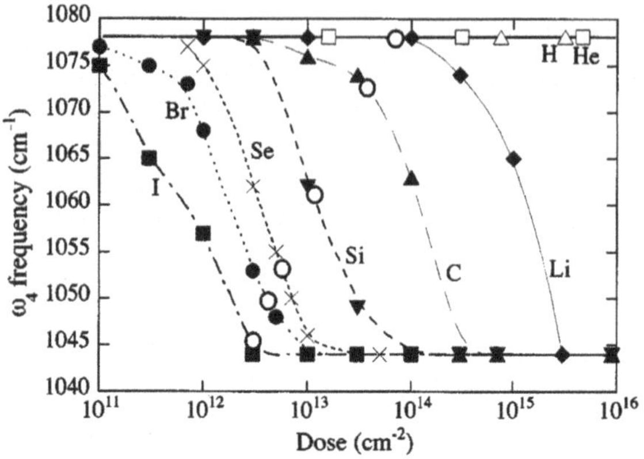

Fig. 5.6 Fluence dependence of the IR ω_4 absorption peak of amorphous SiO_2 for various ions. 10 MeV H, 15 MeV He, 2 MeV Li, 4 MeV C, 30 MeV Si, 35 MeV Se, 67 MeV Br, and 80 MeV I. Open circles correspond to electronic deposition of the Si-O bond energy. Reproduced with permission from [15], ©2000 American Physical Society

the compaction comes to saturation. Closed symbols in Fig. 5.6 show the fluence dependences of compaction monitored by the shift of the IR vibration of SiO_2 (ω_4 mode) for various ions. Open circles indicate the fluences which correspond to the electronic energy deposition $E_{el} \sim E_{Si\text{–}O} = 3.4$ eV/bond. In the case of 80 MeV I ions ($S_e = 10.4$ keV/nm), the saturation of the closed squares attained at almost the same fluence as the open circle: i.e., $E_{el}^{sat} \sim 3.4$ eV/bond, where E_{el}^{sat} denotes the electronic energy deposition for the saturation. However, for 67 MeV Br ions (8.2 keV/nm), the saturation is not attained at the fluence of the open circle (5×10^{12} ions/cm^2), but requires double the fluence (1×10^{13} ions/cm^2), i.e., $E_{el}^{sat} > 3.4$ eV/bond. In the case of 30 MeV Si ions, $E_{el}^{sat} \sim 3.4$ eV is attained at 1×10^{13} ions/cm^2 but the saturation reaches at 1×10^{14} ions/cm^2. With decreasing the S_e of ions, the deviation becomes larger, and approaches to the ratio of $E_{el}^{sat}/E_{n}^{sat} \sim 660$, which was reported by Devine for the low energy cases [2].

According to another viewpoint based on the ion track picture, the fluence dependence of the ω_4 mode frequency $\nu(\Phi)$ is assumed to be described by following relation [16],

$$\nu(\Phi) = \{\nu(0) - \nu(\infty)\}\exp(-\sigma\Phi) + \nu(\infty) \tag{5.3}$$

where $\nu(0)$ and $\nu(\infty)$ denote the frequencies in unirradiated and in the saturation, respectively. From the fitting of Eq. (5.3), the cross-section of the ion tracks $\sigma = \pi R_{track}^2$ is obtained, which roughly coincides with the track size determined from other techniques such as XRD (See e.g. [16]). It should be noted that the track sizes determined by other techniques do not precisely coincide with each other, because different techniques detect different properties of tracks. In this analysis, it is assumed that all the SiO_2 in a track is transformed to the saturated state only by a single ion impact. Figure 5.6 indicates that the saturated compacted state is not attained only by a single ion hit for low S_e SHI irradiation.

Since glasses are formed by quenching of molten phase from high temperatures, the high temperature atomic networks are frozen. The temperature corresponding the atomic distribution in the network is called "fictive temperature", which is much higher than the real temperature of the solid glass. The fictive temperature is determined by IR and Raman spectroscopies. Since compacted silica shows higher fictive temperature than before the irradiation, the compacted phase is considered to be quenched from a very high temperature which was transiently attained by ion irradiation. In the case of keV ion irradiation, the high temperature could be attained by collisional thermal spikes.

The compacted silica shows much higher etching rate to chemicals such as hydrofluoric acid HF. While the original diameter of the (latent) ion tracks are around ten nm or less, the tracks are enlarged by etching and observed by optical microscopy. However, the highly enhanced etching rate in the irradiated regions (tracks) sounds contradictory to the higher density of irradiated silica. Higher density but opened structures in non-thermal equilibrium state are proposed and confirmed by Devine et al. [4]. Raman spectroscopy clearly confirmed that the compacted silica contained lower-membered ring structures such as three- and four-membered rings. This change is confirmed in silica irradiated with swift heavy ions by Awazu et al. [15]. A phenomenological explanation is that the compacted state is regarded as a metastable state induced by irradiation, having higher free energy compared with the ground state. Consequently, the energy barrier to the etching is lower for the metastable state than the ground state [17].

Under the etching treatments, the unirradiated regions are also solved at much lower rate compared with the irradiated ones. The irradiation simply accelerates the etching rate. However, TiO_2 is an exception: While this material is not solved in HF acid in unirradiated state, SHI irradiation induces structural changes which are susceptible to dissolved [18].

It should be noted again that ion irradiation induces the extremely rapid thermal cooling of ~10^{16} K/s for 0.1 keV/u ions and ~10^{15} K/s for 1 MeV/u ions [20], which cannot be attained by other methods. Based on thermal quenching simulations using molecular dynamics (MD), Benyagoub succeeded to explain the fluence dependence of the compacted ion tracks [19]. Starting from the unirradiated density of silica of 2.20 g/cm^3, the ion impact raises the temperature over 5000 K as shown in Fig. 5.7. Depending on the cooling rate, the density follows a certain route in the density-temperature space to cool down. However, the reached density is always higher than the unirradiated value, i.e., compacted. Moreover, when the irradiated regions are hit again, the regions are heated and cool down to the same compacted density, if the cooling rate is the same as before. Under swift heavy ion irradiation, the compacted region increases as

$$f = 1 - \exp(-\sigma \Phi) \tag{5.4}$$

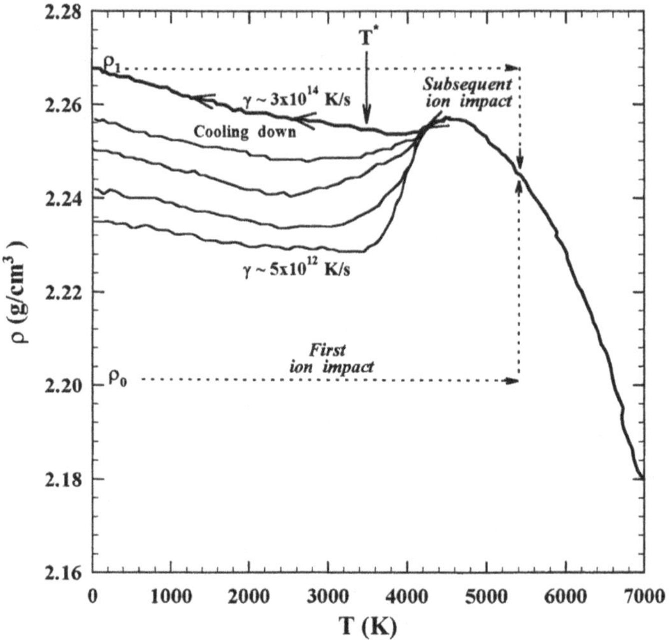

Fig. 5.7 Schematic variation of the density of amorphous silica for different cooling rates deduced from molecular dynamics simulations. Reproduced with permission from [19], ©1998 Elsevier

where σ and Φ denote the cross-section of an ion track and fluence, respectively. The compacted state could be regarded as a high temperature phase, which is metastable at room temperature. The energy deposition by irradiation induces the transformation of the unirradiated phase to the high temperature (compacted) phase and quenching of the high temperature phase down to room temperature.

5.1.4 Microscopic Origin of Compaction and Core/Shell Ion Tracks

A new microscopic picture has been proposed for the compaction induced by swift heavy ion irradiations since 2008 [21, 22]. Swift heavy ion irradiation generates core/shell ion tracks in silica, which have track cores with lower density and shells with higher density than the unirradiated state. The formation of the core/shell ion tracks in SiO_2 was confirmed by small angle X-ray scattering (SAXS) [21, 22]. The compaction which is observed in macroscopic scales, is a spatially averaged effect of the underdense cores and the overdense shells. Figure 5.8 shows the density profiles of ion tracks of various electronic stopping power S_e simulated by the molecular dynamics (MD) [21]. With increasing S_e, the density of the track core decreases in addition to an increase in the density of the shell.

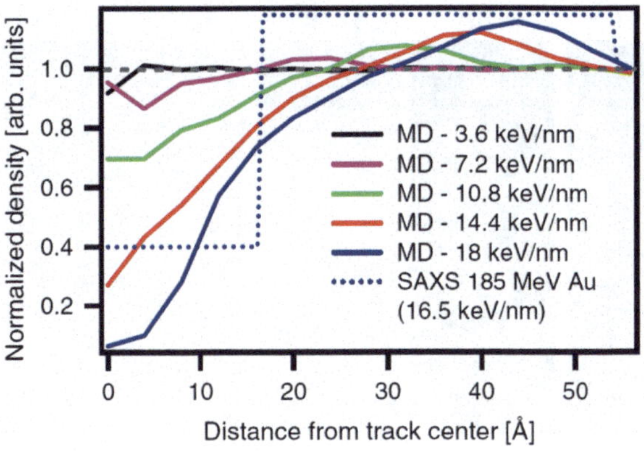

Fig. 5.8 MD simulations of normalized density profiles of ion tracks in SiO_2 under various electronic stopping powers. Reproduced with permission from [21], ©2008 American Physical Society

An important difference between the core/shell ion tracks and the conventional hard-cylinder tracks is the destruction of the previous tracks by a new ion impact. Figure 5.9 shows MD simulations of density profiles of SiO_2 against the successive two ion impacts with the distances of (a) 8 nm and (b) 3 nm. The two successive impacts are independently recorded in the case of 8 nm distance.

However, in the case of 3 nm distance, the previous underdense track core (circles) is overwritten by a part of overdense shell of the newly formed core/shell track. Consequently, with increasing the fluence, all the irradiated surface of SiO_2 has never been completely filled with the underdense state or the overdense state.

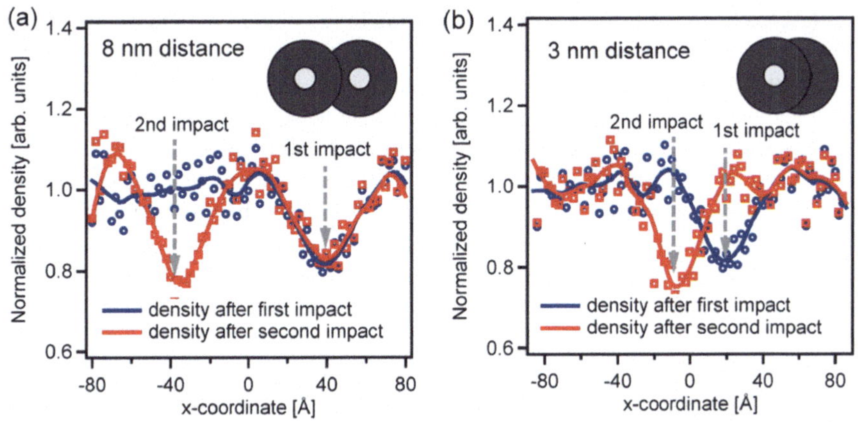

Fig. 5.9 MD simulations of density profiles of successive two ion impacts with the distance of **a** 8 nm and **b** 3 nm. Circles with blue curves for the profiles after the 1st impact and squares with red curves for those after the 2nd impact. Reproduced with permission from [22], ©2011 American Institute for Physics

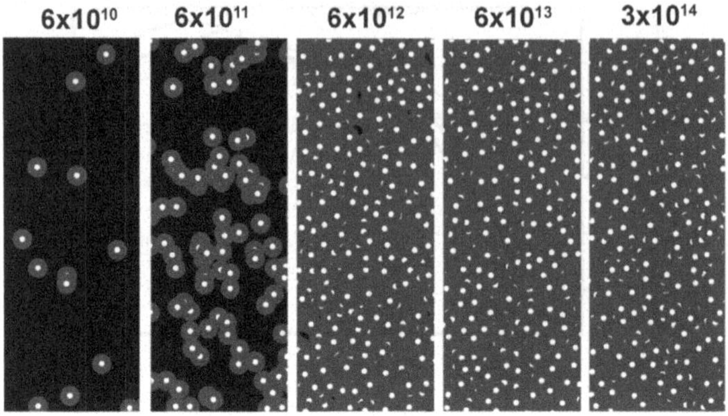

Fig. 5.10 Simulations of random ion impacts of core/shell ion tracks in amorphous SiO_2 for the fluences shown above in the unit of ions/cm^2. Black, brown, and white areas denote unirradiated, high-density, and low-density regions, respectively. Reproduced with permission from [22], ©2011 American Institute of Physics.

Therefore, the static steady state has never been attained but dynamic steady state, where always a new impact partly destroys the previous core/shell tracks and generates a new core/shell track. Macroscopic measurements show that the compaction of SiO_2 reaches a *saturation* at a relatively low fluence. In a microscopic picture, the core/shell tracks have not reached any microscopic steady state. The situation is simulated by MD and shown in Fig. 5.10. Black, gray, and white areas denote unirradiated, high-density shell and low-density core regions, respectively. Up to the fluence of 6×10^{11} ions/cm^2, the region covered by the core/shell tracks increases. The increase of the gray region results in an increase of the macroscopic density, i.e., the compaction. Up to 6×10^{12} ions/cm^2, all the area is covered by gray and white regions. Exceeding this fluence, a ratio of white to ray regions maintains almost the same value, while the positions of the white areas change with fluence. Therefore, the system comes a dynamical steady state. Always the core/shell ion tracks are generated but simultaneously destroy the previous tracks. Even under the dynamical steady state, the macroscopic density comes to a saturation value.

The numerical simulations shown in Fig. 5.10 were experimentally supported by SAXS: If all the regions were occupied by the conventional hard-cylinder tracks, all the regions became homogenous up to 6×10^{12} ions/cm^2. No X-rays were scattered, because the spatial density fluctuation, i.e., tracks, scatters the X-rays. However, the SAXS signal from the core/shell tracks is observed up to 3×10^{14} ions/cm^2 and higher [22].

Saturated areal density of the ion tracks has been experimentally confirmed by TEM observation in crystals of $MgAl_2O_4$ [23] and CeO_2 [24]. Figure 5.11 exhibits the plan-view TEM images in $MgAl_2O_4$ crystal irradiated with 200 MeV Xe ions at four different fluences. Ion tracks are observed as white dots. Figure 5.11c and d were irradiated to 1×10^{13} and 2×10^{14} ions/cm^2, respectively, i.e., (d) was irradiated to 20 times higher fluence than (c). However, both the images show almost the same

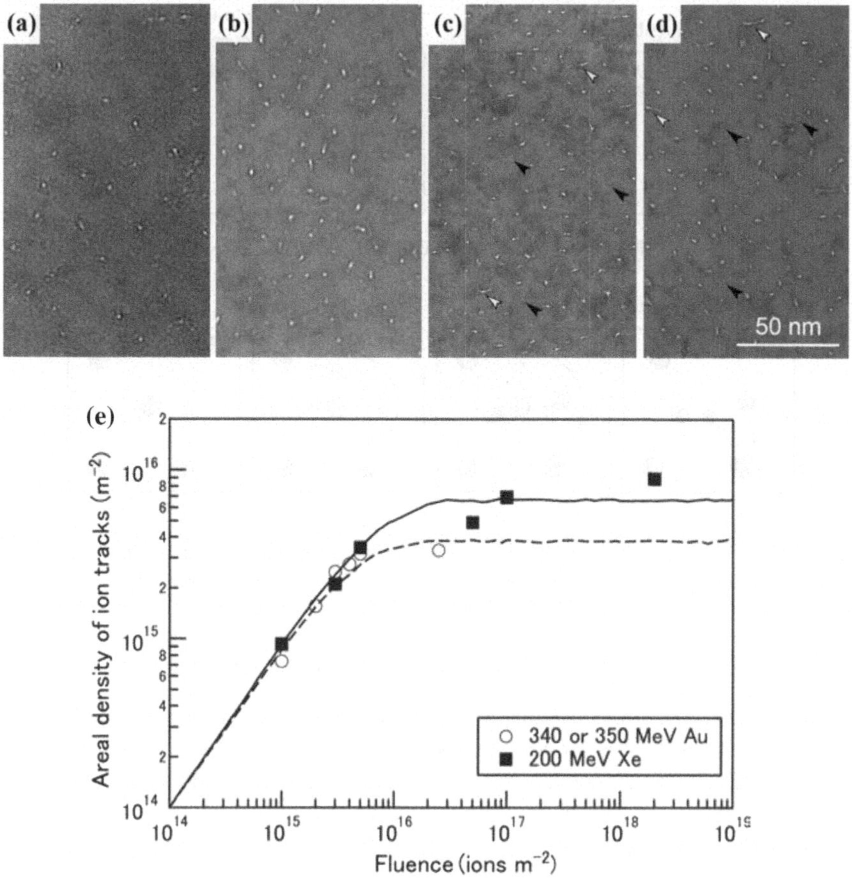

Fig. 5.11 Plan-view TEM images of ion tracks in $MgAl_2O_4$ crystal irradiated with 200 MeV Xe ions to fluence of **a** 3 × 10^{11}, **b** 5 × 10^{12}, **c** 1 × 10^{13}, **d** 2 × 10^{14} ions/cm^2. **e** Fluence dependence of areal density of tracks in the SI units, i.e., m^{-2} not cm^{-2}. Reproduced with permission from [23], ©2011 Carl Hanser Verlag GmbH & Co.KG, Muenchen

track areal density, clearly indicating the saturation of the track areal density. Fluence dependence of the track areal density is shown in Fig. 5.11e, which shows the linear increase of track density at low fluences and saturation exceeding 10^{16} ions/m^2 = 10^{12} ions/cm^2. Yasuda et al. consider that the saturation of the track density is not a consequence of core/shell tracks but a consequence of recrystallization. They assume that tracks of much larger diameters are formed just after impacts but the tracks shrink by recrystallization. Only the cores of the initially-formed larger tracks remain as permanent tracks. Amekura et al. proposed a similar model to explain initial steep degradation and successive stable tolerance of ZnO NPs as shown in Fig. 5.12 [5]. At the low fluence, the number of the tracks linearly increases even with the large recrystallization. At the high fluences, the number of the tracks

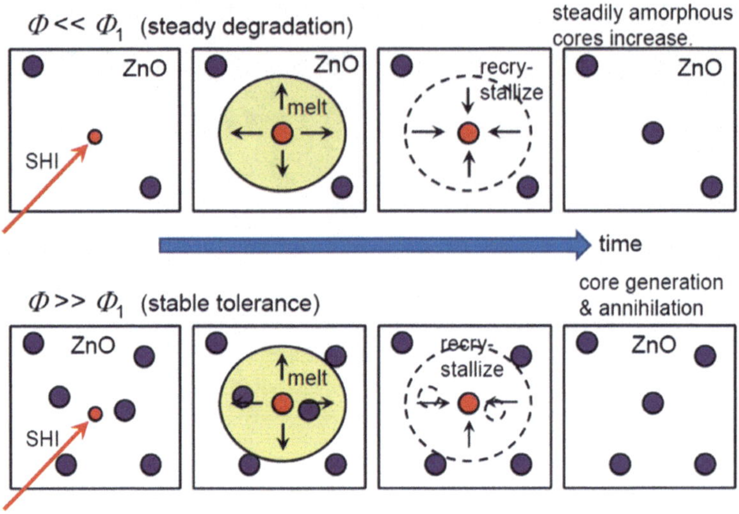

Fig. 5.12 Schematically depicted mechanism proposed for the initial steep degradation at low fluence and successive stable tolerance of ZnO NPs. Much larger areas of melting and recrystallization than the permanent tracks are the key. Reproduced with permission from [5], ©2018 American Institute of Physics

saturates due to the impact-induced recrystallization. The partial recrystallization of tracks in ZnO was also supported with MD simulations by Pakarinen et al. [25].

It should be, however, noted that the mechanism could be different between amorphous SiO_2 and crystalline oxides even if both the materials show the saturated track areal density.

5.1.5 *Inelastic Thermal Spike Model for the Core/Shell Track Formation and Inconsistency*

The formation mechanism of the core/shell ion tracks of SiO_2 had been explained by inelastic thermal spike (i-TS) effect. After the passage of swift heavy ion, the core part of the track is heated and overcomes both the melting and the vaporization temperatures. The shell part is also heated and overcomes the melting temperature but not the vaporization temperature. The outside of the shell regions does not reach the melting temperature and maintains the solid phase. The core part is vaporized and generates high pressure, which presses the molten shell. Therefore, the core is expanded and the shell is compressed. After extremely quick quenching, the core and the shell maintain lower and higher density, respectively.

If this model would be correct, the core radius and the total radius of the tracks determined by SAXS could match with those of the vaporized and the molten regions, respectively. In fact, relatively good agreement was reported between them as shown

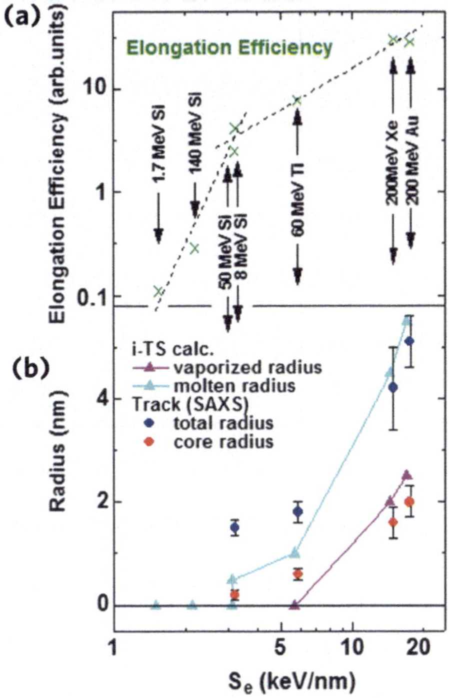

Fig. 5.13 Electronic stopping power S_e dependence of **a** the elongation efficiency of metal nanoparticles of **b** the radii of core/shell ion tracks in amorphous silica determined by SAXS and of corresponding vaporization/melting regions. Reproduced with permission from [26], ©2018 American Physical Society

in Fig. 5.13, in the cases of relatively high S_e irradiations ($S_e > 10$ keV/nm), where the vaporization is induced [21]. However, while the vaporization is no longer induced at $S_e = 6$ keV/nm or lower, still the core/shell tracks are detected as shown in Fig. 5.13 [26]. This is clear counterevidence to the broadly accepted core/shell track formation mechanism by the i-TS effect. The mechanism should be reexamined.

In the i-TS model, melting and vaporization in ion tracks are presumed. However, the vaporization induced in an ion track could be much different from that at the atmospheric pressure, because the vapor is confined in a small volume of the ion track in a solid. High pressure is expected in the track, and such high pressure could suppress the vaporization itself. However, it should be noted that the vaporization of SiO_2 cores was reproduced by the two-temperature molecular dynamics (TT-MD) simulations [26], as shown in Fig. 5.14.

Here shows results of the TT-MD simulations for discrimination of the vaporized state of SiO_2. TT-MD is an extension of the i-TS calculations. However, the discrimination between the vaporized and the molten state was difficult, since the boiling and melting temperatures are changed under high pressure. Instead, the pair correlation function was calculated from simulation cells of SiO_2 before and after the impact of 200 MeV Xe ion and of 50 MeV Si ion [26]. According to the i-TS calculation performed at the atmospheric pressure, vaporization is induced by the 200 MeV Xe ion impact but not by the 50 MeV Si ion impact, as consistent with Fig. 5.13. The results of TT-MD simulations are exhibited in Fig. 5.14. Comparing

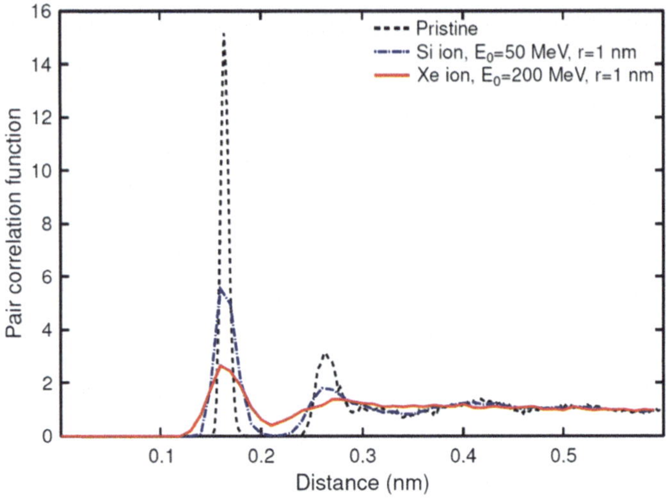

Fig. 5.14 Pair correlation functions for SiO_2 within the cylinder of 1 nm radius before and after 5 ps from the impacts of 50 MeV Si ion and 200 MeV Xe ions. The missing gap between the 1st and 2nd peaks corresponds the vaporization of SiO_2. Reproduced with permission from [26], ©2018 American Physical Society

with the pristine, the correlation peak decreased at 5 ps after the ion impact, but the correlation converged to the unity at large distances. A small but clear difference was observed: a gap was observed between the 1st and 2nd correlation peaks for the pristine and the 50 MeV Si case. The gap was missing for the 200 MeV Xe case. To find conditions for the missing gap, SiO_2 cells were made at various temperatures in computer. After equilibration, the pair correlation function was evaluated. It was found that the missing gap corresponded to the vaporization. Therefore, the vaporization is induced in an ion track of SiO_2 under 200 MeV Xe irradiation but not 50 MeV Si irradiation [26]. Anyway, the vaporization in an ion track in solid was confirmed by TTMD simulations.

5.1.6 Ion Hammering

In 1983, Klaumunzer and Schumacher observed that metallic glass $Pd_{80}Si_{20}$ exhibited large growth of the sample dimensions perpendicular to the ion beam without appreciable change in volume [27] under 285 MeV Kr ion irradiation. This phenomenon was confirmed since then in many amorphous materials [28–30] including silica [31]. Since the samples are expanded perpendicular to the beam and shrunk parallel to the beam as if they were hit by hammers, this phenomenon was named as "ion hammering". Figure 5.15 exhibits deformation of microstructures irradiated with ion beams observed by secondary electron microscopy (SEM). Since the samples (a), (d), (f), and (h) were irradiated with 16 MeV Au, 30 MeV Cu, 4 MeV Xe,

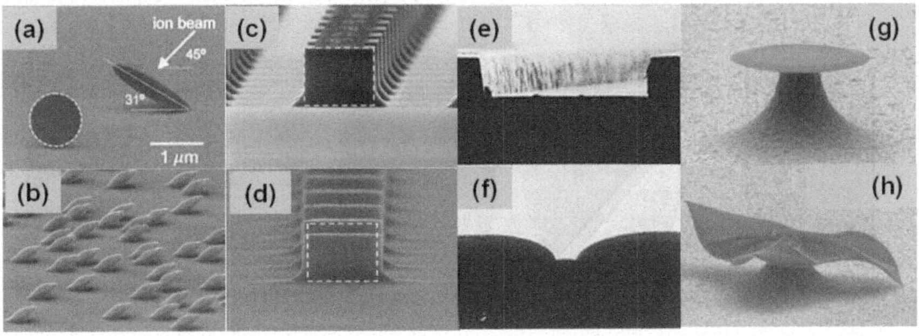

Fig. 5.15 Deformations in microstructures induced by 4–30 MeV ion irradiations observed by SEM, which the ion hammering could contribute: A silica micro-sphere deforms to an oblate spheroid. **a** Many silica micro-spheres deform to the same direction. **b** Crystalline Si micro-pillars, **c** deformed by ion irradiation after amorphization. **d** Five-μm-wide trench in a thermally grown SiO_2 film **e** are deformed. **f** An optical micro-resonator composed of a SiO_2 toroid supported by a Si post **g** shows complex deformation (**h**). Reproduced with permission from [32], ©(2004) Elsevier, [33], ©(2004) American Institute of Physics, [34], ©(2004) Elsevier, and [35], ©(1994) American Institute of Physics

and 30 MeV Cu, respectively, i.e., not purely SHIs, the deformations are not surely ascribed to the pure hammering effect only. However, this figure clearly demonstrates that the high energy ion irradiation is useful to modify the shapes of micro-structures.

As shown in Fig. 5.15a, the incident angle of the ion beam was set to 45°, while the angle of the oblate spheroid was 31°. The hammering with inclined irradiation induces not only the in-plane dilatation and the out-plane shrinkage, but also the rotation due to the off-diagonal components of the strain tensor as shown later in Eq. (5.5). This is evidence that the hammering does not play an important role in the elongation of NPs. Under an inclined SHI irradiation, it was shown that the NPs elongate with the same angle with the incident angle of the ion beam [36], which is different from the hammering.

The deformation mechanism of the hammering is ascribed to the viscous flow with the shear stress relaxation in ion tracks [37], as schematically depicted in Fig. 5.16. When a SHI goes through an amorphous solid, a cylindrical thermal spike is formed along the ion trajectory. If the temperature in the spike does not exceed the flow temperature T^*, above which the shear stress is relaxed, the memory of the original shape is preserved. While the hot region expands (Fig. 5.16(2)), the shape recovers after cooling (Fig. 5.16(1)). However, if the temperature exceeds T^*, the shear stress is steeply relaxed. This is a characteristics of an amorphous material. Consequently, the dilation along the hot spike is relaxed, i.e., the dilated hot region (Fig. 5.16(3)) returns to a flat shape (Fig. 5.16(4)). Cooling down begins from the flat shape but a hot axis. Since hotter region shrinks more, the center part becomes thinner than before the SHI hit. Consequently, the cylindrical spike region shows shrinkage parallel to the ion beam and expansion perpendicular to the beam in order to conserve the volume, i.e., the hammering.

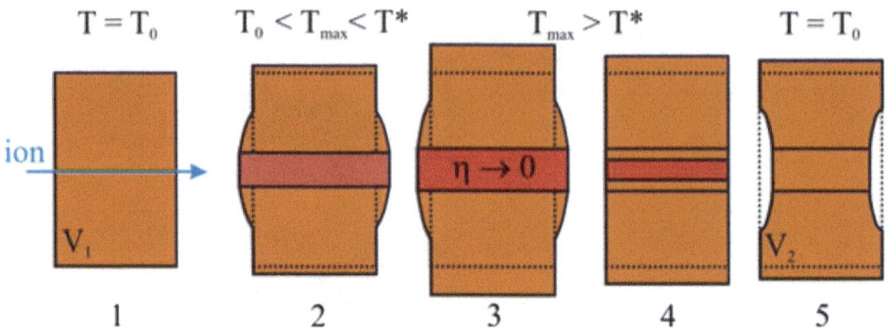

Fig. 5.16 Schematically depicted deformation of ion hammering induced by a cylindrical thermal spike in a solid. (2) If the maximum temperature T_{max} is below the flow temperature T^*, the deformation is elastic and recovered with cooling down. (3–4) If $T_{max} > T^*$, viscous deformation by shear stress relaxation is induced. (5) After cooling, a certain strain remains. Reproduced with permission from [37], ©1995 American Physical Society and [38], ©2016 Springer-Nature

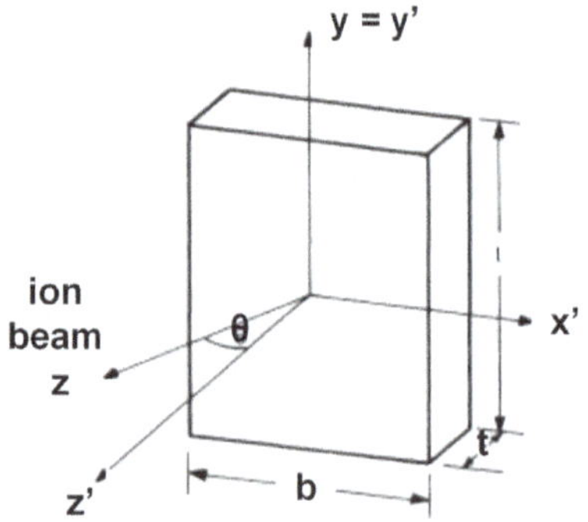

Fig. 5.17 Geometric relationship between a brick-shaped sample and ion beam. The ion beam parallel to the z axis is tilted around the y' axis by an angle θ. ($y = y'$). The width b, length l, and thickness t of the sample are defined as shown

The incident angle θ dependence of the hammering is described with the strain tensor ε, which shows the relative dimensional changes along the corresponding directions, e.g., $\varepsilon_{xx} = \Delta b/b_0$. Here, the z'-, z-, and y-axis coincide with the sample normal, the ion beam direction and the tilt axis, respectively. See Fig. 5.17 [39]. The tensor is given as,

$$\varepsilon = A(S_e, T_{irrad})\Phi \begin{pmatrix} 1 - 3\sin^2\theta & 0 & 3\sin\theta\cos\theta \\ 0 & 1 & 0 \\ 3\sin\theta\cos\theta & 0 & 1 - 3\cos^2\theta \end{pmatrix} \quad (5.5)$$

where A denotes the deformation yield, which is a function of stopping power S_e and irradiation temperature T_{irrad}. Φ denotes the ion fluence [17].

The trace, i.e., the sum of the diagonal components, of the strain tensor always vanishes, indicating the volume conservation. For the incidence angles of $\theta = 0°$ and $45°$, the angle dependence parts read, respectively,

$$\frac{\varepsilon(\theta=0^\circ)}{A\Phi}=\begin{pmatrix}1&0&0\\0&1&0\\0&0&-2\end{pmatrix},\ \frac{\varepsilon(\theta=45^\circ)}{A\Phi}=\begin{pmatrix}-1/2&0&3/2\\0&1&0\\3/2&0&-1/2\end{pmatrix} \tag{5.6}$$

In the case of $\theta = 0^\circ$, zz component is -2, whose absolute value is the double of xx and yy components. The sample shrinks along the z-direction and dilates both in the x and y directions. The former magnitude is twice of the latter. In the case of $\theta = 45^\circ$, the x and z components shrink but the y component dilates. Since the off-diagonal parts appear, the deformation is more complicated, which includes the rotation.

The constitutive equations for the ion-hammering is described [40] as,

$$\frac{d\varepsilon_{ij}}{dt}=A^0_{ij}\frac{d\Phi}{dt}+\frac{1}{2\mu}\frac{d}{dt}\left[\sigma_{ij}-\frac{\nu}{1+\nu}\sigma_{kk}\delta_{ij}\right]+\frac{1}{2\eta}\left[\sigma_{ij}-\frac{1}{3}\sigma_{kk}\delta_{ij}\right] \tag{5.7}$$

where σ_{ij}, μ, and ν are the macroscopic stress tensor, the shear modulus, and Poisson's ratio, respectively. The ion-flux-dependent shear viscosity η is assumed as $1/\eta \propto d\Phi/dt$. The strain rate tensor A^0_{ij} includes the angular dependence shown in Eq. (5.5). The 1st, 2nd, and 3rd terms describe the unconstrained ion-hammering, Hooke's law for isotropic elastic media, and the macroscopic irradiation-induced Newtonian viscous flow.

5.2 Shape Elongation of Nanoparticles

5.2.1 Observations of the Shape Elongation of Nanoparticles

The shape elongation of nanoparticles (NPs) was firstly observed by D'Orleans et al. in 2003 under swift heavy ion (SHI) irradiation [41]. They formed Co NPs with a mean diameter of ~10 nm by implantation with 160 keV Co ions to 300 nm thick SiO_2 layers on Si wafers at elevated temperature of 873 K. The fluence was set to 1×10^{17} ions/cm^2. In the previous paper [42], they studied the implantation temperature dependence of NP diameter. The mean diameter increases with the implantation temperature, as 3.5 nm (77 K), 4.8 nm (295 K), and ~8 nm (873 K). The selection of large NPs was a key to the new world, i.e., the observation of the shape elongation phenomenon, because it was later found that the elongation is induced for only the NPs larger than the ion track diameter [43]. The large NPs were a requisite for the shape elongation of NPs.

The NP samples were irradiated with 200 MeV ^{127}I ions to fluences ranging from 1×10^{11} to 1×10^{14} ions/cm^2 at room temperature. The track diameter is estimated as ~8 nm, i.e., comparable to the NP mean diameter. Furthermore, they reported that the mean diameter of NPs increased under the SHI irradiation up to the fluence of 10^{12} ions/cm^2 with remaining the NPs spherical. At 1×10^{13} ions/cm^2, some NPs

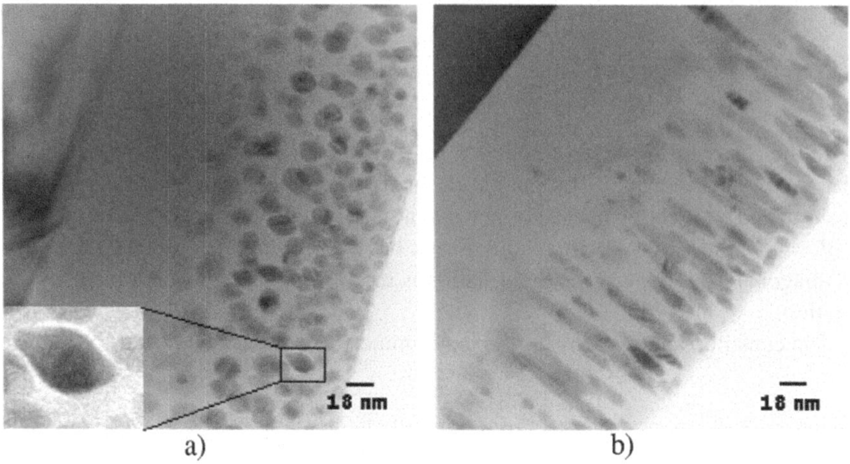

Fig. 5.18 Cross-sectional TEM images of Co NPs fabricated in SiO_2 by 160 keV ion implantation at 873 K, which were then irradiated with 200 MeV ^{127}I ions to **a** 1×10^{13} and **b** 1×10^{14} ions/cm^2. Reproduced with permission from [41], ©2003American Physical Society

exhibited a lemon shape as shown in the inset of Fig. 5.18. At 1×10^{14} ions/cm^2, a more drastic change was induced. NPs were elongated along the same direction with the propagation of the SHI ions. The mean sizes of the NPs changed to 35 nm along the beam direction and 9 nm perpendicular to it.

In the next year of this discovery, the same phenomenon was observed as a different configuration, i.e., in chemically-synthesized self-standing Au-core/silica-shell colloidal NPs as shown in Fig. 5.19 [44]. After irradiation with 30 MeV selenium

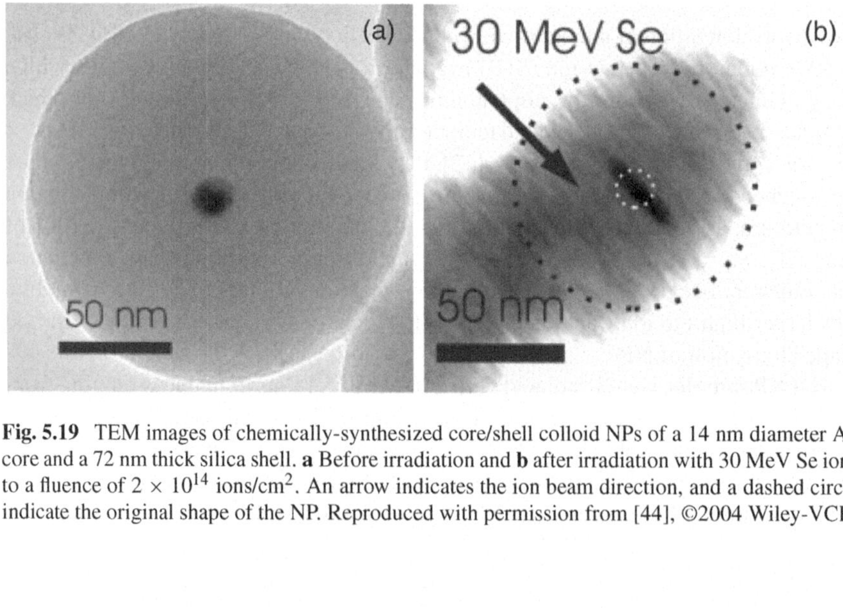

Fig. 5.19 TEM images of chemically-synthesized core/shell colloid NPs of a 14 nm diameter Au core and a 72 nm thick silica shell. **a** Before irradiation and **b** after irradiation with 30 MeV Se ions to a fluence of 2×10^{14} ions/cm^2. An arrow indicates the ion beam direction, and a dashed circle indicate the original shape of the NP. Reproduced with permission from [44], ©2004 Wiley-VCH

ions to a fluence of 2×10^{14} ions/cm^2, the spherical gold core of 14 nm in diameter elongated along the direction of the ion beam to a rod of 6 nm by 54 nm. Simultaneously the silica shell shows expansion perpendicular to the beam and shrinkage parallel to the beam as shown in Fig. 5.19b. Since colloidal silica NPs without the Au cores show the similar shape changes, the deformation of the SiO_2 shells were ascribed to the ion hammering [45].

To elucidate the deformation mechanism, a series of core/shell NPs were formed with different silica shell thickness ranging from 15 and 72 nm and an identical Au core diameter of 14 nm. No measureable deformation of the Au core was confirmed for the shells thinner than 26 nm under 30 MeV Se irradiation. Of course, Au NPs of the same diameter of 14 nm without the silica shell were not deformed under the irradiation. From these observations, Roorda et al. concluded that silica shells play a major role in the deformation of the Au cores, and the cores are deformed by an indirect effect. The driving force for deformation of the Au cores is the ion hammering effect of the silica shells. The silica expands perpendicular to the ion beam due to the hammering effect, and in-plan stress in silica acts on the Au cores. To release it, the core deforms to outer-plan direction, i.e., parallel to the ion beam. This scenario was applied to the elongation of metal NPs embedded in bulk or thin-film silica such as shown in Fig. 5.18.

However, the ion hammering can build up the stress in the order of 100 MPa only, which is too low to induce clear deformation in solid state Au NPs. Therefore, Roorda et al. assumed that radiation-induced softening of Au NPs was simultaneously induced. However, Klaumunzer pointed out that the observed large elongation of NPs was not explained by any known mechanisms of the radiation-induced softening [46].

After then, many researchers supported the synergy model, in which the shape elongation was induced by the synergy effect of the ion hammering (the stress generation) and the transient melting of NPs (the softening of NPs) due to the thermal spike effect. However, the authors don't know who firstly proposed this model. The description of the model is found in Ref. [48]. It should be noted that the recent major model is the thermal pressure model, which will be described later.

The shape elongation phenomenon induced under SHI irradiation has been observed in various elemental metal NPs as summarized in Table 5.2. It seems that the shape elongation is the common phenomenon for elemental metal NPs. Also it was observed in NPs of a solid-solution of $Au_{1-x}Ag_x$ [49] and of intermetallic compound of FePt [50].

Semiconductor Ge NPs show a bimodal behavior [51] as shown in Fig. 5.20: While small NPs show the prolate deformation similar with metal NPs, medium-sized NPs

Table 5.2 Species of nanoparticles which exhibit the ion shaping

Metal	V, Co, Ni, Cu, Zn, Ag, Sn, Pt, Au, Pb, Bi
Semiconductor	Ge [51]
Solid solution	$Au_{1-x}Ag_x$ [49]
Intermetallic compound	FePt [50]

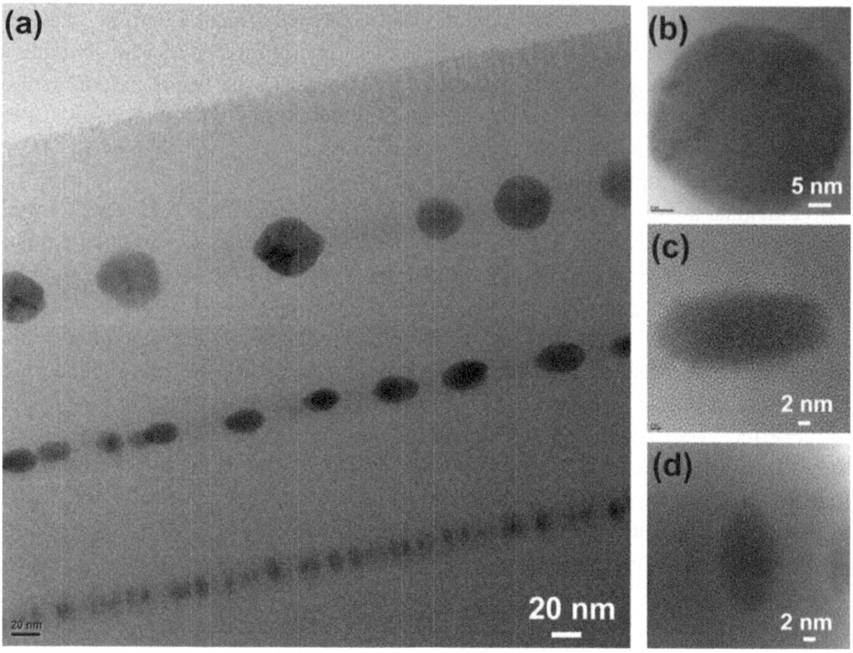

Fig. 5.20 Cross-sectional TEM images of Ge NPs with three different sizes in SiO_2, which were irradiated with 38 MeV I^{7+} to 1×10^{14} ions/cm^2. While the smallest NPs exhibited prolate deformation, the medium-sized ones oblate deformation. The largest ones did not deformed. Reproduced with permission from [51], ©2009 Elsevier

show the oblate deformation probably due to the ion hammering. The largest NPs did not change. Since crystalline Ge NPs is much more sensitive to amorphization under irradiation as mentioned in Chap. 4, the irradiated Ge NPs are probably in amorphous and exhibit the ion hammering. Araujo et al. formed Ge NPs in SiO_2 by 2 MeV Ge ion implantation and thermal annealing at 1100 °C. They have confirmed the shape elongation of Ge NPs under Au irradiation of 27, 54, and 185 MeV Au ions [52].

Amekura et al. irradiated ZnO NPs in SiO_2 with 200 MeV Xe ions and observed the decomposition of ZnO NPs to Zn metal NPs, and the shape elongation of newly formed Zn NPs as shown in Fig. 5.21. It seems that the SHI irradiation induced the partial decomposition of ZnO NPs to Zn and O atoms, the re-aggregation of Zn, and the shape elongation [53].

Recently Dawi et al. reported that NiO NPs in SiO_2 exhibited the shape elongation under 32 MeV Cu^{5+} ion irradiation [54]. However, the result is not easy to be accepted because they reported that the magnetization loops of the NiO NP samples before and after the irradiation in the same paper, which clearly showed a ferromagnetic behavior. Since NiO and Ni are anti-ferromagnetic and ferromagnetic, respectively, the observation clearly indicates the existence of a certain amount of metallic Ni NPs

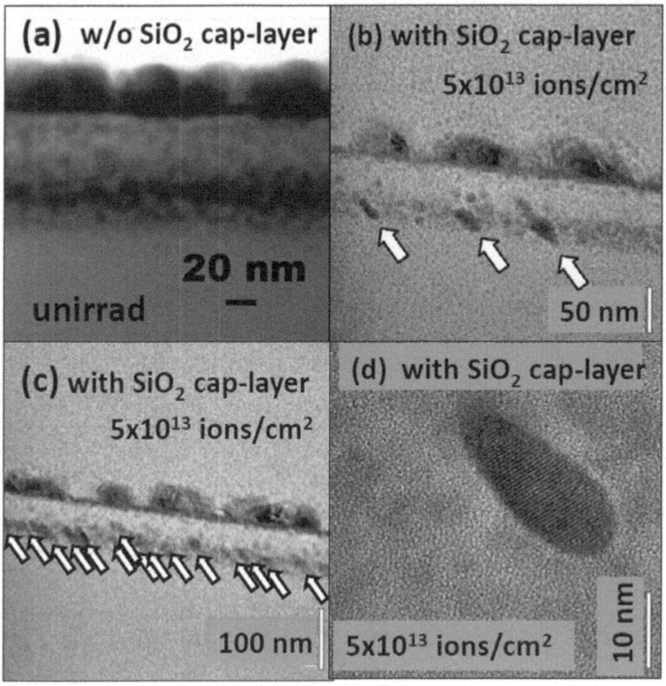

Fig. 5.21 Cross-sectional TEM images of ZnO NPs embedded in SiO_2 which were irradiated with 200 MeV Xe ions. The sample consists larger ZnO NPs in upper layer and smaller ZnO NPs in lower layer. Elongated metallic Zn NPs indicated by arrows appeared in the lower layer. Reproduced with permission from [53], ©2013 American Institute of Physics

in the samples. The observed elongated NPs could be ascribed to metallic Ni but not NiO. The SHI irradiation may decompose NiO NPs to Ni NPs. The decomposed Ni metal NPs may show the shape elongation as usual. This is the same behavior observed in Fig. 5.21 [53].

5.2.2 Minimum Width for the Elongation

Giulian et al. measured major and minor dimensions (D_{minor}, D_{major}) of more than 500 nano-rods from TEM micrographs, and plotted D_{minor} as a function of D_{major}. Each data point in Fig. 5.22 represents the mean value of D_{minor} for a given $D_{\mathrm{major}} \pm 1$ nm; i.e., each data point is not from a NPs but after averaging over many NPs. Before the irradiation, the data points fall on a straight line with a slope of unity as shown by white diamonds, indicating the NPs are spherical. With increasing the fluence, only the NPs larger than the threshold diameter of ~6.5 nm exhibit the elongation. The slope of the data points larger than ~6.5 nm decreases, i.e., $D_{\mathrm{minor}} < D_{\mathrm{major}}$, with increasing

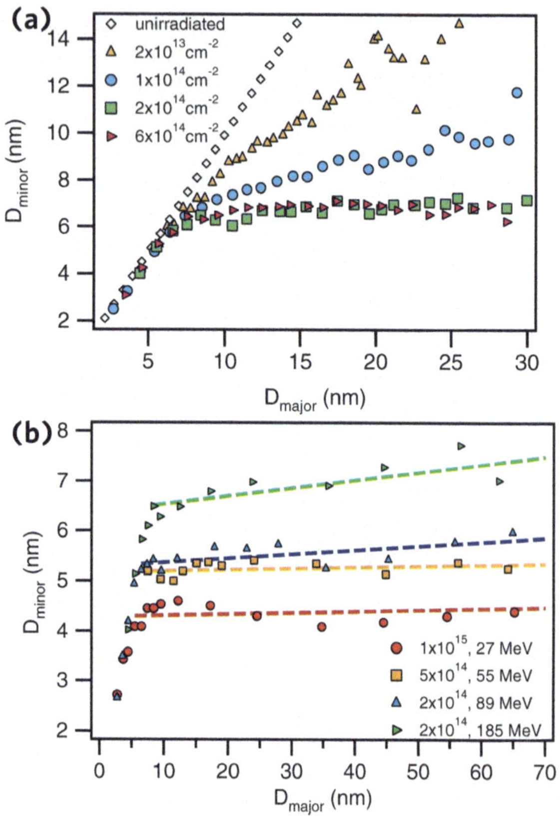

Fig. 5.22 **a** Major and minor dimensions of Pt nano-rods determined by TEM observation are plotted for different fluences of 185 MeV Au ion irradiation. **b** The same plot but different ion energies of Au ions at saturated fluences which are indicated in the figure. Reproduced with permission from [55], ©2008 American Physical Society

the fluence. At the fluence of 2×10^{14} ions/cm^2 and higher, the slope becomes zero; i.e., all the elongated nano-rods exhibit almost the same width. Even increasing the fluence further, the width of nano-rods no longer becomes thinner. Contrary the NPs smaller than the threshold of ~6.5 nm seem to maintain the spherical shapes, with staying on the diagonal line. However, this observation does not mean that smaller NPs than the threshold maintain the size and the spherical shapes during SHI irradiation. Rather it is considered that these small NPs are suffered by vaporization by SHI impacts. See the "particle-in-vacuum model" in Sect. 5.2.3. During these processes, parts of NPs are dissolved into the matrix. The dissolved atoms are again aggregated or absorbed to remaining NPs and nano-rods. The newly aggregated NPs show spherical shapes. The smaller NPs also experience the dynamical processes under SHI irradiation.

The threshold diameter of ~6.5 nm is in the case of 185 MeV Au irradiation, which changes with the ion energy, ion species and of course NP species. Figure 5.22b shows the saturated (D_{minor}, D_{major}) distributions for different energies of Au ions. With decreasing the ion energy from 185 MeV to 27 MeV, the threshold diameter decrease from ~6.5 to ~4 nm.

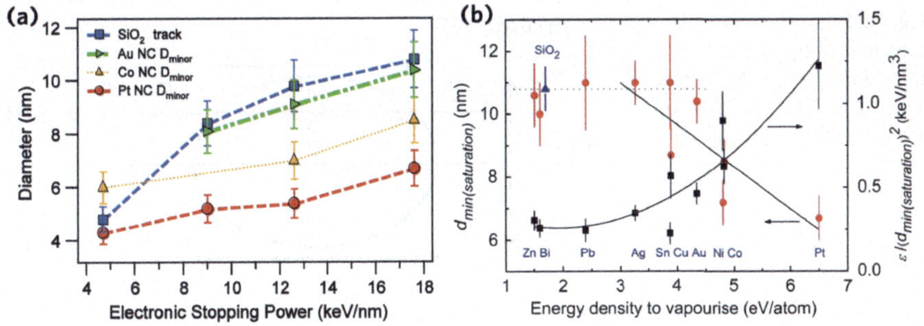

Fig. 5.23 **a** Saturated NP width for Au, Co, Pt NPs and the ion track diameter in SiO_2 are plotted with electronic stopping power. **b** Saturated NP width is plotted by circles with the energy density per atom required for vaporization. Reproduced with permission from [56], ©2011 Elsevier and [43], ©2011 American Physical Society

The fluences for the saturation are noted in the figure. Lower energy irradiation requires higher fluence for the saturation. If readers want to make thinner width nano-rods or to induce elongation of smaller NPs, they should apply lower energy SHI beam, because the saturated width of NPs is thinner. Of course, lower energy SHI beams exhibit lower elongation efficiency, i.e., require longer irradiation time.

Saturated widths of Au, Co, and Pt nano-rods are plotted with the electronic stopping power in Fig. 5.23a. For comparison, the diameter of molten tracks in SiO_2 is also plotted. The saturated width of Au nano-rods is very close to the diameter of ion tracks. The saturated width of Co and Pt nano-rods are smaller but could be limited by the track diameter. Data of saturated nano-rod widths of more NP species are plotted with the energy density for vaporization in Fig. 5.23b. Lower melting point NPs, i.e., Zn, Bi, Pb, Ag, Sn, Au, in this case, show the saturated widths very close to the track diameter in SiO_2, while higher melting point NPs, i.e., Ni, Co, Pt, in this case, show somewhat smaller saturated widths than the ion track diameter.

Rizza et al. studied the minimum width for elongation based on TEM observation and discussed the fragmentation process in terms of the Rayleigh instability. However, the minimum width estimated from the Rayleigh instability was ~15 nm, which was twice larger than the minimum width observed (~8 nm) for the shape elongated nano-rods [49]. It should be reminded that the fragmentation and dissolution of nano-rods are different phenomena. It is important to clarify which process is dominant. It seems that both the dissolution and fragmentation of NPs are simultaneously induced [56].

5.2.3 *Particle-in-Vacuum Model*

The saturation phenomenon is partly understood by a following simple phenomenological model "particle-in-vacuum model" as shown in Fig. 5.24. In this model, a

Fig. 5.24 A schematically depicted configuration for calculating energy deposition in a prolate NPs from passing through swift heavy ion

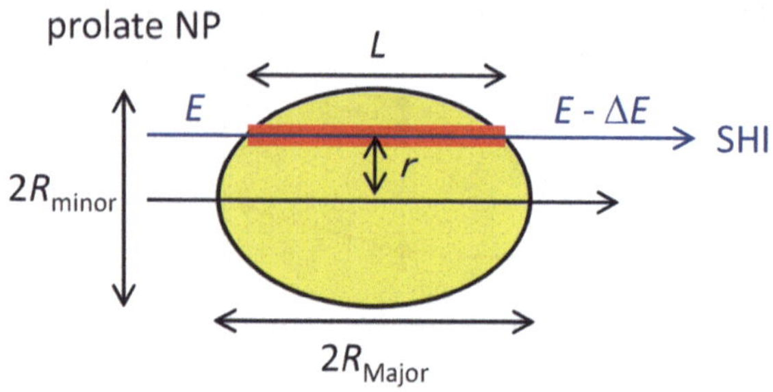

prolate NP embedded in a perfect thermal insulator is presumed. As known, a spherical NP is a special case of prolate NP. When SHI has passed through along the major axis (i.e., on-axis impact), the energy E_0,

$$E_0 = 2R_{\text{major}}\left(\frac{dE}{dx}\right) = 2R_{\text{major}}S_e^{\text{NP}} \tag{5.8}$$

is deposited in the NP, where $2R_{\text{major}}$, and S_e^{NP} denote the major axis of the prolate NP and the electronic stopping power in the NP, respectively. In the case of off-axis impact with the impact distance r, i.e., the distance between the impact point and the axis, the path-length L along which a SHI passes through in the NP is given as,

$$L(r) = 2R_{\text{major}}\sqrt{1 - \left(\frac{r}{R_{\text{minor}}}\right)^2} \tag{5.9}$$

When ions come with random impact distances, the mean path length L_{mean} is given after averaging as,

$$L_{\text{mean}} = \iint L(r)rdrd\theta \Big/ \iint rdrd\theta = \tfrac{4}{3}R_{\text{major}} \tag{5.10}$$

and the mean energy deposition E_{mean} is given as,

$$E_{\text{mean}} = L_{\text{mean}}S_e^{\text{NP}} \tag{5.11}$$

The deposited energy E is assumed to be equally shared by all the atoms consisting the elongated NP, but not to be leaked to the outside of the NP. Consequently, each atom in the elongated NP has the deposited energy of E_{mean}/N,

$$\frac{E_{\text{mean}}}{N} = \frac{L_{\text{mean}}S_e^{\text{NP}}}{nV} = \frac{S_e^{\text{NP}}}{\pi n}\frac{1}{R_{\text{minor}}^2} \tag{5.12}$$

where $N = nV$ denotes the number of atoms forming the NP, V, the volume of the NP, n, the atomic number density in the NP, R_{minor}, the semi-minor axis.

An important feature is that the deposited energy per atom, E_{mean}/N, is inversely proportional to R^2_{minor}, and does not depend on R_{major}. The deposited energy per atom in the NP are converted to the temperature of the NP transiently reached by thermal spikes, using T-dependent specific heat and the latent heats at melting and vaporization. For example, when E/N is higher than the melting energy but lower than the vaporization energy, the transient maximum temperature T is determined from E_{mean}/N using following relationship,

$$E_{\text{mean}}/N = \int_{T_{\text{irrad}}}^{T_m} C_s(T)dT + L_{\text{melt}} + \int_{T_{\text{im}}}^{T} C_m(T)dT \tag{5.13}$$

where T_{irrad}, T_m, $C_s(T)$, $C_m(T)$, and L_{melt} denote irradiation temperature, melting temperature, the specific heat in the solid phase and in molten phase, and the latent heat of melting, respectively.

At first, spherical NPs are considered: In case, the R_{minor} in Eq. (5.12) is replaced by the radius of NPs. The Eq. (5.12) indicates that the deposited energy per atom, i.e., the maximum temperature of the NP, which increases with decreasing the radius of NPs even under the same impact with SHI. While much larger NPs are always in solid phases even under the same thermal spike effect, NPs smaller than a certain size show the transient molten transformation. Further decrease in the radius finally induces the vaporization transition in a NP. NPs having the minimum size for elongation $R_0{}^{\text{th}}$ or smaller experience the vaporization transition, which cancels the elongation.

The saturation width of the elongated NPs, $2R^{\text{sat}}_{\text{minor}}$, at high fluences is also explained from Eq. (5.12): The maximum temperature of NPs is determined only by the semi-minor axis, R_{minor}, which is independent of R_{major}, as shown in Eq. (5.12). This implies that NPs having the same R_{minor} have the same maximum temperature, irrespective of R_{major}. The NPs having $R^{\text{sat}}_{\text{minor}}$ came to the boundary where no shorter R_{minor} exists. If an elongated NP has shorter R_{minor} than $R^{\text{sat}}_{\text{minor}}$, the NP will be vaporized with the next SHI impact, which results in the fragmentation or dissolution of NPs.

While the above-mentioned particle-in-vacuum model well explains the existence of the minimum size for elongation, $2R_0{}^{\text{th}}$, and of the saturation of the minor axis, $2R^{\text{sat}}_{\text{minor}}$, following points are inconsistent with the observations.

The particle-in-vacuum model assumed all the energy deposited by a single ion was shared within the NP only, and stays always within the NP. On the contrary, the 3D thermal spike calculation [57] on an Au NP in SiO_2 showed that a certain amount of the deposited energy is transferred to SiO_2 region close to the NP boundary, i.e., the "outer boundary heating" scenario described in 5.3.1. Consequently, the energy is shared within larger volume, resulting in more dilute energy than the particle-in-vacuum model. From Eq. (5.12), the saturated minor axis $2R^{\text{sat}}_{\text{minor}}$ is calculated as

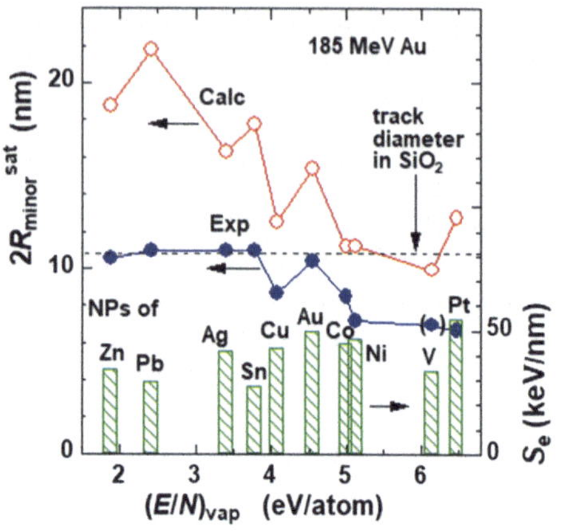

Fig. 5.25 Saturated minor axes for various NP species from experimental (closed circles) [43] and calculated by the particle-in-vacuum model (open circles). S_e in various NPs are shown by bars

$$R_{\text{minor}}^{\text{sat}} = \sqrt{\frac{S_e}{\pi n} \frac{1}{(E/N)_{\text{vap}}}} \tag{5.14}$$

The Eq. (5.14) indicates that lower (*E*/*N*) for vaporization, larger the $2R_{\text{minor}}^{\text{sat}}$ is. However, Ridgway et al. showed that the R_0^{th} of different NP-species with low and medium MPs were rather constant, which were comparable with the track radius in SiO_2 matrix, as shown in Fig. 5.23b. Their data were re-plotted in Fig. 5.25 with a simple prediction from the particle-in-vacuum model, i.e., Eq. (5.14).

The experimental $R_{\text{minor}}^{\text{sat}}$ of high MP species decreased with increasing the *E/N*, which indicated that the rule $R_{\text{minor}}^{\text{sat}} \propto (E/N)_{\text{vap}}^{-1/2}$ held in the very limited region. However, the experimental data are rather constant in the low and medium $(E/N)_{\text{vap}}$ values, while Eq. (5.14) predicts that $R_{\text{minor}}^{\text{sat}}$ increases with decreasing $(E/N)_{\text{vap}}$. Some modification on the particle-in-vacuummodel or completely new model is necessary.

Rizza et al. observed that NPs larger than the melting criterion of the particle-in-vacuum model also exhibit the shape elongation, and ascribed to the elongation due to the surface melting [58]. In these larger NPs, the deposited energy from a SHI impact is not enough for melting of whole the volume of NP. Even a NP in an incomplete melting form, i.e., molten surface layer and a solid core, can contribute the elongation. This is also the consequence of the "outer boundary heating" scenario described in 5.3.1.

5.2.4 Mass Non-conservation of Nano-Rods

Dawi et al. [59] deposited chemically synthesized monodisperse colloidal Au NPs in spherical shapes on the surface of oxidized Si substrates, and covered them by SiO_2 layer of 150 nm by reactive sputtering. With adjusting the immersion time to the Au NP solution, samples with different NP areal density were prepared. Degree of the elongation was monitored from the width of the Au peak of Rutherford backscattering spectrometry (RBS) using 2 MeV $^4He^+$ with a scattering angle of 120°. (The RBS method is not adequate to detect small elongation such as the elongation threshold as pointed out in Ref. [60], but convenient for detecting large elongation as discussed here.) As shown in Fig. 5.26, the FWHM of the Au peak increases with the fluence up to 2×10^{14} ions/cm^2, irrespective of the particle density. Above the fluence, the lowest density sample (2.5×10^9 NPs/cm^2) shows saturation in FWHM. The saturation is ascribed to the fact that the minor widths of all the monodispersed NPs reach to 'the minimum width' (see Sect. 3.2.2). Since the initial diameters, consequently the volume, of all the NPs were the same irrespective of the NP areal density, and since the elongation speed is consider to be the same between the samples, all the samples should reach to 'the minimum width' at the same fluence, irrespective of the NP areal density. However, higher density samples show further increase of the FWHM, indicating the mass transfer between the NPs. Probably some NPs were destroyed by the irradiation. Au atoms from the destroyed NPs transferred to other NPs for further elongation.

This behavior was more clearly confirmed by Rizza et al. [49] using TEM observation as summarized in Fig. 5.27. Around 2×10^{14} ions/cm^2, the width D_{minor} of the nano-rods reaches the minimum critical value as shown in Fig. 5.27a, irrespective of the NP density. During further irradiation, some nano-rods may have the width smaller than the critical value. However, they are vaporized and (partly) dissolved into the matrix. The dissolved atoms are transferred to other nano-rods to increase the mass (volume) of the nano-rods and to maintain further elongation. As shown in Fig. 5.27b, the normalized volume of NPs starts to increase exceeding the fluence of 2×10^{14} ions/cm^2, but saturates around 5×10^{14} ions/cm^2. However, the saturated volume is higher for the higher NP density.

Applying X-ray absorption near-edge structures (XANES) measurements, Ridgway et al. evaluated the fraction of Co atoms in metallic and oxidized (isolated) environments for Co NPs of 12 nm diameter in SiO_2, irradiated with Au ions of 9,

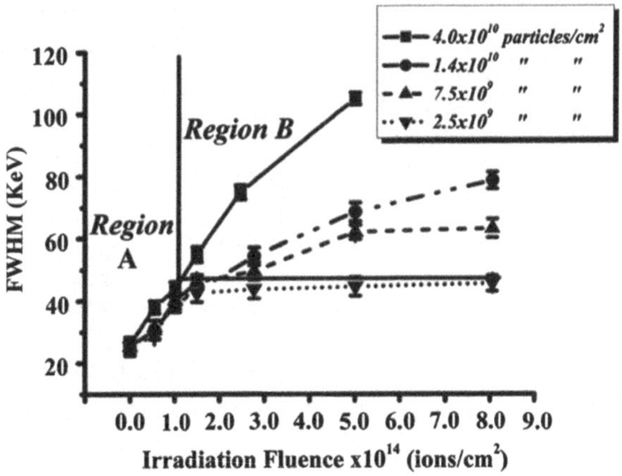

Fig. 5.26 Fluence dependence of FWHM of Au RBS peak from Au NPs (15 nm diameter) in SiO_2 irradiated with 54 MeV Ag^{8+} ions, for different particle areal densities. Reproduced with permission from [59], ©2011 American Institute of Physics

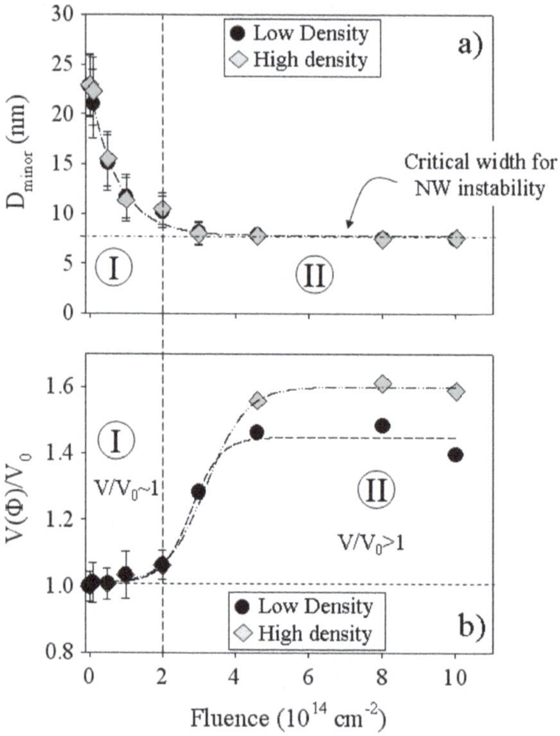

Fig. 5.27 Fluence dependent evolution of **a** the minor axis and of **b** normalized NP volume for low and high NP densities. Reproduced with permission from [49], ©(2011) IOP

27, 89, and 185 MeV as shown in Fig. 5.28 [56]. The oxidized environment means Co atoms whose neighborhoods are not occupied by other Co atoms but by O atoms, i.e., isolated Co atoms, while the metallic environment means the neighborhood are occupied by other Co atoms. Before the irradiation, the metallic fraction was ~100%.

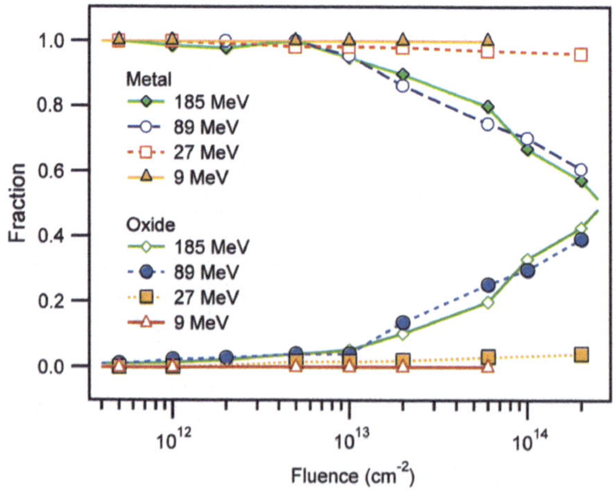

Fig. 5.28 Fraction of Co atoms in metallic and oxidized (isolated) environments detected by XANES for Co NPs of 12 nm diameter in SiO_2, irradiated with Au ions. The ion energy is shown in the figure. Reproduced with permission from [56], ©2009 Elsevier

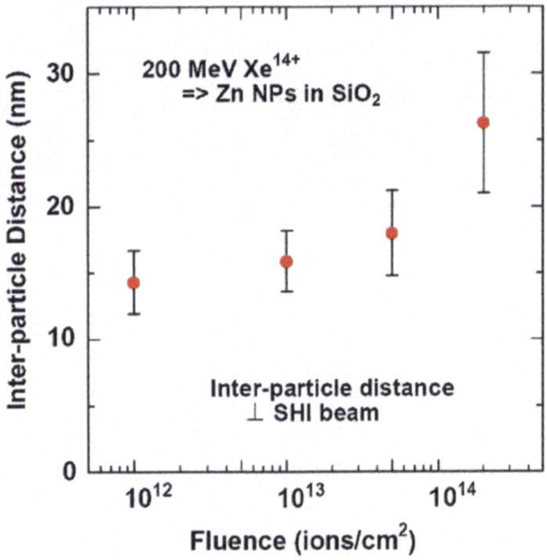

Fig. 5.29 Fluence dependence of the mean inter-particle distance between Zn NPs, in the plane perpendicular to 200 MeV Xe ion beam, determined from the interference peak of SAXS. Zn NPs were formed by 60 keV Zn^+ implantation. Reproduced with permission from [47], ©2014 Wiley-VCH Verlag

With increasing the fluence, the oxidized (isolated) fraction gradually increases for ions of 89 and 185 MeV, indicating progressive dissolution of Co NPs into the matrix. At the fluence of 2×10^{14} ions where the minimum width is attained for most of NPs, the oxidized (isolated) fraction increases to almost 50% for 89 and 185 MeV. Contrary, the dissolution of NPs is very small for 27 MeV and lower ions, indicating that the dissolution is induced by electronic excitation but not the ballistic mixing.

Amekura et al. studied the small angle X-ray scattering (SAXS) from Zn NPs in SiO_2, which were formed by 60 keV Zn^+ implantation and irradiated with 200 MeV Xe ions at different fluences [47]. Since the NPs are formed by low energy keV ion implantation, the inter-particle distance is much smaller compared with NPs formed by MeV implantation. In fact, the NPs showed strong interference peak, from which the mean inter-particle distances were evaluated. It was found that the inter-particle distance increases with the fluence as shown in Fig. 5.29. With increasing the fluence, the dissolution of NPs is progressed and some NPs are missing. Since SAXS detects NPs only but did not detect the dissolved atoms, the missing of NPs due to the dissolution is observed as the increase of the inter-particle distance.

5.2.5 Initial Nanoparticle Size Dependence on the Elongation

Rizza et al. fabricated various sizes of monodispersed colloid Au NPs in spherical shapes by chemical synthesis, and embedded them in SiO_2, which were irradiated with 74 MeV Kr ions. As shown in Fig. 5.30, the morphological evolution of the NPs strongly depends on the initial size of the NPs [58]:

(i) Satellite NPs (<10 nm) remain spherical in shape even after the irradiation. This is because the vaporization is induced for NPs smaller than ~10 nm when they are impacted with SHI.
(ii) NPs in the range of 10–30 nm transform continuously into nano-rods.

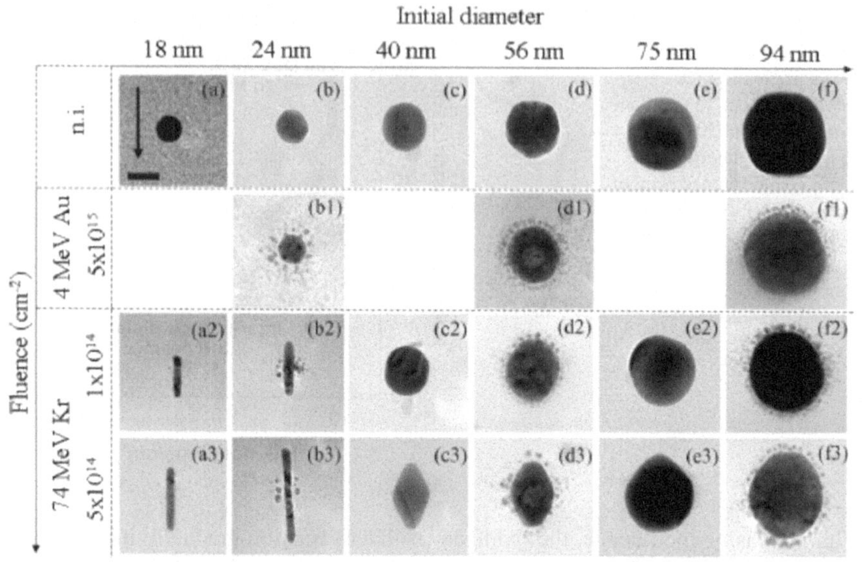

Fig. 5.30 Initial size dependent TEM images of Au NPs irradiated with 74 MeV Kr ions to the fluences of 0, 1×10^{14}, and 5×10^{14} ions/cm^2. To form satellite NPs, the samples (**b**), (**d**), and (**f**) were irradiated with 4 MeV Au ions prior to the Kr irradiation. Reproduced with permission from [58], ©2012 American Physical Society

(iii) NPs in the range of 30–70 nm require longer time for deformation, and transform to facetted NPs as shown in Fig. 5.30(c3). This is because only the surface layers are heated to melt but the cores maintain in a solid phase.
(iv) Larger NPs (>70–80 nm) are not noticeably deformed. It should be noted that a bulk Au no longer transforms to a molten phase by SHI irradiation, because of high heat conductivity and low electron-lattice coupling.

Figure 5.31a exhibits fluence dependence of the major axis of NPs normalized by the initial diameters D_o of 29, 35, 46, and 95 nm. With increasing the initial diameter, the normalized deformation $L(\Phi)/D_o$ becomes smaller. Noted that the results of smaller NPs than 29 nm are not shown in Fig. 5.31a. The normalized maximum deformation $L(\Phi = 5 \times 10^{14}$ ions/cm$^2)/D_o$ is plotted against the initial diameter in Fig. 5.31b. The maximum length increases with the initial diameter from 15 to 30 nm, but turns to decrease after showing a peak around 30 nm. The increasing behavior up to 30 nm is understood by following model.

Since the fluence of 5×10^{14} ions/cm^2 is high enough for NPs with initial diameters of 15–30 nm to attain the saturated minimum width of the nano-rods. The initial volume of NP is $(4\pi/3)R_o^3$ and the saturated volume of the nano-cylinder is $\pi\left(R_{\mathrm{minor}}^{\mathrm{sat}}\right)^2 L_{\max}$, where $R_{\mathrm{minor}}^{\mathrm{sat}}$ denotes the saturated minimum radius of the nano-rods. As described in the previous sub-section, the volume (mass) of NPs is not always preserved. However, in the case that the particle density is low enough, the mass conservation can be assumed, i.e., $(4\pi/3)R_o^3 = \pi\left(R_{\mathrm{minor}}^{\mathrm{sat}}\right)^2 L_{\max}$. Consequently, the

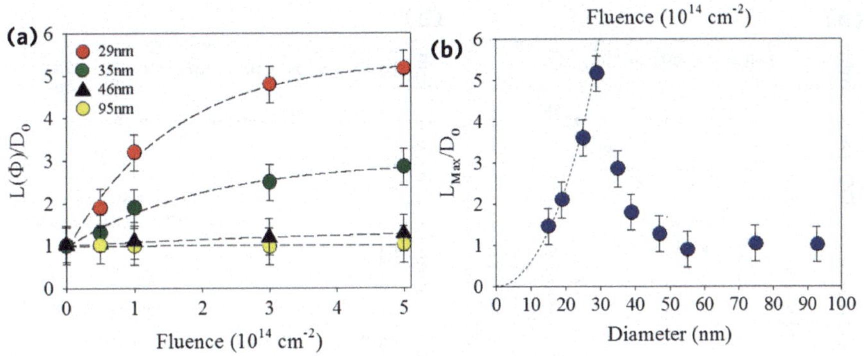

Fig. 5.31 **a** Evolution of the normalized major axis $L(\Phi)/D_o$ with the fluence Φ for $D_o = 29$, 35, 46, and 95 nm, where D_o is the initial NP diameter. **b** The initial NP diameter dependence of the normalized maximum deformation length, obtained at a fluence of 5×10^{14} ions/cm^2. Reproduced with permission from [58], ©2012 American Physical Society

normalized major axis $L/2R_o$ of NP increases proportional to the square of the initial radius R_o,

$$\frac{L_{\max}}{2R_0} = \frac{2}{3\left(R_{\text{minor}}^{\text{sat}}\right)^2} R_0^2 \tag{5.15}$$

which is indicated by a curve in Fig. 5.30b. Larger than 30 nm, the NPs do not follow this relation, indicating that the deformation does not reach the saturated state or that NPs are not fully in a molten state.

5.2.6 Scaling Law for the Elongation

Amekura et al. evaluated degree of the elongation of NPs using the polarization angle dependent anisotropic optical absorption, i.e., optical linear dichroism (OLD) [16], for various SHI irradiations [61]. Fluence dependences of the elongation determined from the OLD for various ion irradiations are exhibited in Fig. 5.32a. High S_e ions, e.g., 200 MeV Xe, induce larger elongation at lower fluence. Lower S_e ions require higher fluence for a given elongation. Here, the elongation efficiency is defined by the inversed value of the fluence, at which the given ΔOD of 0.1 is induced. The elongation efficiencies are estimated for various ions from this figure and plotted with nuclear and electronic stopping powers, as shown in Fig. 5.32b, c, respectively [61]. While the elongation efficiency increased with S_e, non-correlated dependence was observed between the elongation efficiency and the nuclear stopping S_n, confirming that the shape elongation is driven by the electronic energy deposition process.

Some readers might consider it is a matter of course that not S_n but S_e plays an important role in the shape elongation of NPs because this phenomenon is induced

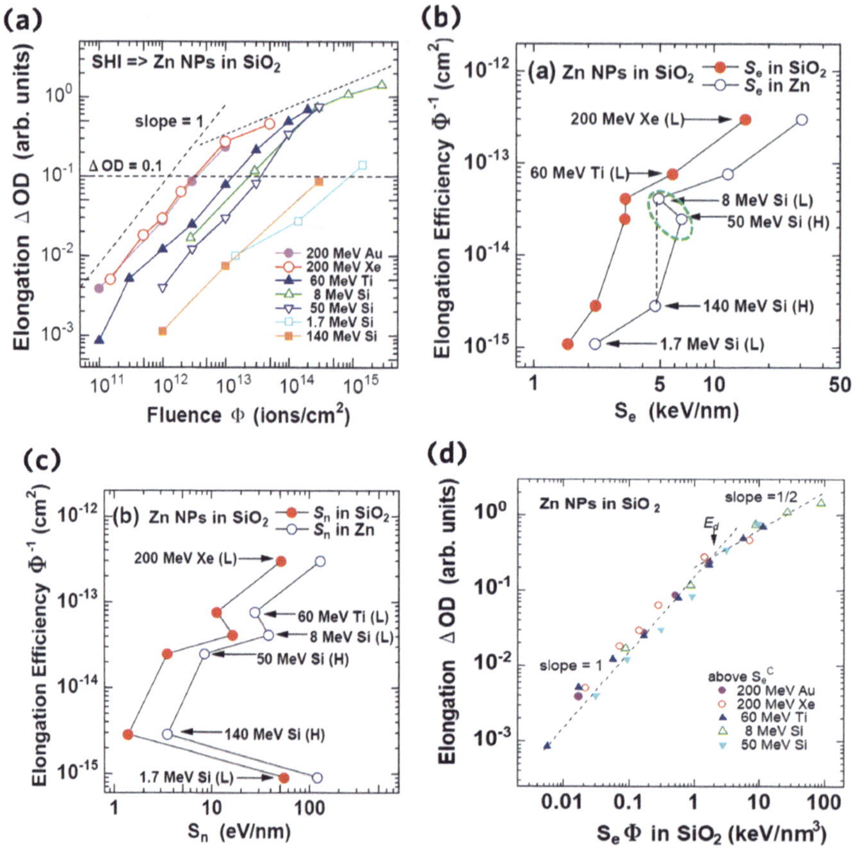

Fig. 5.32 **a** Fluence dependence of the elongation of Zn NPs in SiO_2 determined by optical linear dichroism (OLD) for various SHI irradiations. **b**, **c** The elongation efficiency of each beam was determined from **a** and plotted against S_e and S_n, respectively. **d** All the data points in this figure (**a**) except having S_e less than 2.5 keV/nm are plotted with the electronic deposition energy $S_e\Phi$ in SiO_2. All of the data fall on the same line, i.e., the elongations are well normalized by the deposited energy $S_e\Phi$. Reproduced with permission from [61], ©2014 IOP

under swift heavy ion irradiation where $S_e \gg S_n$. However, it was not trivial. Here we show a phenomenon where not S_e but S_n plays an important role, which is even induced under up to 200 MeV Xe ion irradiation. Lattice constant changes of yttria-stabilized zirconia (YSZ), which were induced by various ions from 7.3 MeV Fe ions to 2.2 GeV Au ions, are plotted in Fig. 5.33 against (a) the electronic and (b) the nuclear energy deposition. When the data is plotted against the nuclear energy deposition, the data points fall around an identical curve. However, it is not the case for the electronic energy deposition. While the lattice constant change is induced under 200 MeV Xe ions, the process is governed by S_n.

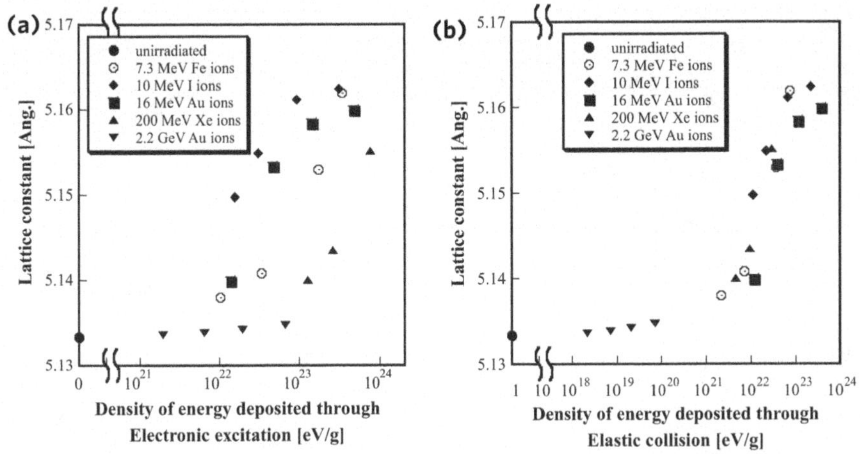

Fig. 5.33 Changes in the lattice constant of yttria-stabilized zirconia (YSZ) irradiated with 7.3 MeV Fe, 10 MeV I, 16 MeV Au, 200 MeV Au, and 2.2 GeV Au ions. The data points are plotted in **a** the electronic energy deposition and **b** the nuclear energy deposition. The nuclear energy deposition governs the data better. Reproduced with permission from [62], ©2016 Elsevier

Another question is which plays more important role, S_e in SiO_2 or S_e in metal NPs. While we have not had a conclusion yet, S_e in SiO_2 exhibits better correlation as shown in Fig. 5.32b [61].

The fluence dependences shown in Fig. 5.32a, except $S_e < 2.5$ keV/nm, are plotted with the deposited energy dependences, i.e., $S_e\Phi$, in Fig. 5.32d. All the data measured at different ions/energies fall on the same line, except ions with $S_e < 2.5$ keV/nm. This observation indicates that the shape elongation induced by different ion beams is normalized using the electronic energy deposition $S_e\Phi$.

5.3 Mechanism of Shape Elongation

5.3.1 Inelastic Thermal Spike

As described in Sect. 5.2.1, one of the candidate mechanisms is the synergy effect of ion hammering and melting of NPs by thermal spike effect. Another major candidate is the thermal pressure effect, which is recently more supported. Since both the models assume the melting of the NPs and of the matrix by the inelastic thermal spike (i-TS) effect, the concept of the i-TS model is briefly described here:

When a swift heavy ion (SHI) goes through a solid, the solid is suffered by high density electronic excitation but almost no direct atomic displacements. Then the excitation is transferred from the electronic sub-system to the atomic sub-system. In the i-TS model, these excitation-relaxation processes are described by the electronic temperature T_e and the lattice temperature T_a, via a pair of coupled heat-diffusion equations:

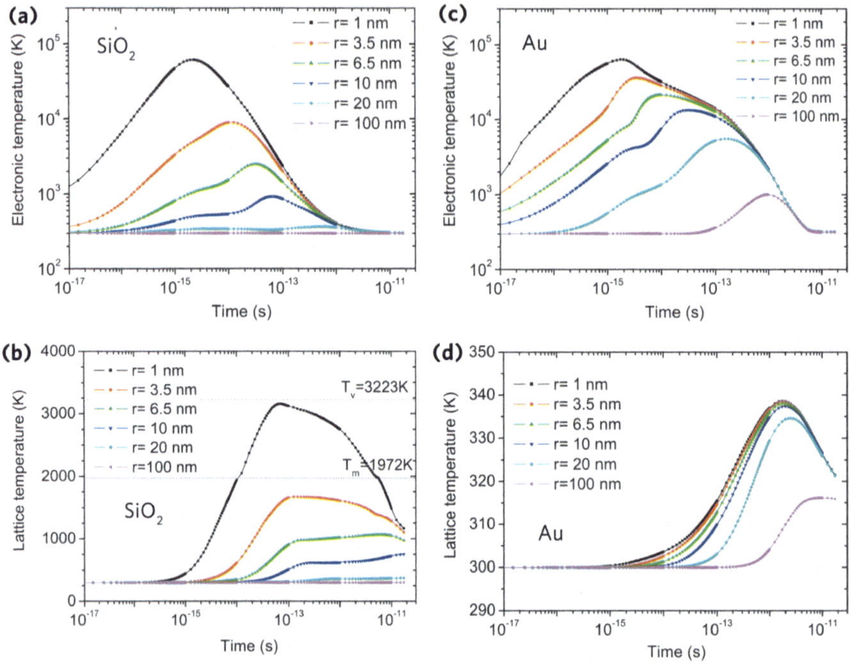

Fig. 5.34 Time evolution of electronic (**a** and **c**) and lattice (**b** and **d**) temperatures in a bulk SiO_2 (**a** and **b**) and a bulk Au (**c** and **d**) at various distances from the impact point of 110 MeV Br ions, calculated by the inelastic thermal spike model. Reproduced with permission from [63], ©2008 American Physical Society

$$C_e(T_e)\frac{\partial T_e}{\partial t} = \nabla \cdot (K_e(T_e)\nabla T_e) - g(T_e - T_a) + A \tag{5.16}$$

$$C_a(T_a)\frac{\partial T_a}{\partial t} = \nabla \cdot (K_a(T_a)\nabla T_a) + g(T_e - T_a) \tag{5.17}$$

where C_i, K_i, T_i ($i = e, a$), g, and A denote the specific heat, thermal conductivity, temperature of the electronic ($i = e$) and atomic ($i = a$) subsystems, electron-lattice coupling, and source term, respectively. $C_a(T_a)$ and $K_a(T_a)$ are often inputted from experimental data. As shown in Eq. (5.16), only the electronic sub-system has a source term A, i.e., a SHI excites the electronic sub-system only, but not the atomic sub-system. The excitation is transferred via the second terms in Eqs. (5.16) and (5.17), i.e., the terms including the electron-lattice coupling g. The thermal flow is proportional to $T_e - T_a$, which finally induces the thermal equilibrium between the electronic and atomic sub-systems.

Figure 5.34 exhibits time evolution of T_e and T_a, in a bulk SiO_2 and a bulk Au at various distances from the impact point of a SHI. Regarding T_e, both SiO_2 and Au reach the highest temperature of several 10^4 K at $r = 1$ nm. In the case of SiO_2, the highest temperature steeply decreases with r, e.g., below 1×10^4 K at

3.5 nm and ~1×10^3 K at 10 nm. Contrary, the highest temperature decay in space is slower for Au, where ~1×10^3 K is reached even at 100 nm away from the impact point. This difference is explained from much higher thermal conductivity of Au and much lower electron-lattice coupling g. Regarding to T_a, much clearer difference is observed. While SiO_2 is heated to 3000 K at the maximum, Au is heated only to 340 K, i.e., an increment of 40 K only. Materials which are not influenced by SHI irradiation are called ‘insensitive’. Au is a typical ‘insensitive’ material. Since a big difference between the thermal conductivities of Au and SiO_2 are known, we here comment on the electron-lattice coupling g, which is estimated by femto-second laser irradiation measurements or the free-electron theory. The g value of Au and amorphous SiO_2 are reported as 2.3×10^{10} and 1.25×10^{13} W cm^{-3} K^{-1} [57]. The amorphous SiO_2 has 500 times larger g than bulk Au.

Wang et al. evaluated the values of the electron-lattice coupling g for various elemental metals shown in Table 5.3 [64]. The g value relates with the electric conductivity $\sigma_e(T_e)$ and the electronic thermal conductivity $K_e(T_e)$ as,

$$g = \frac{\pi^4 (k_B n_e v)^2}{18 L \sigma_e(T_e) T_e},$$
$$g = \frac{\pi^4 (k_B n_e v)^2}{18 K_e(T_e)} \tag{5.18}$$

where n_e, v, L denote the electron density, the sound velocity, and the Lorentz number, respectively. Since $n_e = z\, n_{atom,}$ where z and n_{atom} denote the valence number of the electrons and atomic density, respectively. With comparisons to the experiments, $z = 2$ was applied.

The mean energy deposition Q is given as $Q = 0.63\, S_e/\pi\lambda^2$, where the 63% of S_e is deposited on the electron of a cylinder of radius λ. The ratio $\eta = Q/\Delta H$f, determines whether the metal is ‘sensitive’ ($\eta > 1.3$) or ‘insensitive’ ($\eta < 0.7$).

The quantity called the electron-phonon mean free path λ is the diffusion length of the energy on the electrons before its transfer to the atoms, which related with the electron-lattice coupling g as,

$$\lambda = \sqrt{D_e \tau_{ep}} = \sqrt{\frac{D_e C_e}{g}} \tag{5.19}$$

where C_e, D_e, and τ_{ep} denote the specific heat, the thermal diffusivity of the electrons, and the electron-phonon interaction time ($\tau_{ep} = C_e/g$), respectively. Since λ has the universal relationship with the band gap energy in non-metallic materials [65], it could be a more fundamental quantity than g. However, the universal relationship was recently slightly revised for semiconductors [66].

While a bulk Au is $\eta = 0.18$, i.e., ‘insensitive’, Au NPs are affected by SHI irradiation. Figure 5.35 exhibits time evolution of (a) electronic and (b) atomic temperatures of 2-dimensional (2D) Au NP surrounded by SiO_2 [57]. After a SHI impacting the center of the NP, high electronic temperature regions exceeding 3000 K (indicated

Table 5.3 List for various elemental metals, of the estimated values of the electron-lattice coupling g with assuming the valence of $z = 2$, the electron-phonon mean free path λ, the energy required to melt ΔH_f, and the ratio $\eta = Q/\Delta H_f$, where $Q = 0.63 S_e/\pi\lambda^2$, i.e., the mean energy density deposited on the electrons of a cylinder of radius λ

Metal	ΔH_f (J cm^{-3})	S_e (TRIM91) (keV nm^{-1})	g ($\times 10^{11}$) (W cm^{-3} K^{-1})	λ (10^{-7} cm)	η	S_e effect
Be	9368	23	293	3.92	5.2	Yes
Mg	2270	20	6.82	21.6	0.61	No
Al	3275	28	8.14	20.9	0.63	No
Ti	6701	42	92.8	6.14	5.4	Yes
V	8907	52	66.4	7.56	3.3	Yes
Cr	9075	63	94.0	6.51	5.3	Yes
Mn	7042	63	444	2.98	32	Yes
Fe	10977	70	49.8	8.97	2.6	Yes
Co	12199	75	34.5	10.9	1.7	Yes
Ni	10529	77	40.5	10.1	2.2	Yes
Cu	6895	73	4.94	28.5	0.42	No
Ga	1061	46	19.6	13.1	8.1	Yes
Zr	4873	48	35.0	9.55	3.5	Yes
Nb	9074	63	15.0	15.2	0.97	No
Pd	7616	81	13.9	16.4	1.3	Yes
Ag	4118	70	1.26	53.1	0.19	No
Sn	1184	45	3.69	28.7	15	Yes
W	14011	93	12.4	17.0	0.74	No
Pt	9003	109	10.3	18.9	1.1	No
Au	4443	99	0.91	62.5	0.18	No
Pb	1109	55	1.56	43.3	0.85	No
Bi	1136	50	8.20	18.4	4.2	Yes
U	3149	95	12.4	16.3	3.7	Yes

by yellow) expand from the center as shown in Fig. 5.35a. After all the parts of the NP become in high electronic temperature around ~10^{-15} s, gradually the high electronic temperature is also conducted to the SiO_2 region through the Au–SiO_2 boundary. Exceeding 10^{-12} s, the electronic temperature turns to decay. Contrary, an increase in the lattice temperature begins not from the Au NP but SiO_2 region close to the Au–SiO_2 boundary. Then the lattice temperature of the NP was heated from outside of the NP, while the electronic temperature increases from the axis of the NP ("the outer boundary heating"). As described before, all the parts of NP turn to melt when the NP is smaller than 30 nm, while the core of the NP does not turn to

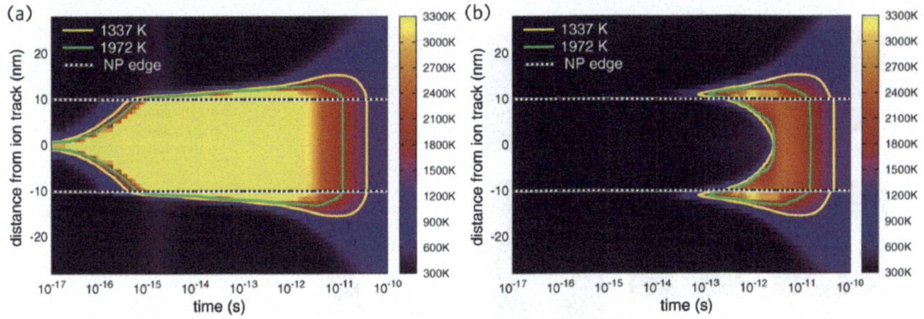

Fig. 5.35 **a** Electronic and **b** atomic temperature evolution of a 2-dimensional (2D) gold NP (diameter 20 nm) after an impact of SHI at the center of the NP. Reproduced with permission from [57], ©2012 Institute of Physics

melt for the NPs larger than 30 nm, because the NP is heated from the outside. The outer boundary heating is a consequence of the large difference of electron-lattice coupling between Au and SiO_2.

5.3.2 Two-Temperature Molecular Dynamics

The i-TS model calculates temperature profiles but not atomic displacements. Recently Helsinki group has combined the i-TS model and molecular dynamics (MD) and formulated the two-temperature molecular dynamics (TT-MD). They succeeded in calculation of atomic motions induced by SHI irradiation, e.g., the shape elongation of NPs [26, 67]. In the TT-MD, the atomic system is evolved by MD Eq. (5.21) while the electronic system is evolved by the heat diffusion Eq. (5.20), i.e.,

$$C_e(T_e)\frac{\partial T_e}{\partial t} = \nabla \cdot (K_e \nabla T_e) - g(T_e - T_a) + A \tag{5.20}$$

$$m_i \frac{\partial v_i}{\partial t} = F_i(t) + \xi m_i v_i \tag{5.21}$$

where $F_i(\mathrm{t})$ is the force on an atom i from the MD interatomic potential, and the second term of (5.21) comes from the acceleration of atoms by high electronic temperature via the electron-lattice coupling. See Ref. [68] for details.

This method was applied to an Au NP in SiO_2 irradiated with SHIs, and has succeeded to reproduce the shape elongation phenomenon as shown in Fig. 5.36. One SHI impact transformed a spherical crystalline Au NP to a weakly elongated amorphous Au NP ("1") shown in Fig. 5.36a. However, the second impact to the amorphous Au NP did not induce further elongation. If the amorphous NP is artificially crystallized, the second impact induced further elongation. In this simulation,

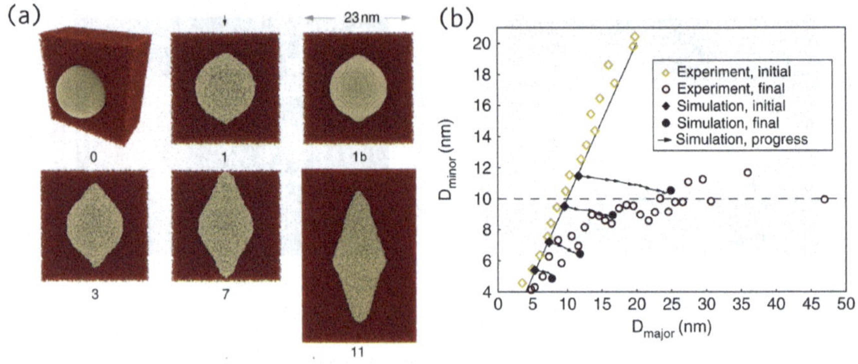

Fig. 5.36 MD simulations of elongation of Au NP in SiO_2. **a** Each image was taken 50 ps after ion impact. The numbers in the figure indicate the number of ion impacts. Images 1 and 1b are amorphous Au NP just after one ion impact and after the intentional crystallization, respectively. **b** Dimensions of NPs before and after ion irradiation. Closed diamonds and closed circles denote the simulation results. Each arrow represents one impact. Reproduced with permission from [67], ©2014 Tayler & Francis

artificial crystallization was carried out after every impact. Since amorphous Au NPs are not experimentally observed after SHI irradiation, the crystallization of the impacted Au NPs is induced in time scale longer than the maximum TT-MD simulation time (100 ps). Preliminary simulation in the order of nano-seconds indicated that the amorphous NP evolved to a polycrystalline state. It is clear from the experimental ion flux that a characteristic time scale between the ion impacts on an individual NP is in the order of seconds. Therefore, the crystallization of Au NPs before the next impact is an acceptable presumption.

The simulation cells are initially stress-free (except for the transient thermal stresses introduced by the ion itself), i.e., the in-plan stress due to the ion hammering effect is not included. Hence the current simulations show that the first stages of the elongation can be explained by thermal expansion and flow of matter from the nanocluster into the track core of the core/shell track in SiO_2, and is not a consequence of the ion hammering effect or any diffusion-like processes.

Figure 5.36b exhibits dimensions of NPs before and after an ion impact. The results are inconsistent with the principle of 'the minimum width for elongation' as indicated in Fig. 5.22a. This principle states that the NPs smaller than 'the minimum width' should be spherical, which is inconsistent with the numerical result. (In this case, the minimum width seems 10 nm.) However, experimental results inconsistent with 'the principle of the minimum width for elongation' have been reported [69] and also plotted in the figure as open circles. Further confirmation of the minimum width principle is necessary.

Further results of MD simulations are exhibited in Fig. 5.37a [26, 70], which indicates that the formation of core/shell ion track in SiO_2 coincides with the elongation of NP. Blue and green curves indicate relative density of SiO_2 in the innermost cylinder of 1 nm radius along the ion trajectory (blue) and that in the second shell of 1 nm

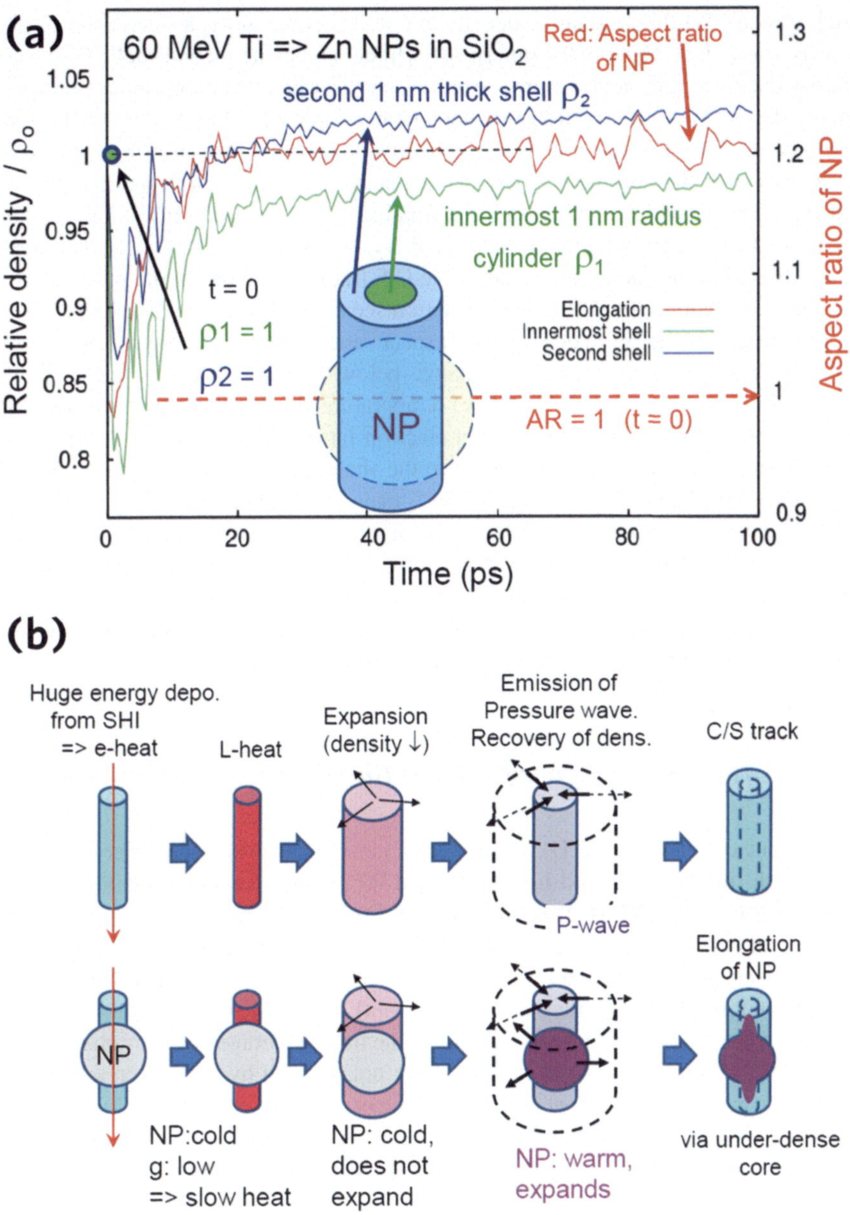

Fig. 5.37 **a** Time evolution of track formation and elongation in a Zn NP in SiO_2 irradiated with 60 MeV Ti ions. Blue and green curves indicate relative density of SiO_2 in the innermost cylinder of 1 nm radius along the ion trajectory (green) and in second shell of 1 nm thick (blue). The core/shell track formation is visible. A red curve indicates the aspect ratio of Zn NP. The elongation is coincident with the core/shell track formation. **b** Schematically depicted formation processes of a C/S track and NP elongation. Reprint from Amekura et al. [70]

thick (green). While the relative density at $t = 0$ was the unity, a steep drop of the density was induced both in the innermost cylinder and in the second shell. The steep density drop, i.e., the steep volume expansion, is ascribed to a consequence of huge energy deposition from the SHI, as schematically depicted in Fig. 5.37b. After a few picoseconds, both the densities turned to recover, probably after emitting a pressure wave. However, the saturated density in the innermost cylinder was lower than that in the second shell, which resulted in the formation of the core/shell (C/S) ion track in SiO_2. Well coincident with the density recovery, the aspect ratio of the NP increased and saturated in the same time scale of the C/S track formation. In the processes of the shape elongation, the C/S track formation seems to play an important role.

In contrast to the SiO_2 density, the aspect ratio of NP was unity up to ~1 ps, i.e., does not experience any steep change below 1 ps, as shown as red curve in Fig. 5.37a. This is because the (lattice) temperature rise is much slower in the NP than the SiO_2 matrix, due to much lower electron-lattice coupling of NP. After ~1 ps, the NP becomes warm, and susceptible to the shape change. Consequently, the NP experienced the late shrinkage stage of SiO_2 matrix only but not the early expansion stage of the matrix, because the NP was hot in the late stage but cold and stiff in the early stage. The NP in the shrinkage stage of SiO_2 matrix results in the elongation of NP through the underdense cores of the core/shell ion tracks.

The elongation could happen when NPs of low electron-lattice coupling are dispersed in a matrix of high electron-lattice coupling [70].

5.3.3 Criticisms on Synergy Model Based on Ion Hammering

Optical linear dichroism (OLD) spectroscopy, i.e., optical absorption difference measured with linearly polarized light of 0° and 90°, detects small elongation of NPs more sensitively than TEM. This is because OLD detects averaged signal over macroscopic numbers of NPs. As shown in Fig. 5.38a, the OLD signal decreases *linearly* with decreasing the fluence down to 1×10^{11} ions/cm^2 under 200 MeV Xe ion irradiation. Judging from the track radius (~4.5 nm) of 200 MeV Xe ions in SiO_2, the fluence of 1×10^{11} ions/cm^2 corresponds to the track coverage of the irradiated area of 5% only. Therefore, most of the NPs have not been hit by SHI. Even some NPs are hit, the impact was only once. NPs hit twice or more are negligible. This was also confirmed by simulations shown in Ref. [16]. This fact indicates that the elongation of NP is induced by only a single impact of SHI, while multiple impacts increases the degree of the elongation.

This observation is inconsistent with the synergy model of hammering and NP melting. A single impact of SHI to a NP may induce the melting of NP, but it probably does not induce the build-up of the strong in-plain stress enough for the elongation. Furthermore, it is known that SiO_2 shows compaction by a single impact of SHI, rather than the hammering. The crossover of the compaction to the hammering is shown as the fluence dependence of the relative strain $\Delta L/L$ in Fig. 5.38b. While the crossover is expected at 0.51×10^{13} ions/cm^2, the linear fluence dependence of

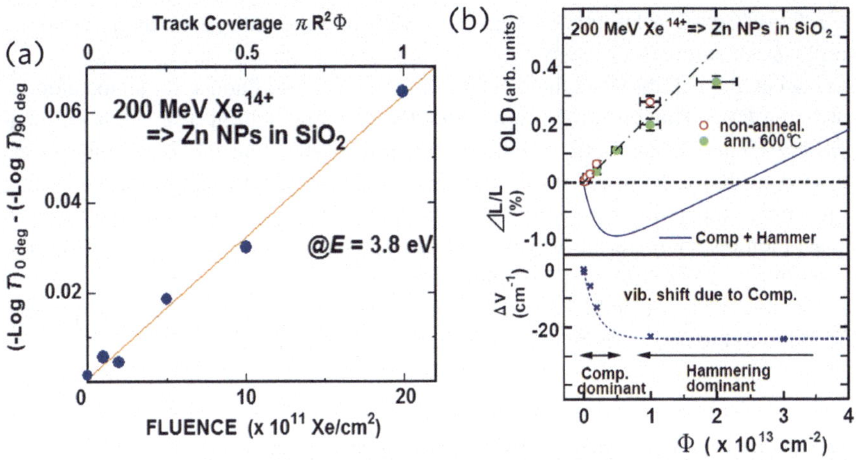

Fig. 5.38 **a** Fluence dependence of optical linear dichroism (OLD) signal from Zn NPs in SiO_2 irradiated with 200 MeV Xe ions. The OLD signal is proportional to the mean-aspect-ratio −1. **b** Comparison among the fluence dependences of the OLD in non-annealed and annealed sample at 600 °C, of the relative deformation $\Delta L/L$ of SiO_2 due to compaction and hammering, and of the IR vibration shift of SiO_2 due to the compaction. Reproduced with permission from [16], ©2011 American Physical Society and [60], ©2017 American Institute of Physics

OLD is also confirmed at the same fluence region. If the hammering would play an important role in the shape elongation of NPs, deviation from the linear relation could be observed. It should be noted that the fluences where the crossover was observed were more than one order higher than the minimum fluences where the linear relation was confirmed.

Since the Zn NPs were formed by ion implantation of 60 keV, a criticism was raised that SiO_2 near the NPs were fully compacted by the 60 keV implantation before the SHI irradiation. Consequently, only the hammering but not the compaction was induced by SHI irradiation. To answer this criticism, the same samples but annealed at 600 °C were irradiated with SHI under the same conditions. In a past literature, we have studied the annealing behavior of SiO_2 samples compacted with 60 keV implantation, which shows that the compaction is annealed out at 600 °C [71]. The results of the annealed samples are exhibit in Fig. 5.38b as closed circles, which fall on the same line with the non-annealed samples, excluding the influences of the compaction induced by 60 keV implantation.

While it is still on the way, we are studying the irradiation temperature dependence of the shape elongation of NPs, since the hammering exhibits strong irradiation temperature dependence [31]. If the hammering was involved, the shape elongation could show a strong irradiation temperature dependence.

Recently, Slablab and co-workers [36] fabricated nano-rods elongated with the angles of 30°, 45°, and 60° from the surface normal, with selecting the same angles of incidence. Also these observations are inconsistent with the hammering, where the relation between the incident angle of SHIs and the elongation angle is not so simple, as shown in Sect. 5.1.6 .

Dawi et al. [72] observed the existence of the threshold fluence for shape elongation using RBS. However, it is known that the depth resolution of the conventional RBS is ~10 nm. Even if NPs exhibit elongation shorter than 10 nm, it may not be detected by RBS. To prove it, the RBS spectra were numerically simulated using Rump code under the same conditions as the experiments reported. As shown in the supplementary material of Ref. [60], the calculated spectra clearly showed the insensitivity of RBS for small elongation of NPs.

The observed threshold fluences for the NP elongation by RBS is a spurious effect due to the insensitivity of the conventional RBS method.

5.4 Optical Properties of Elongated Nanoparticles

5.4.1 Optical Linear Dichroism

Figure 5.39 exhibits optical absorption spectra of Zn NPs in SiO_2 irradiated with 200 MeV Xe^{14+} ions at an incident angle of 45° to five different fluences, detected by linearly polarized light with three different polarization angles of 0°, 45°, and 90°. The 0° angle was defined when the polarization plane includes the major axes of NPs [16].

Since a spherical NP does not have a special direction, the absorption spectrum does not change with rotating the polarization angle from 0 to 90°. This is the case of 0 Xe/cm^2 in the figure, i.e., of unirradiated. While each real NP is not perfectly spherical, the deviation of the directions and the magnitudes are random. Consequently, after the averaging of macroscopic number of NPs, which is automatically

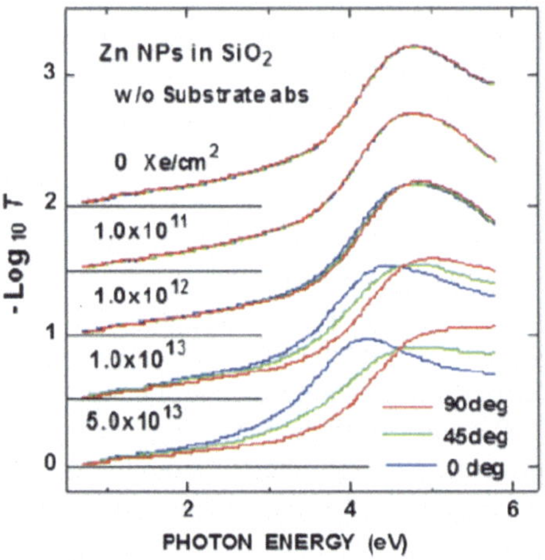

Fig. 5.39 Optical absorption spectra of Zn NPs in SiO_2 irradiated with 200 MeV Xe^{14+} ions at an incident angle of 45° to five different fluences, detected with linearly polarized light of the polarization angle of 0°, 45°, and 90°. Reproduced with permission from [16], ©2011American Physical Society

carried out during the macroscopic optical absorption measurements, the spectra do not depend on the polarization angle.

With the SHI irradiation, the metal NPs elongate toward the direction parallel to the SHI beam, i.e., the same direction with each other. Triply-degenerated surface plasmon resonance (SPR) modes in a spherical NP split into doubly-degenerated transverse (T) modes and a non-degenerated longitudinal (L) mode. In prolate spheroids, the L mode shifts to lower energy but the T modes shift to higher energy with elongation. However, the latter shift is smaller compared with the former, because the T modes are doubly degenerated. The absorption spectra depend on the polarization angle. The optical absorption depending on the polarization angle of linearly polarized light is called 'linear dichroism'. With increasing the fluence, the aspect ratio of NPs increases and the absorption deviation between different polarizations become higher as shown in Fig. 5.39. Optical linear dichroism of Ag NPs elongated by SHIs is also reported in Ref. [73].

An advantage of this system is that NPs are elongated toward the same direction, i.e., parallel to the direction of the SHI beam, which can be controlled. While each elongated NP has small microscopic anisotropy only, a large macroscopic anisotropy is realized since all the elongation axes are aligned to the same direction. All the small microscopic anisotropies sum up over macroscopic number of the nano-rods. While nano-rods can be formed by other methods, e.g., wet-chemical synthesis, the alignment of the major axes of nano-rods of macroscopic number is not easy in the most of cases. In the present method, the SHI beam does.

Anisotropic absorption of a NP in the quasi-static limit is described by Rayleigh theory [74]. Unfortunately, the shape elongation requires NPs diameters larger than ion track diameter, and NPs elongate longer than several tens nm. The assumption of the quasi-static limit is not sure. Certain corrections should be necessary for the Rayleigh theory.

According to the Rayleigh theory, the extinction cross-section C_{ext} is described by

$$C_{\mathrm{ext}} = k\mathrm{Im}(\alpha) + \frac{k^4}{6\pi}|\alpha|^2 \tag{5.22}$$

where $k = 2\pi/\lambda$ the wavenumber and α the polarizability. The first and the second terms describe the absorption and the scattering cross sections, respectively. A spheroid of prolate type ($a > b = c$) is assumed, where a and b denote the major and minor semi-axes, respectively. The polarizability against the field parallel to the principal axis i ($i = a, b$) is given by

$$\alpha_i = \left(\frac{4\pi}{3}ab^2\right)\frac{\varepsilon_{\mathrm{NP}} - \varepsilon_d}{\varepsilon_d + L_i(\varepsilon_{\mathrm{NP}} - \varepsilon_d)}, \ (i = a, b) \tag{5.23}$$

where $\varepsilon_{\mathrm{NP}}$, ε_d, and L_i denote (complex) dielectric functions of NPs, of matrix, and the geometrical depolarization factors, respectively. The depolarization factor L_i ($i = a, b$) is given by

$$L_a = \frac{1-e^2}{e^2}\left\{\frac{1}{2e}\ln\left(\frac{1+e}{1-e}\right)-1\right\}$$
$$L_b = \frac{1}{2}(1-L_a) \tag{5.24}$$

5.4.2 Birefringence

According to the complex refractive index theory, the imaginary part of the refractive index corresponds to the extinction constant, which relates with the photo-absorption. The imaginary part of the linear dichroism, i.e., the polarization angle dependent absorption, describe the (linear) birefringence, i.e., the polarization angle dependent refractive index. The elongated Zn NPs exhibiting the linear dichroism as shown in Fig. 5.39 also exhibit the birefringence [75].

The birefringence of Zn NPs in SiO_2 was confirmed by the following method: First, a pair of linear polarizers was set in the crossed Nicols configuration, i.e., the relative angle between the polarizers was set to $\theta = 90°$. Shown as the bottom spectrum in Fig. 5.40, no light is transmitted through this configuration (spectrum 'No sample' in the figure). When a silica sample including elongated Zn NPs is inserted between the polarizers, a certain amount of light transmits through the crossed Nicols configuration except $\alpha = 0°$ or $90°$, i.e., when the major axes of the elongated NPs are parallel to either of the polarizers.

Polarization plane of light incident to the sample is determined by the 1st polarizer. The 2nd polarizer is set perpendicular to it. Because of the birefringence nature of the elongated NPs, the polarization plane is rotated after going through the sample. In this case, the polarization transmitted through the sample is no longer perpendicular to the

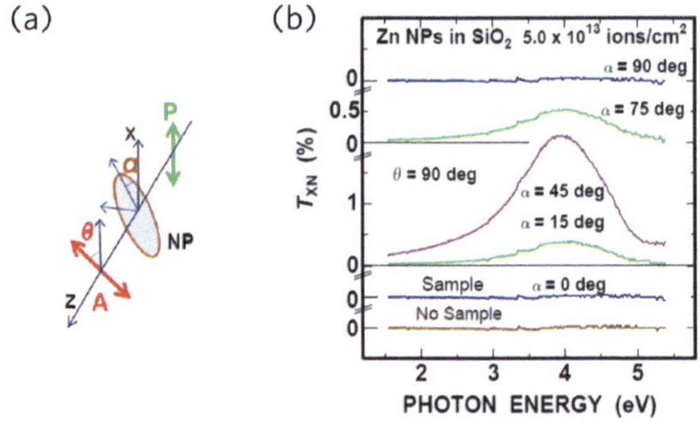

Fig. 5.40 **a** Configuration between two polarizers and the elongated NPs. α is the angle between the polarizer axis and the major axes of the elongated NPs. θ is the angle between the polarizer and the analyzer, which is always set to 90°. **b** The optical transmission spectra through the crossed Nicols configuration. Reproduced with permission from [75], ©2014 The Optical Society of America

2nd polarizer. The rotation of the polarization is due to the different light velocities of $\alpha = 0°$ mode and 90° mode, both of which are excited by any polarization angle except parallel to $\alpha = 0°$ or 90°. When the incident polarization is parallel to $\alpha = 0°$ or 90°, only one of the two is excited. No rotation of the polarization is induced, i.e., no transmission through the Crossed Nicol configuration.

The birefringence from Ag nano-rods in SiO_2, elongated by SHI irradiation is also reported in Ref. [76].

5.4.3 *Possible Application to UV Polarizer of Nanometric Thickness*

One of the possible applications of this system (Zn NPs in SiO_2 elongated by SHIs) is an optical polarizer of the absorption type working from the near-UV to deep-UV region. The same application has already been proposed for elongated Ag NPs [76]. However, the UV SPR absorption band of Zn NPs is wider than that of Ag NPs, which results in a wider working photon-energy range. The performance of the polarizer is characterized by the extinction ratio, i.e., the ratio of transmittance of one polarization to that of the orthogonal polarization. In this application, a higher extinction ratio is desirable which produces strong absorption for one polarization but simultaneously high transmission for the orthogonal polarization at the same photon energy.

An extinction ratio of ~5 dB was observed at 4 eV in the sample irradiated with a fluence of 5.0×10^{13} ions/cm^2. While an extinction ratio of ~5 dB would be small, this was attained from an NP layer of ~75 nm thick in the present sample [16]. Table 5.4 compares the extinction ratio and layer thickness of the present sample with some literature data, which have been referred to as exhibiting "giant birefringence" [76]. While the extinction ratio of our sample is lower than the other samples listed in Table 5.4, the extinction ratio of our sample was attained with the layer thickness that 1–3 orders of the magnitude thinner than the others. If the thickness of the NP layer could be increased to 250 nm, e.g., using higher energy ion implantation or some other fabrication method, the extinction ratio would be expected to increase to 17 dB. According to Ref. [76], an extinction ratio of 20 dB is a typical commercial value used in the 1.55 μm band. We consider that our sample has much higher potential

Table 5.4 Comparison of the extinction ratios and the required thicknesses of films for polarizer applications from past literature. Reproduced with permission from [75], ©2014 The Optical Society of America

Authors	Extinction ratio (dB)	Thickness (μm)	Refs.
Kuenzner et al.	17	100	[77]
Genereux et al.	9	235	[78]
Muskens et al.	40	15	[79]
Amekura et al.	5	0.075	[75]

because a comparable extinction ratio is attainable with a sample thickness a few orders of magnitude thinner.

Regarding to the methods to form thicker NP layer, various candidates are expected: A simple extension of the present method is higher energy implantation with, e.g., 250 keV Zn ions, which introduces Zn ions up to the depth of 250 nm. Simultaneous multi-energy implantation is better recommended to form a layer with a homogenous concentration along the depth. Other methods, e.g., sputtering co-deposition, could be applicable. A limitation from the SHI elongation method is that NPs should be larger than the track diameter [43]. In the case of 200 MeV Xe ions, only NPs larger than ~9 nm in diameter can be elongated. The upper limit could be several tens nm. The elongation of NPs by SHI irradiation strongly depends on the NP sizes, but not on the NP concentration. This method is applicable to much more dilute NPs.

Another limitation which comes from SHI irradiation is weak tolerance against vacuum degradation, while the SHI irradiation may strongly enhance degasifying from samples, e.g., polymers. This can be disadvantageous for nano-composites formed by wet chemical processes and so on. Since the projectile range of 200 MeV Xe ions is 20 μm long, the elongation can be induced whole the NP layer in SiO_2 up to, say, 18 μm thick.

5.4.4 *Second-Harmonic Generation Microscopy*

Gold nano-rods (NRs) in SiO_2 elongated by 92 MeV Xe ion irradiation with four different incident angles of 0°, 30°, 40°, and 60° from the surface normal were observed by the second-harmonic generation (SHG) microscopy. The results are shown in Fig. 5.41 [36]. The fundamental frequency light generated from a mode-locked femto-second laser (pulse width 140 fs, wavelength 1060 nm) was focused on a spot of ~350 nm in diameter, which is comparable to the length of NRs. The backscattered SHG signal was collected by appropriate optical filters and a cooled photomultiplier in the photon-counting mode. Not only the linear polarization (LP) but also the cylindrical vector beams of the radial polarization (RP) and the azimuthal polarization (AP) were applied. The SHG signal was not detected in the AP configuration, as predicted by theory. The experimental results shown in Fig. 5.41 were well reproduced by calculations.

5.4.5 *Electron Energy Loss Mapping of Elongated Nanoparticles*

While this is not an optical property of the elongated NPs, nowadays, electronic excitation in the visible photon energy range is possible by scanning TEM electron

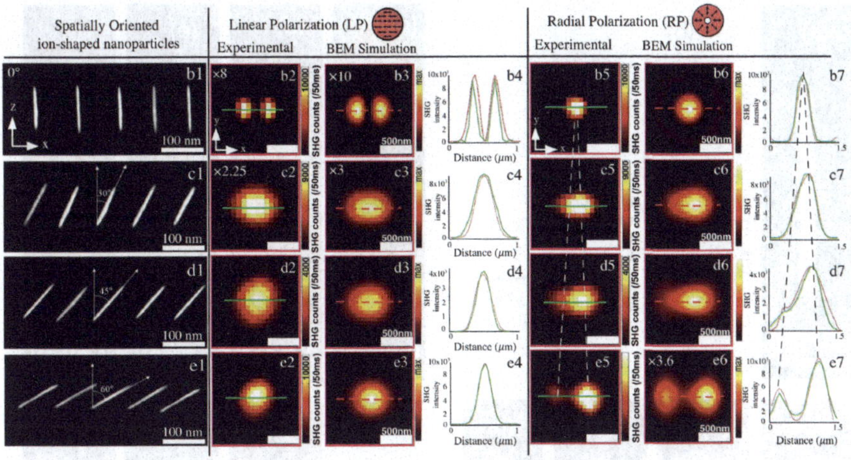

Fig. 5.41 Second-harmonic generation images of oriented nano-rods formed by SHI irradiation with four different incident angles of 0, 30, 45, and 60°, excited by linear polarization (b2–e2, b3–e3) and radial polarization (b5–e5, b6–e6). Experimental (b2–e2, b5–e5) and numerical simulations (b3–e3, b6–e6). Reproduced with permission from [36], ©2016 Nature-Springer

energy loss spectroscopy (STEM-EELS) mapping. Since the objective is concerning the surface plasmon resonances (SPRs) of the elongated NPs, this is described here. The STEM-EELS mapping was carried out to visualize the SPRs of individual ion-shaped Au nano-rods (NRs) embedded in SiO_2. With scanning e-beam in TEM, EELS at each position of a NR was detected. Figure 5.42 shows the results of Au NR with a length of 92.6 nm and a width of 7.8 nm, i.e., an aspect ratio of 11.9. As shown in (b), the NR has four resonances at 0.86, 1.27, 1.66, and 2.40 eV. Spatial mappings at the four energies are shown in (c)–(f). The corresponding simulation results are shown in (k)–(n), with the energies at 0.67, 1.17, 1.53, and 2.60 eV. Similar results of different aspect ratios are shown in the original paper [80].

5.5 Some Comments Related to Track Formation

As already mentioned, ion tracks, i.e., modified regions of cylindrical shaped with diameters ranging from nm to tens of nm and lengths more than μm, are observed after SHI irradiation in some materials. At the moment, the most accepted model of the track formation is the inelastic thermal spike (i-TS) model, where the tracks are formed as transiently-molten regions along the ion paths due to the localized thermal energy. Here we comment on three topics concerning track formation.

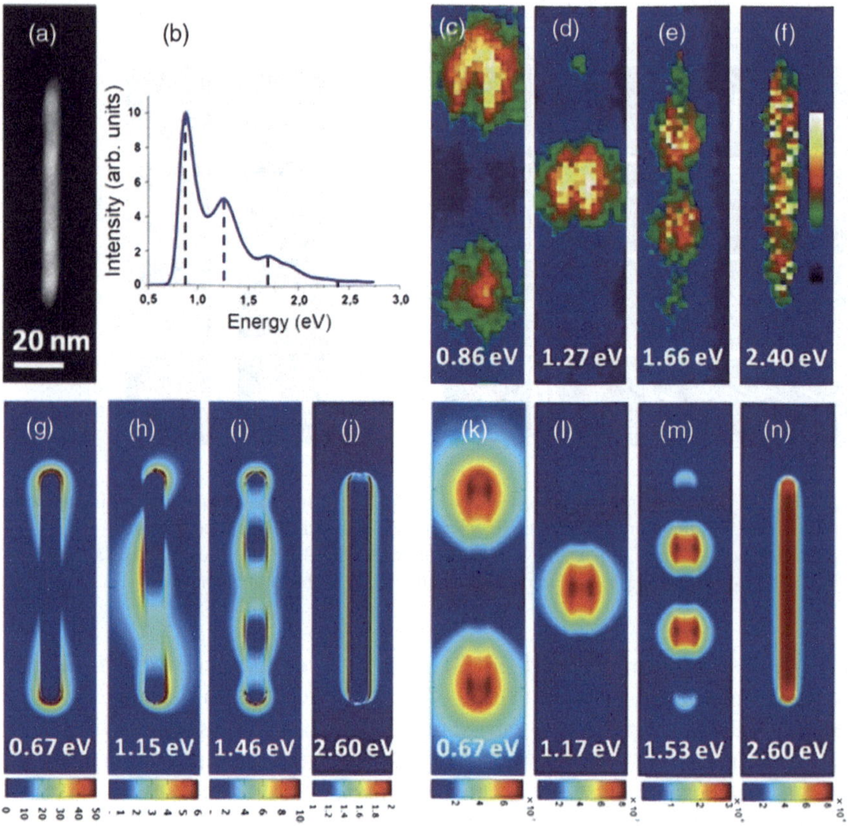

Fig. 5.42 Electron energy loss spectroscopy (EELS) mapping of a Au nano-wire fabricated by SHI irradiation. **a** STEM-HAADF image, **b** EELS spectrum, **c**–**f** experimental EELS maps, **g**–**j** Simulated electric field enhancement, **k**–**n** Simulated EELS maps. Reproduced with permission from [80], ©2018 American Physical Society

5.5.1 Primary Ionization Rate and Velocity Effect Free Theory

With increasing the energy of ions, the electronic stopping power S_e increases with proportional to the velocity in the low energy region, but turns to decrease in the Bethe-Bloch region, with exhibiting the Bragg peak. Consequently, the same S_e value is attained at two different energies: One is lower and the other is higher than the Bragg peak, i.e., at low and high velocity, respectively. Unfortunately, irradiation effects, e.g., the track radius, are different at the low and high velocities while at the same S_e value. This is called "velocity effect". The δ-rays generated by high velocity ions have higher mean energy, which penetrate deeper. Consequently, the excitation density becomes lower.

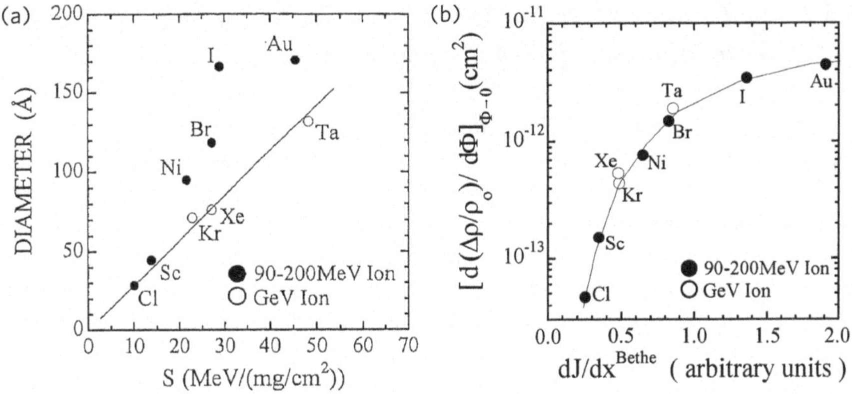

Fig. 5.43 Diameters of ion tracks in $EuBa_2Cu_3O_y$ formed by irradiation with various swift heavy ions are plotted with **a** the electronic stopping power S_e and **b** the primary ionization rate dJ/dx. Reproduced with permission from [81], ©2000 The Physical Society of Japan

Figure 5.43a shows S_e dependence of track diameters in $EuBa_2Cu_3O_y$, formed by various ions in the low (90–200 MeV) and the high (GeV) energies. The low velocity ions (Ni, Br, I, Au) exhibit larger track diameters than the high velocity ions (Cl, Sc, Kr, Xe, Ta) [81]. This observation may indicate that the S_e is not a good measure of the irradiation effects by SHIs. To improve this situation, a revised concept of S_e, e.g., the restricted energy loss, was proposed [82].

On the other hand, Iwase, Ishikawa, and coworkers proposed that the primary ionization rate $\mathrm{d}J/\mathrm{d}x$ explains the experimental results better than S_e [81]. The same data in Fig. 5.43a were replotted in Fig. 5.43b with $\mathrm{d}J/\mathrm{d}x$. All the data fall on an identical curve, indicating the superiority of the $\mathrm{d}J/\mathrm{d}x$.

The electronic energy loss S_e and the primary ionization rate $\mathrm{d}J/\mathrm{d}x$ are similar quantities with each other, and represented following relations:

$$S_e = N \sum_n E_n \sigma_n$$

$$\frac{dJ}{dx} = N \sum_n \sigma_n \tag{5.25}$$

where N denotes the number density of target atoms, σ_n the cross-section of n-th electron having energy of E_n. Since valence (core) electrons have large (small) σ_n but low (high) E_n, both the valence and the core electrons contribute to S_e. However, mostly the valence electrons contribute to $\mathrm{d}J/\mathrm{d}x$, because of larger σ_n.

Anyhow, the primary ionization rate $\mathrm{d}J/\mathrm{d}x$ could be a better measure of irradiation effects induced by SHI irradiation. The primary ionization rate $\mathrm{d}J/\mathrm{d}x$ plays an important role in the Coulomb explosion model.

5.5.2 *Self-trapped Exciton Model Versus Exciton Mott Transition and Thermal Spike Heating*

Some point-defect formations or atomic movements induced by UV/visible photo-excitation are known. A famous example is the photography: Photo-images are recorded using photo-films with dispersed micro-crystals of silver halides. Illumination of blue/purple light (exceeding the bandgap of AgBr, e.g., 2.7 eV at 4.2 K) induces the formation of fine Ag NPs in silver halides, which are later enlarged by chemical treatments for imaging. Another examples are point defect formation in alkali halides under the above-band-gap photo-excitation [83]. In case, the defects are formed via the non-radiative decays of self-trapped excitons (STEs). Since usually excitons are formed by electrons in the conduction band and positive holes in the valence bands, they have an extended nature. However, excitons are localized (self-trapped) because of strong coupling between (mostly) holes and lattice. The self-trapped holes are called "small polarons", which interact with acoustic phonons and strongly localized. The holes (electrons) interacted with optical phonons are called "large polarons", which have an extended nature, i.e., are not localized, but having heavier mass [84].

While the atomic displacements induced by light illumination in alkali halides are not common in solids, it is relatively common in molecules, which is called "photochemical reactions". A hydrogen molecule (ion) is an example as shown in Fig. 5.44, in which the energies of the bonding and anti-bonding orbits are plotted against the distance between the two hydrogen atoms. In the bonding state, the minimum energy is attained at the inter-nuclear distance $R \sim 2.5\ a_B$, where a_B denotes the Bohr radius. If the electronic excitation is induced in this molecule by, e.g., light illumination, the

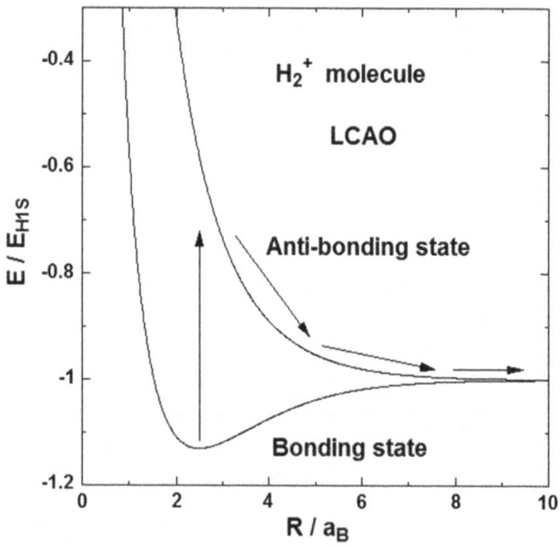

Fig. 5.44 Total energy of a H_2^+ molecule is plotted with the separation between the two H atoms, for the bonding and the anti-bonding orbitals. a_B denotes Bohr radius

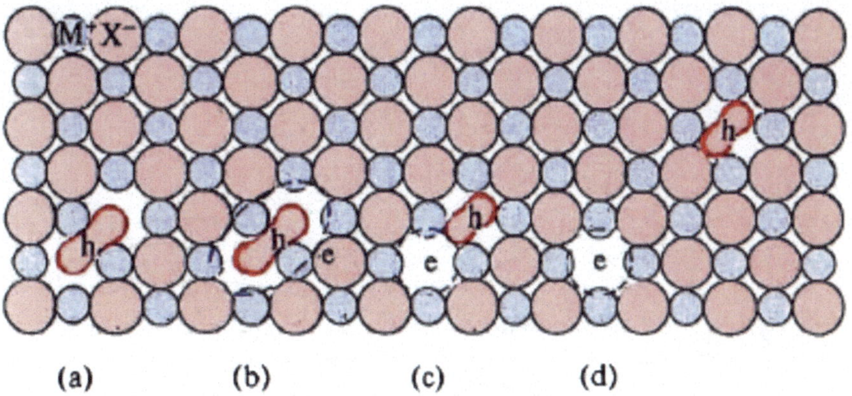

Fig. 5.45 Schematically depicted {100} plane of the NaCl-type alkali halide, with small and large circles for alkali and halide atoms, respectively. **a** V_K center, **b** V_K center binding an electron, a model of the on-center STE, **c** the off-center STE, and **d** a pair of F center and H center

molecule is excited to the anti-bonding orbital. In search for the energy minimum along the anti-bonding curve, the distance R increases toward the infinity. Therefore, the bond-breaking of H_2^+ molecule is induced by electronic excitation.

A difficulty to induce photo-chemical reactions in solids is the extended nature of electron (hole) wave-functions. In order to break a certain bond by electronic excitation, the excitation energy should localize around the objective bond. However, light illumination induces excitation to states whose wave-functions are extended. A certain localization mechanism is necessary. A localized wave-function of a point defect could help. Another candidate is a STE, where the self-trapping localizes the exciton wave function, which had been extended all over the crystal. One of the roles of STEs in the photo-chemical reactions in solids is to localize carrier or an exciton at a certain site and convert the localized electronic energy to atomic motions.

Figure 5.45 schematically depicts atomic configurations of STEs and point defects in {100} plane in the NaCl-type alkali halide [83], where smaller and larger circles represent alkali and halogen atoms, respectively. When a positive hole localizes, the hole is shared by two adjacent halogen atoms along [110] direction. The two halogen atoms come closer with each other to trap the hole, forming a halogen dimmer called V_K center (a). A V_K center binding an electron was a model of on-center STE (b). Because of the pseudo-Jahn Teller effect in the excited state, a symmetry breaking is induced and the off-center STE (c) is formed. The stability of the on-center STEs depends on the alkali halide species. Simultaneously the off-center STE is regarded as a pair of point defects, i.e., an F center and an H center in the nearest neighbor. With further migration of the H center along [110] direction, a stable pair of F center and H center is formed. Figure 5.46 shows a schematic view of the lowest adiabatic potential surface of the STE in NaCl with assuming the luminescent STE (STE_L) is in the off-center configuration ($Q_2 = 0$). Here Q_1 and Q_2 denote the interaction coordinates of hole self-trapping and of the off-center displacement, i.e., the displacement of

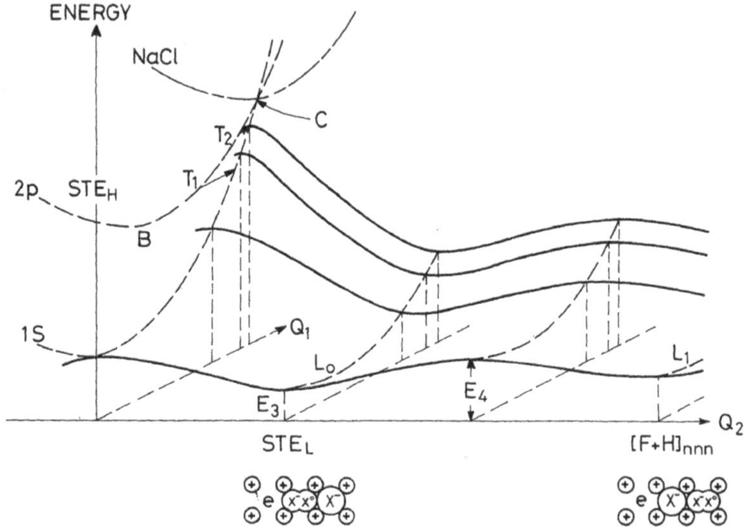

Fig. 5.46 Schematically depicted adiabatic energy surfaces of NaCl for self-trapping and non-radiative defect formation, with Q_1 and Q_2 coordinates which represent the hole self-trapping and the off-center displacement, respectively. Reproduced with permission from [85], ©1987 The Physical Society of Japan

the center of the halogen molecule X_2^- from the midpoint of the V_K center. After self-trapping along the Q_1 coordinate, the state moves along the Q_2 coordinate as shown in Fig. 5.45c, d.

In fact, it is confirmed from luminescence spectra the formation of STEs during the SHI irradiation. An interesting observation is that the materials showing STEs, e.g., both amorphous [31] and crystalline [11] SiO_2 (2 keV/nm) and alkali halides such as LiF (4 keV/nm), exhibit much lower threshold S_e for the track formation than the other materials not showing STEs, e.g., MgO (20 keV/nm) [12]. However, the number of the examples are too limited. Itoh and Stoneham raised only two materials, SiO_2 and LiF, which surely show the STEs and low S_e thresholds, while they categorized $Y_3Fe_5O_{12}$ and $LiNbO_3$ as "probably yes" for STE. While some materials were added by Agullo-Lopez et al. [86], still the examples are too limited.

Itoh estimated the number of STEs $\rho(r)$ at a distance r from the ion trajectory with dividing the deposited energy density $D(r)$ reported by Katz et al. [87] by the electron-hole pair formation energy W, which is 2–3 times of the bandgap energy [88]. He presumed that a track is registered if the average concentration of STEs within a radius is larger than the concentration of SiO_2 molecules, and reproduced the S_e dependence of track radius reported by Meftah et al. [89].

However, STE concentration comparable to the atom concentration is impossible. Here we recall the concept of "the exciton Mott transition", which has been studied since 1960s. Figure 5.47 exhibits an old example of exciton phase diagram in Ge [90]. When the density of free-excitons increases so that the exciton Bohr orbitals overlap

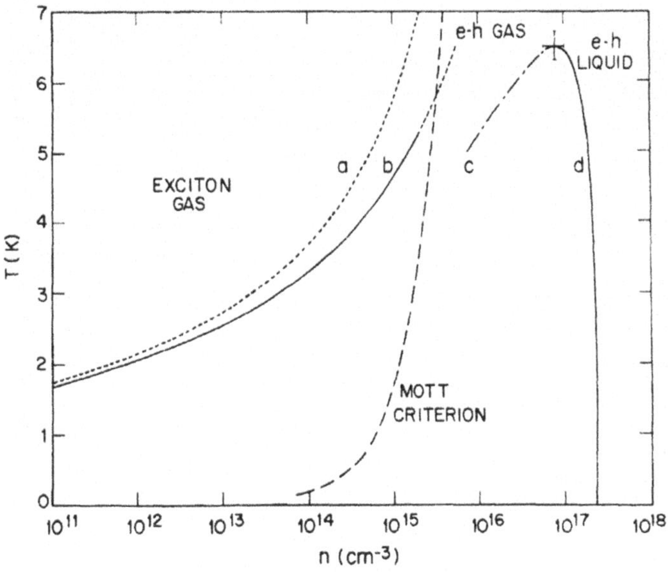

Fig. 5.47 A phase diagram plotted in e-h pair density and temperature for excitons and e-h plasma/droplet in Ge. Mott transition was shown by a broken line. Reproduced with permission from [90], ©1976 American Physical Society

with each other, the Coulomb attraction between the electron and hole is partially shielded. Consequently, the excitons are no longer stable and could be dissociated to electron-hole (e-h) plasma or electron-hole droplets. In the case of Ge, the exciton Mott transition, i.e., the transition from exciton gas to e-h plasma, is induced at the exciton density of 10^{15}–10^{16} cm^{-3}, which is much lower than the SiO_2 molecule density of ~10^{22} cm^{-3}. The much low Mott density of Ge is partly due to the large dielectric constant of ε_0 ~ 16, which results in large exciton Bohr radius. Furthermore, the indirect nature of the band gap of Ge is appropriate for accumulating excitons and observing the exciton Mott transition. The transition is observed also in direct-gap materials such as CuCl [91]. The Mott density was ~5 × 10^{19} cm^{-3} in CuCl, which was still lower than the STE density of ~10^{22} cm^{-3} proposed by Itoh for SiO_2 [88]. For reference, the exciton Bohr radius in CuCl is as small as 0.7 nm.

Of course, the above discussion is for free excitons. However, STEs in many materials including alkali halides consist of a loosely bound electron with a self-trapped hole. If some STEs are formed within the distance of the exciton Bohr radius, they become unstable due to the shielding by the loosely bound electrons. Therefore, the STE concentration comparable to the SiO_2 molecule density is not attained.

Figure 5.48 shows temperature dependence of STE luminescence intensity and the F center absorption. With increasing the temperature, the potential barrier E_4 in Fig. 5.46 is easily overcome. Consequently, the luminescence intensity decreases while the F center absorption increases, i.e., the defect formation efficiency increases

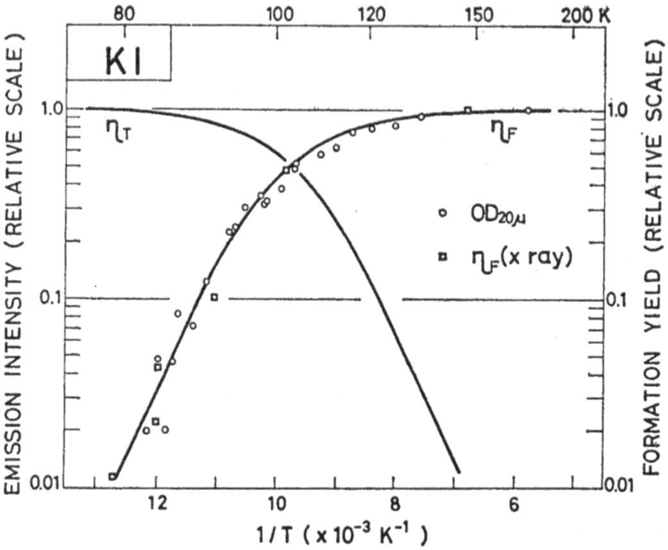

Fig. 5.48 Temperature dependence of the thermal conversion from STE luminescence (η_T) to the F center formation yield (η_F) in KI. Reproduced with permission from [95], ©1975 The Physical Society of Japan

with temperature. Agullo-Lopez and coworkers developed the synergy model of the thermal spike and the point defect formation via non-radiative decays of STEs to $LiNbO_3$ [92], TiO_2 [93], and NaCl [94]. They described the exciton decay as,

$$\frac{dN_{\mathrm{STE}}}{dt} = -N_{\mathrm{STE}}(r, z, t)\left[\nu_1 \exp\left(-\frac{\varepsilon_{\mathrm{STE}\to\mathrm{FH}}}{k_B T(r, z, t)}\right) + \frac{1}{\tau}\right] \tag{5.26}$$

where $\varepsilon_{\text{STE-FH}}$ denotes the potential barrier for the defect conversion, ν_1 a frequency factor, τ the radiative lifetime, and $T(r, z, t)$ the temperature evolution in the spike. The defect concentration is given as,

$$n_d(r, z) = \int_0^\infty dt N_{\mathrm{STE}}(r, z, t)\nu_1 \exp\left(-\frac{\varepsilon_{\mathrm{STE}\to\mathrm{FH}}}{k_B T(r, z, t)}\right) \tag{5.27}$$

In this model, STE concentration are not presumed to be high as the SiO_2 molecule concentration. Rather, the amorphization is assumed to be induced at a critical defect concentration lower than 1 dpa.

However, they described that (free) excitons are formed after electron thermalization and then self-trapped [92]. Since the binding energy of the free-exciton is not so high, the (free) excitons could be dissociated by thermal spike heating. It should be noted that $\varepsilon_{\text{STE-FH}}$ in Eq. (5.22) is not the binding energy of the excitons. This dissociation is induced for free-excitons before forming STE. Because of the extended nature, the dissociation of free-excitons does not result in the defect formation. This effect is not included in their model.

5.5.3 Recrystallization of Ion Tracks

With combining Monte Carlo simulation code TREKIS for the electron kinetics induced by SHI impact and classical Molecular Dynamics (MD) LAMMPS code for atomic dynamics, Rymzhanov et al. evaluated the ion track formation in three different oxides having different crystal structures, i.e., MgO, Al_2O_3, and $Y_3Al_5O_{12}$ (YAG), under 167 MeV Xe ion irradiation. The snapshots of MD cell and corresponding X-ray powder diffraction (XRD) patterns are shown in Fig. 5.49. The electron kinetics and the energy transfer to the atomic system were almost the same among the three oxides. Consequently, approximately the same sizes of amorphous tracks were formed around 1 ps after the ion passage. However, big differences were recognized in the atomic dynamics, particularly in the recrystallization behaviors.

In MgO, the amorphous track steeply reduces the size. Around 20 ps, the XRD pattern recovered to that before the ion impact. Almost the perfect recrystallization except some point defects was induced in MgO, probably due to the simplest crystalline structure of the rock-salt type. Contrary, the amorphous tracks are not recrystallized in YAG, due to the most complex crystalline structure of the Garnet type. In Al_2O_3, partial recrystallization was induced and the track shrinks the size but does not disappear. The XRD signal is partly recovered but the pattern was slightly different from the pristine. The calculated results are consistent with the experiments under 167 MeV Xe irradiation. The tracks were not observed in MgO, while the amorphous tracks were observed in YAG. In Al_2O_3, thin and discontinuous crystalline tracks were observed.

In the i-TS model, the criterion of the track formation was whether the melting is induced or not. However, the possibility of drastic recrystallization should be taken into account. Similar recrystallization is numerically confirmed in ZnO by Pakarinen et al. [25]. TEM images reported by Yasuda et al. (Fig. 5.11) [23] were interpreted by the recrystallization.

5.6 Elongation by Cluster Irradiation—Toward Lower Energy and More Accessibility

One of the problems of the shape elongation phenomenon for industrial application is that this method requires SHIs, i.e., very high energy ions, e.g., 200 MeV Xe ions. Acceleration to SHIs requires the world-class big accelerator facilities. The facilities available for materials science applications is quite limited, e.g., a few facilities in Japan. Since a few to several MeV C_{60} ions provide high electronic energy deposition

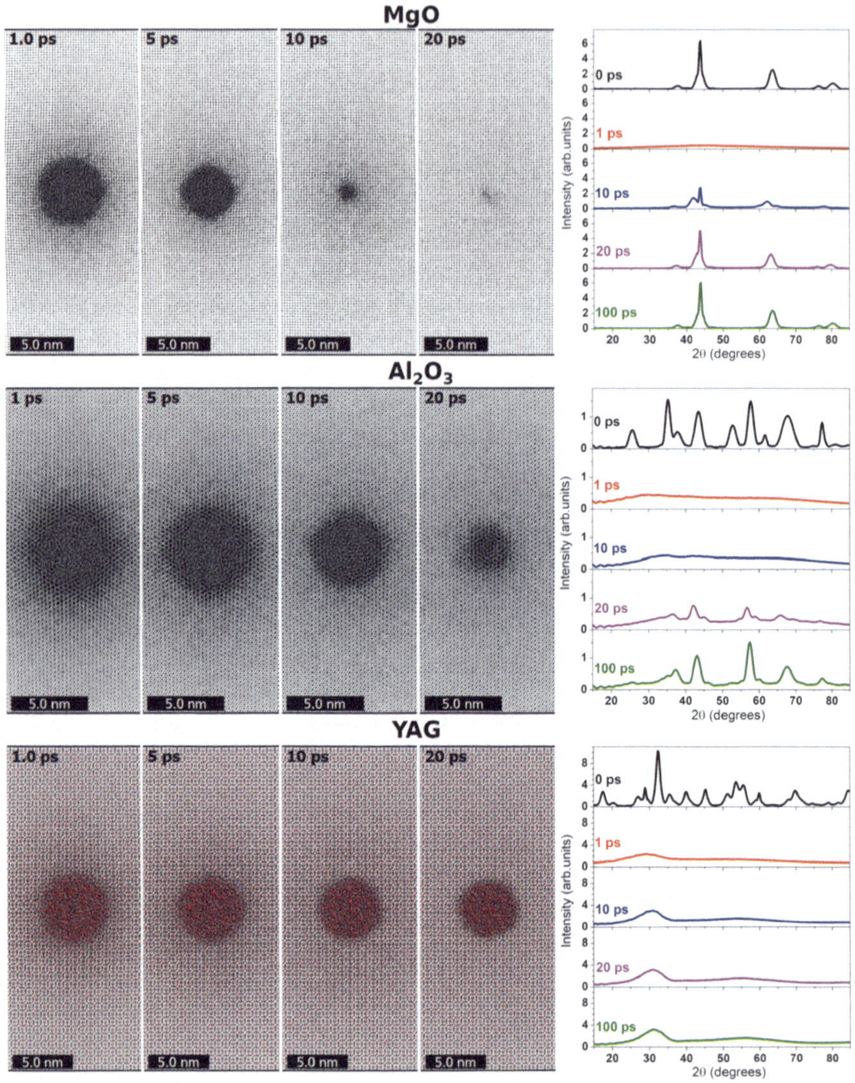

Fig. 5.49 Snapshots of MD cells (left panels) and simulated X-ray powder diffraction patterns (right panels) of MgO, Al_2O_3, and YAG at different times after the passage of 167 MeV Xe ion. Reproduced with permission from [96], ©2019 Springer-Nature

comparable to SHIs, they could be used as substitution of SHIs. The acceleration energy of a few MeV is not high. It is easily attainable in quite common accelerator facilities with the terminal voltage of ~1–2 MV, which are found at many places even in Japan. If the newly-developed high flux C_{60} ion sources can be combined with the commonly-used ~1 MV accelerator facilities, the high-density electronic excitation which are comparable to SHIs would be easily accessible in quite many facilities, i.e., more than 20 facilities in Japan.

As shown in Table 5.5, 6 MeV C_{60}^+ ions provides S_e of 15.5 keV/nm in SiO_2, which is comparable to that of 200 MeV Xe^{14+} ions. A difficulty of the shape elongation experiment using the C_{60} ions was due to low ion flux of the C_{60} ion beam, since the elongation requires high fluences of ~10^{13} ions/cm^2 or more. Impractical irradiation time was required. However, recently a high flux (negative) C_{60} ion source has been developed in QST, Takasaki [98]. Using the newly-developed high-flux ion-source, we have carried out irradiation of Au NPs in SiO_2 with 1–6 MeV C_{60} ions [97]. Figures 5.50a–d exhibit XTEM images of Au NPs (a), (c) before and (b), (d) after

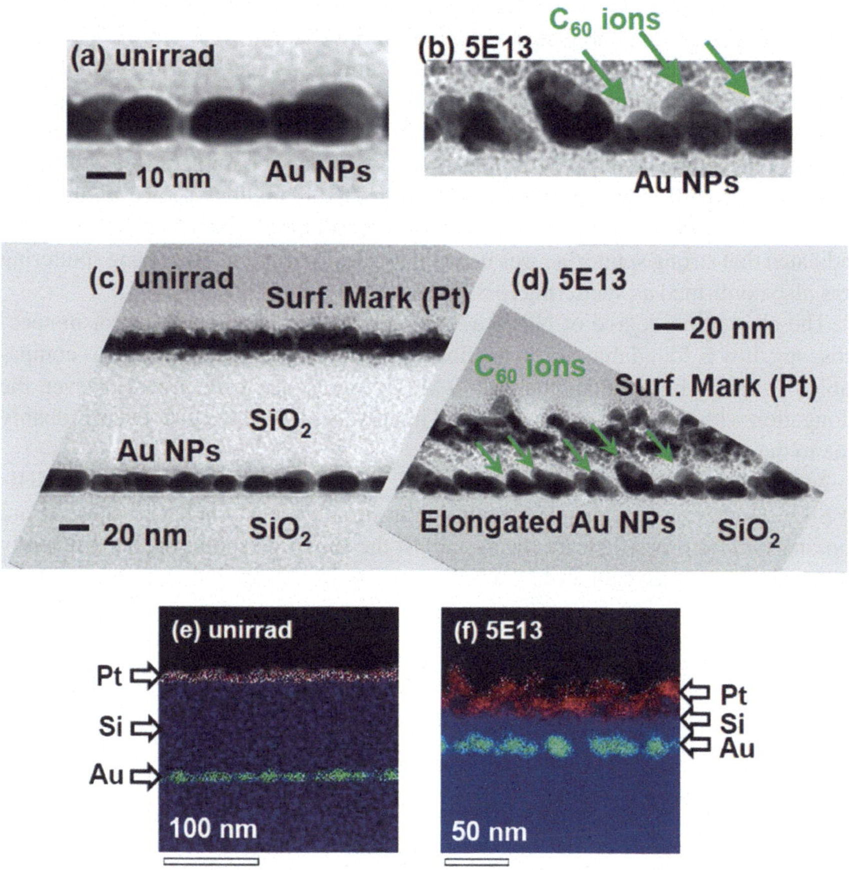

Fig. 5.50 Cross-sectional TEM images of Au NPs embedded in SiO_2 before (**a**), (**c**), (**e**) and after (**b**), (**d**), (**f**) irradiation of 4 MeV C_{60} ions to a fluence of 5×10^{13} C_{60}^+/cm^2. Enlarged images of Au NPs **a**, **b** and overviews of deposited SiO_2 layer on the NPs (**c**), (**d**). A thin layer of Pt was deposited as surface markers. **a**–**d** bright field (BF) images and **e**–**f** STEM-EDS element mappings. In **e**–**f** red, pureple, and green regions correspond to Pt, Si, and Au rich regions. Reproduced with permission from [97], ©2019 Springer-Nature

Table 5.5 Electronic and nuclear stopping powers (S_e, S_n) at the surface of amorphous SiO_2 were calculated. The ion energies per nucleon *E/M*, i.e., the ion vlocties, are also shown.

Ion energy *E* (MeV)	Ion species	S_e (keV/nm)	S_n (keV/nm)	*E/M* (MeV/u)
200	Xe^{14+}	15.0	0.051	1.50
6	C_{60}^{+}	15.5	2.66	8.33×10^{-3}
4	C_{60}^{+}	12.7	3.41	5.56×10^{-3}

4 MeV C_{60} ions to a fluence of 5×10^{13} C_{60}^{+}/cm^2. The Au NPs were fabricated by multiple depositions; First, a gold thin film of 3 nm thick was deposited on a bulk SiO_2. The Au film was transformed to NPs by rapid thermal annealing, then the Au NPs were covered by SiO_2 of 100 nm thick by sputtering deposition. While the Au NPs show oblate spheroidal shapes before the irradiation as shown in Fig. 5.50a, they elongate to the angle of 45° from the sample surface since the sample was irradiated with an incident angle of 45°. Figures 5.50c, d show the overview of the same samples. Before the sample thinning for TEM observation, the surface of the SiO_2 layer was covered by thin Pt layer as the surface marker. Before the C_{60} irradiation, Au NPs were covered by SiO_2 layer of 100 nm thick. However, the SiO_2 layer thickness decreased to ~30 nm after the irradiation. This observation clearly indicated that strong sputtering was induced by C_{60} irradiation. The strong sputtering was also confirmed as sputtering loss by RBS, and optical absorption.

The elongation degree of NPs was evaluated by optical linear dichroism spectroscopy. It was found that the elongation efficiency of 4 MeV C_{60} ions was comparable or slightly higher than that of 200 MeV Xe ions at low fluence. However, the elongation with the C_{60} ions saturated and finally decreased at high fluence, probably due to the strong sputtering.

MeV C_{60} ion irradiation induces high electronic deposition comparable to SHIs in SiO_2. However, the enhanced sputtering is simultaneously induced. Now we are looking for matrices which are strong against the enhanced sputtering but efficiently induce the shape elongation of NPs [99].

References

1. L. Skuja, Optically active oxygen-deficiency-related centers in amorphous silicon dioxide. J. Non-Cryst. Solids **239**, 16 (1998)
2. R.A.B. Devine, Ion implantation-induced and radiation-induced structural modifications in amorphous SiO_2. J. Non-Cryst. Solids **152**, 50 (1993)
3. D.L. Griscom, Optical properties and structure of defects in silica glass. J. Ceram. Soc. Jpn. **99**, 923 (1991)
4. R.A.B. Devine, Macroscopic effects of radiation in amorphous SiO_2. Nucl. Instrum. Meth. Phys. Res. B **91**, 378 (1994)
5. H. Amekura, N. Okubo, F. Ren, N. Ishikawa, Swift heavy ion irradiation to ZnO nanoparticles: steep degradation at low fluences and stable tolerance at high fluences. J. Appl. Phys. **124**, 145901 (2018)
6. H. Amekura, N. Kishimoto, Effects of high-fluence ion implantation on colorless diamond self-standing films. J. Appl. Phys. **104**, 063509 (2008)

7. A. Oliver, J.C. Cheang-Wong, A. Crespo, L. Rodríguez-Fernández, J.M. Hernández, E. Muñoz, R. Espejel-Morales, E' and B_2 center production in amorphous quartz by MeV Si and Au ion implantation. Mater. Sci. Eng. B **78**, 32 (2000)
8. I. Jozwik-Biala, J. Jagielski, B. Arey, L. Kovarik, G. Sattonnay, A. Debelle, S. Mylonas, I. Monnet, L. Thomé, Effect of combined local variations in elastic and inelastic energy losses on the morphology of tracks in ion-irradiated materials. Acta Mater. **61**, 4669 (2013)
9. D. Schauries, B. Afra, T. Bierschenk, M. Lang, M.D. Rodriguez, C. Trautmann, W. Li, R.C. Ewing, P. Kluth, The shape of ion tracks in natural apatite. Nucl. Instrum. Meth. Phys. Res. B **326**, 117 (2014)
10. A. Meftah, F. Brisard, J.M. Costantini, E. Dooryhee, M. Hage-Ali, M. Hervieu, J.P. Stoquert, F. Studer, M. Toulemonde, Track formation in SiO_2 quartz and the thermal-spike mechanism. Phys. Rev. B **49**, 12457 (1994)
11. B. Afra, M.D. Rodriguez, C. Trautmann, O.H. Pakarinen, F. Djurabekova, K. Nordlund, T. Bierschenk, R. Giulian, M.C. Ridgway, G. Rizza, N. Kirby, M. Toulemonde, P. Kluth, SAXS investigations of the morphology of swift heavy ion tracks in α-quartz. J. Phys.: Conden. Mat. **25**, 045006 (2013)
12. N. Itoh, D.M. Duffy, S. Khakshouri, A.M. Stoneham, Making tracks: electronic excitation roles in forming swift heavy ion tracks. J. Phys.: Conden. Mat. **21**, 474205 (2009)
13. M. Lang, J. Lian, J. Zhang, F. Zhang, W.J. Weber, C. Trautmann, R.C. Ewing, Single-ion tracks in $Gd_2Zr_{2-x}T_xO_7$ pyrochlores irradiated with swift heavy ions. Phys. Rev. B **79**, 224105 (2009)
14. P.D. Townsend, P.J. Chandler, L. Zhang, *Optical Effects of Ion Implantation* (Cambridge University Press, 1994)
15. K. Awazu, S. Ishii, K. Shima, S. Roorda, J.L. Brebner, Structure of latent tracks created by swift heavy-ion bombardment of amorphous SiO_2. Phys. Rev. B **62**, 3689 (2000)
16. H. Amekura, N. Ishikawa, N. Okubo, M.C. Ridgway, R. Giulian, K. Mitsuishi, Y. Nakayama, Ch. Buchal, S. Mantl, N. Kishimoto, Zn nanoparticles irradiated with swift heavy ions at low fluences: optically-detected shape elongation induced by nonoverlapping ion tracks. Phys. Rev. B **83**, 205401 (2011)
17. S. Klaumunzer, Ion tracks in quartz and vitreous silica. Nucl. Instrum. Methods Phys. Res. B **225**, 136 (2004)
18. K. Nomura, T. Nakanishi, Y. Nagasawa, Y. Ohki, K. Awazu, M. Fujimaki, N. Kobayashi, S. Ishii, K. Shima, Structural change induced in TiO_2 by swift heavy ions and its application to three-dimensional lithography. Phys. Rev. B **68**, 064106 (2003)
19. A. Benyagoub, S. Klaumunzer, M. Toulemonde, Radiation-induced compaction and plastic flow of vitreous silica. Nucl. Instrum. Meth. Phys. Res. B **146**, 449 (1998)
20. M. Nastasi, J.W. Mayer, J.K. Hirvonen, *Ion-Solid Interaction: fundamentals and Applications* (Cambridge University Press, Cambridge, UK, 1996)
21. P. Kluth, C.S. Schnohr, O.H. Pakarinen, F. Djurabekova, D.J. Sprouster, R. Giulian, M.C. Ridgway, A.P. Byrne, C. Trautmann, D.J. Cookson, K. Nordlund, M. Toulemonde, Fine structure in swift heavy ion tracks in amorphous SiO_2. Phys. Rev. Lett. **101**, 175503 (2008)
22. P. Kluth, O.H. Pakarinen, F. Djurabekova, R. Giulian, M.C. Ridgway, A.P. Byrne, K. Nordlund, Nanoscale density fluctuations in swift heavy ion irradiated amorphous SiO_2. J. Appl. Phys. **110**, 123520 (2011)
23. K. Yasuda, T. Yamamoto, M. Etoh, S. Kawasoe, S. Matsumura, N. Ishikawa, Accumulation of radiation damage and disordering in $MgAl_2O_4$ under swift heavy ion irradiation. Int. J. Mater. Res. **102**, 1082 (2011)
24. S. Takaki, K. Yasuda, T. Yamamoto, S. Matsumura, N. Ishikawa, Structure of ion tracks in ceria irradiated with high energy xenon ions. Prog. Nucl. Energy **92**, 306 (2016)
25. O.H. Pakarinen, F. Djurabekova, K. Nordlund, Density evolution in formation of swift heavy ion tracks in insulators. Nucl. Instrum. Meth. Phys. Res. B **268**, 3163 (2010)
26. H. Amekura, P. Kluth, P. Mota-Santiago, I. Sahlberg, V. Jantunen, A.A. Leino, H. Vazquez, K. Nordlund, F. Djurabekova, N. Okubo, N. Ishikawa, Vaporlike phase of amorphous SiO_2 is not a prerequisite for the core/shell ion tracks or ion shaping. Phys. Rev. Mater. **2**, 096001 (2018)

27. S. Klaumünzer, G. Schumacher, Dramatic growth of glassy $Pd_{80}Si_{20}$ during heavy-ion irradiation. Phys. Rev. Lett. **51**, 1987 (1983)
28. S. Klaumünzer, C. Li, G. Schumacher, Plastic flow of borosilicate glass under bombardment with heavy ions. Appl. Phys. Lett. **51**, 97 (1987)
29. M.-D. Hou, S. Klaumunzer, G. Schumacher, Dimensional changes of metallic glasses during bombardment with fast heavy ions. Phys. Rev. B **41**, 1144 (1990)
30. A. Hedler, S.L. Klaumunzer, W. Wesch, Amorphous silicon exhibits a glass transition. Nat. Mater. **3**, 804 (2004)
31. A. Benyagoub, S. Loeffler, M. Rammensee, S. Klaumunzer, Ion-beam-induced plastic deformation in vitreous silica. Radiat. Eff. Defect Solids **110**, 217 (1989)
32. T. van Dillen, E. Snoeks, W. Fukarek, C.M. van Kats, K.P. Velikov, A. van Blaaderen, A. Polman, Anisotropic deformation of colloidal particles under MeV ion irradiation. Nucl. Instrum. Meth. Phys. Res. B **175–177**, 350 (2001)
33. T. van Dillen, M.J.A. de Dood, J.J. Penninkhof, A. Polman, S. Roorda, A.M. Vredenberg, Ion beam-induced anisotropic plastic deformation of silicon microstructures. Appl. Phys. Lett. **84**, 3591 (2004)
34. T. van Dillen, A. van Blaaderen, A. Polman, Shaping colloidal assemblies. Mater. Today **7**, 40 (2004)
35. E. Snoeks, A. Polman, C.A. Volkert, Densification, anisotropic deformation, and plastic flow of SiO_2 during MeV heavy ion irradiation. Appl. Phys. Lett. **65**, 2487 (1994)
36. A. Slablab, T.J. Isotalo, J. Mäkitalo, L. Turquet, P.-E. Coulon, T. Niemi, C. Ulysse, M. Kociak, D. Mailly, G. Rizza, M. Kauranen, Fabrication of ion-shaped anisotropic nanoparticles and their orientational imaging by second-harmonic generation microscopy. Sci. Rep. **6**, 37469 (2016)
37. H. Trinkaus, A.I. Ryazanov, Viscoelastic model for the plastic flow of amorphous solids under energetic ion bombardment. Phys. Rev. Lett. **74**, 5072 (1995)
38. W. Wesch, T. Steinbach, M.C. Ridgway, Swift Heavy Ion Irradiation of Amorphous Semiconductors. in *Ion Beam Modification of Solids*, Eds. by W. Wesch, E. Wendler (Springer, 2016), p. 403
39. S. Klaumünzer, A. Benyagoub, Phenomenology of the plastic flow of amorphous solids induced by heavy-ion bombardment. Phys. Rev. B **43**, 7502 (1991)
40. T. van Dillen, E. van der Giessen, P.R. Onck, A. Polman, Size-dependent ion-beam-induced anisotropic plastic deformation at the nanoscale by nonhydrostatic capillary stresses. Phys. Rev. B **74**, 132103 (2006)
41. C. D'Orleans, J.P. Stoquert, C. Estournès, C. Cerruti, J.J. Grob, J.L. Guille, F. Haas, D. Muller, M. Richard-Plouet, Anisotropy of Co nanoparticles induced by swift heavy ions. Phys. Rev. B **67**, 220101 (2003)
42. C. D'Orleans, C. Cerruti, C. Estournes, J.J. Grob, J.L. Guille, F. Haas, D. Muller, M. Richard-Plouet, J.P. Stoquert, Irradiations of implanted cobalt nanoparticles in silica layers. Nucl. Instrum. Meth. Phys. Res. B **209**, 316 (2003)
43. M.C. Ridgway, R. Giulian, D.J. Sprouster, P. Kluth, L.L. Araujo, D.J. Llewellyn, A.P. Byrne, F. Kremer, P.F.P. Fichtner, G. Rizza, H. Amekura, M. Toulemonde, Role of thermodynamics in the shape transformation of embedded metal nanoparticles induced by swift heavy-ion irradiation. Phys. Rev. Lett. **106**, 095505 (2011)
44. S. Roorda, T. van Dillen, A. Polman, C. Graf, A. van Blaaderen, B.J. Kooi, Aligned gold nanorods in silica made by ion irradiation of core-shell colloidal particles. Adv. Mater. **16**, 235 (2004)
45. E. Snoeks, A. van Blaaderen, T. van Dillen, C.M. van Kats, M.L. Brongersma, A. Polman, Colloidal ellipsoids with continuously variable shape. Adv. Mater. **12**, 1511 (2000)
46. S. Klaumunzer, Modification of nanostructures by high-energy ion beams. Nucl. Instrum. Meth. Phys. Res. B **244**, 1 (2006)
47. H. Amekura, K. Kono, N. Okubo, N. Ishikawa, Shape elongation of embedded Zn nanoparticles induced by swift heavy ion irradiation: A SAXS study. Phys. Status solidi (b) **252**, 165 (2015)

48. C. Harkati Kerboua, J.M. Lamarre, M. Chicoine, L. Martinu, S. Roorda, Elongation of gold nanoparticles by swift heavy ion irradiation: surface plasmon resonance shift dependence on the electronic stopping power. Thin Solid Films **527**, 186 (2013)
49. G. Rizza, F. Attouchi, P.E. Coulon, S. Perruchas, T. Gacoin, I. Monnet, L. Largeau, Rayleigh-like instability in the ion-shaping of Au–Ag alloy nanoparticles embedded within a silica matrix. Nanotechnology **22**, 175305 (2011)
50. J.C. Pivin, F. Singh, O. Angelov, L. Vincent, Perpendicular magnetization of FePt particles in silica induced by swift heavy ion irradiation. J. Phys. D **42**, 025005 (2009)
51. B. Schmidt, K.H. Heinig, A. Mueklich, C. Akhmadaliev, Swift-heavy-ion-induced shaping of spherical Ge nanoparticles into disks and rods. Nucl. Instrum. Meth. Phys. Res. B **267**, 1345 (2009)
52. L.L. Araujo, R. Giulian, D.J. Sprouster, C.S. Schnohr, D.J. Llewellyn, B. Johannessen, A.P. Byrne, M.C. Ridgway, Structural properties of embedded Ge nanoparticles modified by swift heavy-ion irradiation. Phys. Rev. B **85** (2012)
53. H. Amekura, N. Okubo, N. Ishikawa, D. Tsuya, K. Mitsuishi, Y. Nakayama, U.B. Singh, S.A. Khan, S. Mohapatra, D.K. Avasthi, Swift heavy ion irradiation of ZnO nanoparticles embedded in silica: radiation-induced deoxidation and shape elongation. Appl. Phys. Lett. **103**, 203106 (2013)
54. E.A. Dawi, A.A. Karar, F.H.P.M. Habraken, Anisotropic deformation of NiO nanoparticles embedded in silica under swift heavy ion irradiation. Nanotechnology **30**, 285603 (2019)
55. R. Giulian, P. Kluth, L.L. Araujo, D.J. Sprouster, A.P. Byrne, D.J. Cookson, M.C. Ridgway, Shape transformation of Pt nanoparticles induced by swift heavy-ion irradiation. Phys. Rev. B **78**, 125413 (2008)
56. M.C. Ridgway, P. Kluth, R. Giulian, D.J. Sprouster, L.L. Araujo, C.S. Schnohr, D.J. Llewellyn, A.P. Byrne, G.J. Foran, D.J. Cookson, Changes in metal nanoparticle shape and size induced by swift heavy-ion irradiation. Nucl. Instrum. Meth. Phys. Res. B **267**, 931 (2009)
57. Ch. Dufour, V. Khomenkov, G. Rizza, M. Toulemonde, Ion-matter interaction: the three-dimensional version of the thermal spike model. Application to nanoparticle irradiation with swift heavy ions. J. Phys. D **45**, 065302 (2012)
58. G. Rizza, P.E. Coulon, V. Khomenkov, C. Dufour, I. Monnet, M. Toulemonde, S. Perruchas, T. Gacoin, D. Mailly, X. Lafosse, C. Ulysse, E.A. Dawi, Rational description of the ion-beam shaping mechanism. Phys. Rev. B **86**, 035450 (2012)
59. E.A. Dawi, G. Rizza, M.P. Mink, A.M. Vredenberg, F. Habraken, Ion beam shaping of Au nanoparticles in silica: particle size and concentration dependence. J. Appl. Phys. **105**, 074305 (2009)
60. H. Amekura, N. Okubo, D. Tsuya, N. Ishikawa, Counterevidence to the ion hammering scenario as a driving force for the shape elongation of embedded nanoparticles. AIP Adv. **7**, 085304 (2017)
61. H. Amekura, S. Mohapatra, U.B. Singh, S.A. Khan, P.K. Kulriya, N. Ishikawa, N. Okubo, D.K. Avasthi, Shape elongation of Zn nanoparticles in silica irradiated with swift heavy ions of different species and energies: scaling law and some insights on the elongation mechanism. Nanotechnology **25**, 435301 (2014)
62. K. Nakano, H. Yoshizaki, Y. Saitoh, N. Ishikawa, A. Iwase, XRD study of yttria stabilized zirconia irradiated with 7.3 MeV Fe, 10 MeV I, 16 MeV Au, 200 MeV Xe and 2.2 GeV Au ions. Nucl. Instrum. Meth. Phys. Res. B **370**, 67 (2016)
63. K. Awazu, X. Wang, M. Fujimaki, J. Tominaga, H. Aiba, Y. Ohki, T. Komatsubara, Elongation of gold nanoparticles in silica glass by irradiation with swift heavy ions. Phys. Rev. B **78**, 054102 (2008)
64. Z.G. Wang, C. Dufour, E. Paumier, M. Toulemonde, The S_e sensitivity of metals under swift-heavy-ion irradiation: a transient thermal process. J. Phys. Conden. Mat. **6**, 6733 (1994)
65. M. Toulemonde, W. Assmann, C. Dufour, A. Meftah, F. Studer, C. Trautmann, Experimental phenomena and thermal spike model description of ion tracks in amorphisabel inorganic insulators. Mat. Fys. Medd. Dan. Vid. Selsk. **52**, 263 (2006)

66. M. Sall, I. Monnet, F. Moisy, C. Grygiel, S. Jublot-Leclerc, S. Della-Negra, M. Toulemonde, E. Balanzat, Track formation in III-N semiconductors irradiated by swift heavy ions and fullerene and re-evaluation of the inelastic thermal spike model. J. Mater. Sci. **50**, 5214 (2015)
67. A.A. Leino, O.H. Pakarinen, F. Djurabekova, K. Nordlund, P. Kluth, M.C. Ridgway, Swift heavy ion shape transformation of au nanocrystals mediated by molten material flow and recrystallization. Mater. Res. Lett. **2**, 37 (2014)
68. H. Vázquez, E.H. Åhlgren, O. Ochedowski, A.A. Leino, R. Mirzayev, R. Kozubek, H. Lebius, M. Karlušic, M. Jakšic, A.V. Krasheninnikov, J. Kotakoski, M. Schleberger, K. Nordlund, F. Djurabekova, Creating nanoporous graphene with swift heavy ions. Carbon **114**, 511 (2017)
69. P. Kluth, R. Giulian, D.J. Sprouster, C.S. Schnohr, A.P. Byrne, D.J. Cookson, M.C. Ridgway, Energy dependent saturation width of swift heavy ion shaped embedded Au nanoparticles. Appl. Phys. Lett. **94**, 113107 (2009)
70. H. Amekura, P. Kluth, P. Mota-Santiago, I. Sahlberg, V. Jantunen, A.A. Leino, H. Vazquez, K. Nordlund, F. Djurabekova, On the mechanism of the shape elongation of embedded nanoparticles, unpublished
71. H. Amekura, N. Kishimoto, Implantation of 60 keV copper negative ion into thin SiO_2 films on Si: thermal stability of Cu nanoparticles and recovery of radiation damage. J. Appl. Phys. **94**, 2585 (2003)
72. E.A. Dawi, A.M. Vredenberg, G. Rizza, M. Toulemonde, Ion-induced elongation of gold nanoparticles in silica by irradiation with Ag and Cu swift heavy ions: track radius and energy loss threshold. Nanotechnology **22**, 215607 (2011)
73. A. Oliver, J.A. Reyes-Esqueda, J.C. Cheang-Wong, C.E. Roman-Velazquez, A. Crespo-Sosa, L. Rodriguez-Fernandez, J.A. Seman, C. Noguez, Controlled anisotropic deformation of Ag nanoparticles by Si ion irradiation. Phys. Rev. B **74**, 245425 (2006)
74. C.F. Bohren, B.R. Huffman, *Absorption and scattering of light by small particles* (Wiley, New York, 1983)
75. H. Amekura, N. Okubo, N. Ishikawa, Optical birefringence of Zn nanoparticles embedded in silica induced by swift heavy-ion irradiation. Opt. Express **22**, 29888 (2014)
76. J.A. Reyes-Esqueda, C. Torres-Torres, J.C. Cheang-Wong, A. Crespo-Sosa, L. Rodriguez-Fernandez, C. Noguez, A. Oliver, Large optical birefringence by anisotropic silver nanocomposites. Opt. Express **16**, 710 (2008)
77. N. Künzner, D. Kovalev, J. Diener, E. Gross, V.Y. Timoshenko, G. Polisski, F. Koch, M. Fujii, Giant birefringence in anisotropically nanostructured silicon. Opt. Lett. **26**, 1265 (2001)
78. F. Genereux, S.W. Leonard, H.M. van Driel, A. Birner, U. Gösele, Large birefringence in two-dimensional silicon photonic crystals. Phys. Rev. B **63**, 161101(R) (2001)
79. O.L. Muskens, M.T. Borgstrom, E.P.A.M. Bakkers, J.G. Rivas, Giant optical birefringence in ensembles of semiconductor nanowires. Appl. Phys. Lett. **89**, 233117 (2006)
80. M. Kobylko, P.-E. Coulon, A. Slablab, A. Fafin, J. Cardin, C. Dufour, A. Losquin, M. Kociak, I. Monnet, D. Mailly, X. Lafosse, C. Ulysse, E. Garcia-Caurel, G. Rizza, Localized plasmonic resonances of prolate nanoparticles in a symmetric environment: experimental verification of the accuracy of numerical and analytical models. Phys. Rev. Appl. **9**, 064038 (2018)
81. N. Ishikawa, A. Iwase, Y. Chimi, O. Michikami, H. Wakana, T. Kambara, Defect production induced by primary ionization in ion-irradiated oxide superconductors. J. Phys. Soc. Jpn. **69**, 3563 (2000)
82. E.V. Benton, On latent track formation in organic nuclear charged particle track detectors. Radiat. Eff. **2**, 273 (1970)
83. K.S. Song, R.T. Williams, *Self-Trapped Excitons*, 2nd edn. (Springer, Berlin, Heidelberg, 1996)
84. M. Ueta, H. Kanzaki, K. Kobayashi, Y. Toyozawa, E. Hanamura, *Excitonic Processes in Solids* (Springer, Berlin, Heidelberg, 1986)
85. M. Hirai, Y. Suzuki, H. Hattori, T. Ehara, E. Kitamura, Picosecond laserphotolysis on photo-induced defect formation process in KI crystal. J. Phys. Soc. Jpn. **56**, 2948 (1987)
86. F. Agulló-López, A. Climent-Font, A. Muñoz-Martin, A. Zucchiatti, Alternative approaches to electronic damage by ion-beam irradiation: Exciton models. Phys. Status Solidi (a) **213**, 2960 (2016)

87. R. Katz, K.S. Loh, L. Daling, G.-R. Huang, An analytic representation of the radial distribution of dose from energetic heavy ions in water, Si, LiF, NaI, and SiO_2. Radiat. Eff. Defects Solids **114**, 15 (1990)
88. N. Itoh, Self-trapped exciton model of heavy-ion track registration. Nucl. Instrum. Meth. Phys. Res. B **116**, 33 (1996)
89. A. Meftah, F. Brisard, J.M. Costantini, M. Hage-Ali, J.P. Stoquert, F. Studer, M. Toulemonde, Swift heavy ions in magnetic insulators: a damage-cross-section velocity effect. Phys. Rev. B **48**, 920 (1993)
90. G.A. Thomas, A. Frova, J.C. Hensel, R.E. Miller, P.A. Lee, Collision broadening in the exciton gas outside the electron-hole droplets in Ge. Phys. Rev. B **13**, 1692 (1976)
91. M. Nagai, R. Shimano, M. Kuwata-Gonokami, Electron-hole droplet formation in direct-gap semiconductors observed by mid-infrared pump-probe spectroscopy. Phys. Rev. Lett. **86**, 5795 (2001)
92. F. Agulló-López, A. Mendez, G. García, J. Olivares, J.M. Cabrera, Synergy between thermal spike and exciton decay mechanisms for ion damage and amorphization by electronic excitation. Phys. Rev. B **74**, 174109 (2006)
93. A. Rivera, M.L. Crespillo, J. Olivares, R. Sanz, J. Jensen, F. Agulló-López, On the exciton model for ion-beam damage: the example of TiO_2. Nucl. Instrum. Meth. Phys. Res. B **268**, 3122 (2010)
94. A. Rivera, J. Olivares, G. Garcia, F. Agulló-López, Swift heavy ion damage to sodium chloride: synergy between excitation and thermal spikes. J. Phys.: Conden. Mat. **24**, 085401 (2012)
95. T. Karasawa, M. Hirai, Color center formation in KI and NaCl crystals by pulsed electron beam. J. Phys. Soc. Jpn. **39**, 999 (1975)
96. R.A. Rymzhanov, N. Medvedev, J.H. O'Connell, A. Janse van Vuuren, V.A. Skuratov, A.E. Volkov, Recrystallization as the governing mechanism of ion track formation. Sci. Rep. **9**, 3837 (2019)
97. H. Amekura. K. Narumi, A. Chiba, Y. Hirano, K. Yamada, D. Tsuya, S. Yamamoto, N. Okubo, N. Ishikawa, Y. Saitoh, C_{60} ions of 1 MeV are slow but elongate nanoparticles like swift heavy ions of hundreds MeV, Sci. Rep. **9**, 14980 (2019)
98. A. Chiba, A. Usui, Y. Hirano, K. Yamada, K. Narumi, and Y. Saitoh, Novel approaches for intensifying negative C_{60} ion beams using conventional ion sources installed on a tandem accelerator. Quan. Beam. Sci. **4**, 13 (2020)
99. Rang Li, K. Narumi, A. Chiba, Y. Hirano, D. Tsuya, S. Yamamoto, Y. Saitoh, N. Okubo, N. Ishikawa, C. Pang, F. Chen, H. Amekura, Matrix-material dependence on the elongation of embedded gold nanoparticles induced by 4 MeV C_{60} and 200 MeV Xe ion irradiation. Nanotech. **31**, 265606 (2020)

Chapter 6
Electrooptic Properties of Dielectric Waveguides

Abstract In this chapter, the electrooptic properties of the ion beam produced optical waveguides are demonstrated, including the ion beam property modification and the electrooptic devices with different designs and functions. Section 6.1 summarizes the modified electrooptic coefficients of dielectrics by diverse ion beam techniques. Section 6.2 introduces electrooptic modulators based on dielectric waveguides. Sections 6.3 and 6.4 overview the up-to-date progress on electrooptic modulators based on thin films and on-chip devices, including freestanding and wafer bonded membranes of $LiNbO_3$.

Electro-optics deals with physics, materials, devices, and systems which the propagation of light and its interaction with materials possessing electrooptic (EO) properties [1–4]. Electro-optics is also a branch of Nonlinear Optics. When an electric field is applied to crystals that particularly lack center of symmetry, it causes a change in the anisotropy (the variation of refractive index of the crystal). In the linear case, i.e., the change in birefringence of the crystal varies linearly with the electric field, the corresponding EO effect is usually called the Pockels effect (can be described in terms of a second-order nonlinear susceptibility) [5]. By applying an additional electric field, the refractive index can be changed, which modulates the light propagation properties (e.g., polarization, phase, or intensity/amplitude). In the nonlinear case, the most used nonlinearity is the Kerr effect, which also called the quadratic EO effect (a third-order nonlinearity). For the Kerr effect, the induced index change is directly proportional to the square of the applied electric field. The components and devices with EO modulation functions are usually constructed on waveguide platforms. The basic geometries typically include Fabry-Perot (straight waveguides), Mach-Zehnder interferometers, Micro-cavity (rings, racetracks, etc.) resonators. For efficient modulation, the required EO crystals must have large EO coefficients, at least on the crystal orientation of applied electric field. In practice, ion beam produced waveguides in $LiNbO_3$ as well as LNOI thin-film waveguides have been investigated for efficient EO modulation. In this chapter, the ion beam modification of EO features, and EO modulators in waveguides and membranes (freestanding or wafer-bonded) will be overviewed.

F. Chen et al., *Ion Irradiation of Dielectrics for Photonic Applications*, Springer Series in Optical Sciences 231, https://doi.org/10.1007/978-981-15-4607-5_6

6.1 Modification of Electrooptic Coefficients of Dielectrics

Efficient EO modulation requires materials possessing large EO coefficients. For the Pockels effect, the changes of refractive index by an applied electric field can be described by the EO tensor r through the relationship [6].

$$\Delta\left(\frac{1}{n^2}\right)_i = \sum_{j=1}^{3} r_{ij} E_j \tag{6.1}$$

where r_{ij} is the EO coefficients, and the typical magnitude is of ~10^{-12} mV^{-1} (pm/V). The most used EO crystals include $LiNbO_3$, $LiTaO_3$, β-BBO, KDP, KTN, $BaTiO_3$, and some semiconductors such as GaAs, ZnSe, and ZnS. $LiNbO_3$ is a widely used EO crystal because of the combination of large EO coefficients, available large-size wafers, and relatively low costs. The largest EO coefficient for $LiNbO_3$ is r_{33}, which is about ~31 pm/V (at wavelength of 633 nm). Since the bandgap of $LiNbO_3$ is ~4 eV, which determines the transparent wavelength range of $LiNbO_3$ to be 350–5200 nm, for ultraviolet wavelength regime, β-BBO is a suitable candidate for EO modulation. So far, by using ion beam processed waveguide structures, EO components or devices have been successfully fabricated in a few crystals. The ion beams influence the EO performances of the waveguides in comparison with the bulks (in most cases clear degradation occurs due to the lattice distortion and defect formation induced by the ion beams), which reflects on the reduction of the EO coefficients [7]. In this sense, the modification of EO coefficients is a factor to evaluate the quality of the ion beam produced waveguides. One of the most efficient methods to decrease the irradiation damage is post-implantation annealing. Thermal annealing at suitable temperatures, will cause the re-crystallization of the damage lattices and remove the defects considerably. As a result, the EO properties of the ion implanted waveguides could be recovered to a great extent. Table 6.1 summarizes the results of the EO coefficients of the ion implanted waveguides in EO crystals ($LiNbO_3$ and β-BBO). As can been seen from the table, after suitable annealing treatments, the ion implanted $LiNbO_3$ waveguides possess well preserved EO properties (80–90%) of the bulks, and the swift heavy ion irradiated waveguides maintain about 40–60% of the bulk EO features [8, 9]. The β-BBO crystal operates in shorter wavelength regimes than $LiNbO_3$, e.g., down to ultraviolet wavelength band. As for β-BBO crystals, the EO coefficient r_{22} of the He^+ ion implanted waveguide was determined to be about 60% of the bulk value, denoting some degradation of the EO properties [10]. In addition, there are reports on the photorefractive properties of the ion implanted waveguides in $KNbO_3$, BTO, which suggests the EO properties were preserved to some extent in the guiding structures [11]; however, there was no directly-measured data on the EO coefficients in those waveguides.

In addition to $LiNbO_3$ waveguides, the thin-film $LiNbO_3$ receives more attentions recently. The $LiNbO_3$ thin film can either be freestanding or bonded on a substrate

Table 6.1 The results of the EO coefficients of the ion implanted optical waveguides in EO crystals

Materials	Fabrication	Waveguide configuration	E-O coefficients	Working wavelength (nm)	Refs.
$LiNbO_3$	He ion implantation	Planar	r_{33} = 9.24 pm/V 30% of the bulk r_{13} = 5.2 pm/V 60% of the bulk	633	Destefanis et al. [12]
	He ion implantation E = 1.5 MeV ~10^{16} ions/cm^2	Channel	r_{13} = 8.13 pm/V 80% of the bulk	633	Reed and weiss [13]
	H ion implantation E = 800 keV 6×10^{16} ions/cm^2	Planar	r_{33} = 28.6 pm/V 90% of the bulk r_{13} = 9.3 pm/V 90% of the bulk	633	Boudrioua et al. [14]
	C ion implantation E = 3.8 MeV 1–6 × 10^{14} ions/cm^2	Planar	r_{33} = 28.8 pm/V 90% of the bulk	633	Bentini et al. [10]
	F ion irradiation E = 20 MeV 4×10^{14} ions/cm^2	Planar	r_{33} = 18.1 pm/V 60% of the bulk	633	Olivares et al. [8]
	Cl ion irradiation E = 46 MeV 2×10^{12} ions/cm^2	Planar	r_{33} = 13.5 pm/V 40% of the bulk	633	Olivares et al. [15]
β-BBO	He ion implantation E = 1.65 MeV 0.5–1 × 10^{16} ions/cm^2	Ridge	r_{22} = 1.7 pm/V 60% of the bulk	257	Poberaj et al. [16]

wafer. Those thin films serve as quasi-2D platforms that could be promising candidate for on-chip device fabrication. Table 6.2 shows the EO properties of $LiNbO_3$ films (freestanding or wafer-bonded) by ion beam cutting. The BCB bonded $LiNbO_3$ membranes only preserve up to 50% of the bulk EO properties, because the BCB polymer serves as the cladding that can be only annealed at moderate temperatures (no more than 320 °C) [17]. The ion sliced freestanding $LiNbO_3$ membranes are with perfect EO properties since high-temperature annealing treatment (>800 °C) is much efficient. This is owing to the entire recovery of the damage lattice structures by the high-temperature thermal treatment. The shortcomings of the freestanding $LiNbO_3$ membranes are the relatively small scale (less than 1 cm^2) and large thickness (6–10 μm, depending on the He ion energy), which cannot be used as platforms for chip-scale devices. The updated version of lithium niobate on insulators (LNOI) uses SiO_2 instead of BCB as the bonded material, which can significantly improve the thermal stability of the wafer up to 500 °C. The recent results obtained from several groups have shown that the EO coefficients of LNOI wafer are with almost the same values of the EO coefficients as those of the bulks (larger than 95% of the bulks), which means the LNOI film is a single crystalline material. This achievement enables further development of low-cost, highly integrated photonic devices in chip scales.

6.2 Electrooptic Modulators on Waveguides

EO modulators in waveguide platforms are more efficient due to the compact geometry and the diffraction-free light propagation [26]. The used EO modulator geometry is straight waveguide (in stripes or ridges, like Fabry-Perot (FP) cavity), Mach-Zehnder, and microring resonators. The straight waveguide resonator is based on FP with the two polished end-faces, which is considered as the simplest EO device for modulation [20]. The applied electric field induces the variation of the refractive index through the EO effect, which therefore changes the phase of the light through the waveguide. It causes the shift of the FP resonator wavelength. The FP straight waveguides can be utilized to further combine with other optical elements, such as polarizers, to achieve amplitude modulation. In practice, to obtain the EO properties, the straight waveguide is observed as light intensity modulation using a cross-polarization setup. The input light (e.g., at a wavelength of 1550 nm) with a +45° linear polarization is coupled into the waveguide (e.g., by objective lens or through a polarization maintaining fiber). The output light passing through a −45° polarizer is collected by a photo-detector. The output signals reflect the modulation of the light phase. The Mach-Zehnder interferometer resonators are based on Y-branched waveguides [27–29]. Typically, one arm of the interferometers is EO modulated. Mach-Zehnder interferometer modulators are combined into integrated circuits and offer high-bandwidth, high-speed EO amplitude and phase responses over multiple-gigahertz frequency range. The microring (or with other modified geometries, e.g., racetracks) resonators are based the so-called whispering gallery mode

Table 6.2 shows the EO properties of $LiNbO_3$ films (freestanding or wafer-bonded) by ion beam cutting

Materials	Fabrication	Configuration	E-O coefficients	Working wavelength (nm)	Refs.
$LiNbO_3$ films	He ion slicing $E = 200$ keV 3.5×10^{16} ions/cm^2	Membrane SiO_2 bonded	$r_{33} = 31$ pm/V 100% of the bulk	633	Rabiei and Gunter [18]
	He ion slicing $E = 3.8$ MeV 6×10^{16} ions/cm^2	Membrane	$r_{33} = 32.2$ pm/V 100% of the bulk $r_{13} = 10$ pm/V 100% of the bulk	1547	Ramadan et al. [19]
	He ion slicing $E = 195$ keV 4×10^{16} ions/cm^2	Membrane BCB bonded	$r_{33} = 15$ pm/V 50% of the bulk	1555	Guarino et al. [17]
	Proton exchange	Membrane SiO_2 bonded	$r_{33} = 29.5$ pm/V 95% of the bulk	1550	Cai et al. [20]
	Focused ion beam	Membrane SiO_2 bonded	$r_{33} = 30.8$ pm/V	1498	Qiu et al. [21]
	Proton exchange	Membrane SiO_2 bonded	$r_{33} = 30.8$ pm/V $r_{13} = 9$ pm/V	~1550	Li et al. [22]
	Ar^+ plasma etching	Membrane SiO_2 bonded	$r_{33} = 30.9$ pm/V	~1550	Wang et al. [23]
	Electron-beam lithography	Membrane SiO_2 bonded	$r_{33} = 30$ pm/V	~1556	Wang et al. [24]
	Electron-beam lithography	Membrane SiO_2 bonded	$r_{33} = 30$ pm/V	~1550	Zhang et al. [25]

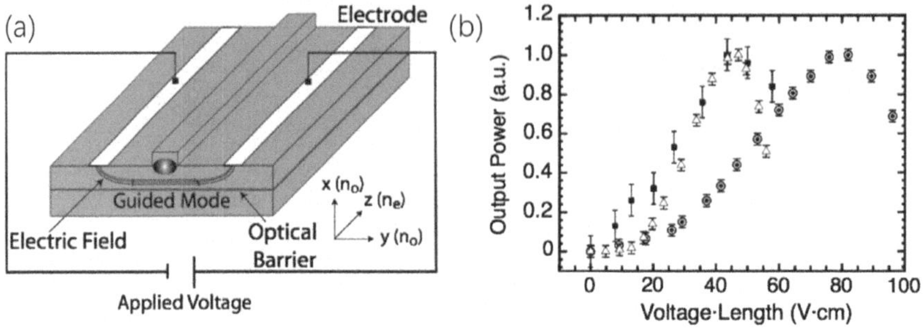

Fig. 6.1 **a** Configuration for the EO amplitude waveguide modulator. **b** Measured half-wave voltage in different modulators at different wavelengths: the ■ points represent the output values measured at 257 nm in a 6.5 mm-long modulator with 16 μm electrode spacing; the Δ points are the values measured at 373 nm in a 6.2 mm-long modulator with 12 μm electrode spacing. Reproduced with permission from [33], ©2007 American Institute of Physics

(WGM) [30–32]. As the additional voltage is applied, the induced refractive index changes modify the effective optical path length of the resonator, leading to the resonance frequency shift. These resonators are with very high Q-factors, allowing high EO modulation tunability in high-bandwidth.

Degl'Innocenti et al. reported on an ultraviolet EO amplitude modulation in β-BBO waveguides, operating in the wavelength range down to 257 nm [33]. The straight ridge waveguides were produced by 1.65 MeV He^+ ion implantation combined with following Ar^+ ion beam etching. The propagation losses were ~10 dB/cm in the UV regime. Figure 6.1 shows (a) the configuration for the EO amplitude waveguide modulator and (b) the measured half-wave voltage in different modulators at different wavelengths. The half-wave voltage V_π can be obtained by the following relationship

$$V_\pi = \frac{\lambda d}{n_o^3 r_{22}(\Gamma_{\mathrm{TE}} + \Gamma_{\mathrm{TM}})L} \tag{6.2}$$

where λ is the operating wavelength of the light, d is the space between two electrodes, $\Gamma_{\mathrm{TE}} + \Gamma_{\mathrm{TM}}$ is the overlap integrals between the applied electric field and the optical field distributions of the TE and TM mode, respectively, and L is the electrode length. Here n_o and r_{22} are the corresponding refractive index and EO coefficient in the modulator geometry, respectively. A half voltage—length product ($V_\pi \cdot L$) of 43 ± 4 V cm at wavelength of 257 nm in the EO modulator was measured.

In telecommunication systems, the most used EO component is Mach-Zehnder interferometer. This geometry possesses two arms for light propagation, which allows EO modulation on one or two arms. Fluck et al. constructed an EO modulator based on an ion-implanted $KNbO_3$ planar waveguide with two 100 μm wide gold electrodes deposited on the top surface separated by 100 μm [34]. In a Mach-Zehnder interferometer configuration, a modulation depth of 20 dB and a half-wave voltage of 18 V at a wavelength of 515 nm was obtained.

Bentini et al. proposed a prototype of integrated scanning Fourier transform micro-interferometer with Mach-Zehnder interferometer geometry based on C or O implanted $LiNbO_3$ waveguides [35]. Figure 6.2 shows the schematic of the operation principles of a scanning Mach-Zehnder micro-interferometer. The micro-device was based on the EO modulation of *x*-cut $LiNbO_3$ channel waveguides. A voltage ramp was applied to one or both arms of the interferometer to produce a proportional linear variation of the refractive index, and therefore the EO modulation resulted in consequent optical path difference between the light travelling along different arms. With an EO modulation under voltage ramp varying from −40 to 40 V, the induced maximum refractive index variation was ~4×10^{-3} corresponding to an optical path difference close to 0.15 mm. By using the integrated micro-interferometer as the major element, compact spectrometer was fabricated, which was successfully used for detecting a certain chemical compound, diluted in fluid media. For example, the gas trace detection by using a calibrated NO_2 optical gas cell showed a sensitivity up to 10 ppb.

Majkic et al. reported on an EO microring resonator based on *z*-cut $LiNbO_3$ waveguides [36]. The resonator consists of a microring and a bus waveguide. A planar waveguide layer was firstly produced by 14.5 MeV fluorine ion irradiation at a fluence of 2.5×10^{14} cm^{-2}. Considerably large refractive index contrast between the waveguide and the cladding was ~0.1 for both extraordinary and ordinary indices (n_e and n_o). The Ar^+ ion etching was performed on the planar waveguide layer with designed photoresist masking to achieve ridge type of structures. The gap between the bus waveguide and microring was sub-μm wide to allow efficient evanescent field coupling of the light fields (see Fig. 6.3 for the scanning electron micrograph of the microring resonator with a bus waveguide [37]). The electrodes were positioned in two vertically separated planes in order to induce the necessary electric field component along the *z*-direction. For the TE polarized light, optical resonances with a modulation depth of 11 dB, separated by a 2.03 nm free spectral range (FSR) were

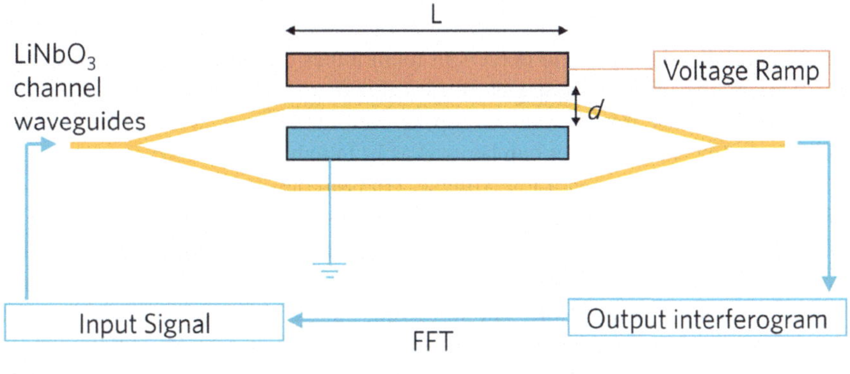

Fig. 6.2 The schematic of a scanning Mach–Zehnder micro-interferometer, which is based on ion implanted *x*-cut $LiNbO_3$ channel waveguides. Reproduced with permission from [35], ©2006 Springer

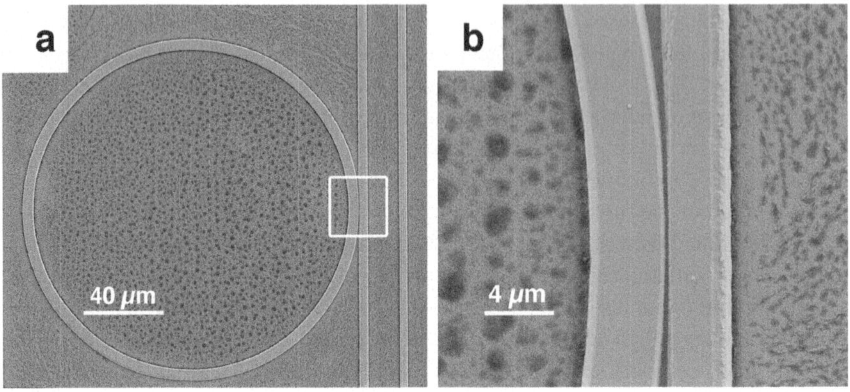

Fig. 6.3 Scanning electron micrograph of a microring resonator and a bus waveguide produced in F ion implanted $LiNbO_3$. **a** The whole ring and the bus waveguide. Ring radius is 80 μm, ridge height is 1.2 μm. **b** Enlarged coupling region, the gap size is ~0.2 μm. Reproduced with permission from [35], ©2006 Springer

observed at ~1.56 μm wavelength band. Additional applied voltage of 10 V onto the resonator results in shifts the resonance by 100 pm, which determines the microring's EO tunability to be 10 pm/V. This value corresponds to frequency tunability of 1.23 GHz/V at 1.55 wavelength band.

6.3 Electrooptic Modulators on Membrane

Ion crystal slicing followed by the lift-off technique enables fabrication of freestanding $LiNbO_3$ membranes with thickness of ~10 μm. The thickness is determined by the energy of ions (e.g., 3.8 MeV He^+). High-quality crystalline membranes depend on the post-lift-off annealing treatment at high temperature (e.g., ~800 °C), which can eliminate the damages and defects in membranes during the implantation process. This is one of the advantages of the $LiNbO_3$ membrane over ion implanted $LiNbO_3$ waveguides. Based on the well-preserved EO features in comparison to the bulks, the $LiNbO_3$ membranes can be used as EO modulation devices.

Izuhara et al. reported on a low-voltage TE/TM converter on ion-sliced lithium niobate membrane [38]. A Ti-indiffused channel waveguide was first produced in a 0.5 mm thick $LiNbO_3$ wafer, supporting both TE and TM modes. The thermally evaporated Cr/Au grating-electrodes were patterned by photolithography and subsequently formed by the lift-off. The following 3.8 MeV He ion implantation and selective acid etching enabled 10 μm membrane formation, and the final annealing at 550 °C for 6 h was applied to recover the EO properties of the thin film. The shift of the mode-conversion wavelength was measured by applying a tuning voltage between the top and bottom electrodes, and the obtained tunability in the $LiNbO_3$

waveguide was 0.26 nm/V, which was three times larger than that of a conventional bulk device.

Ramadan et al. investigated the EO modulation in *z*-cut $LiNbO_3$ freestanding thin film [19]. The measurements of the modulator properties were performed by using optical fiber butt coupling technique. A tunable probe laser was used to measure the wavelength-dependent phase shift between TE and TM modes. It was found that the post annealing processing was significantly important to recover the crystalline properties of the membrane. Figure 6.4a shows the schematic of the EO modulator based on $LiNbO_3$ freestanding membrane. The variation in output power in the modulator film with the $V_\pi \cdot L$ value is shown in Fig. 6.4b. At wavelength of 1550 nm, the $V_\pi \cdot L$ was 12 V·m for 1.1 mm electrode length, whist at 1547 nm, 8 V·m for 1 mm electrode length was obtained. For an ideal bulk EO modulator based on $LiNbO_3$

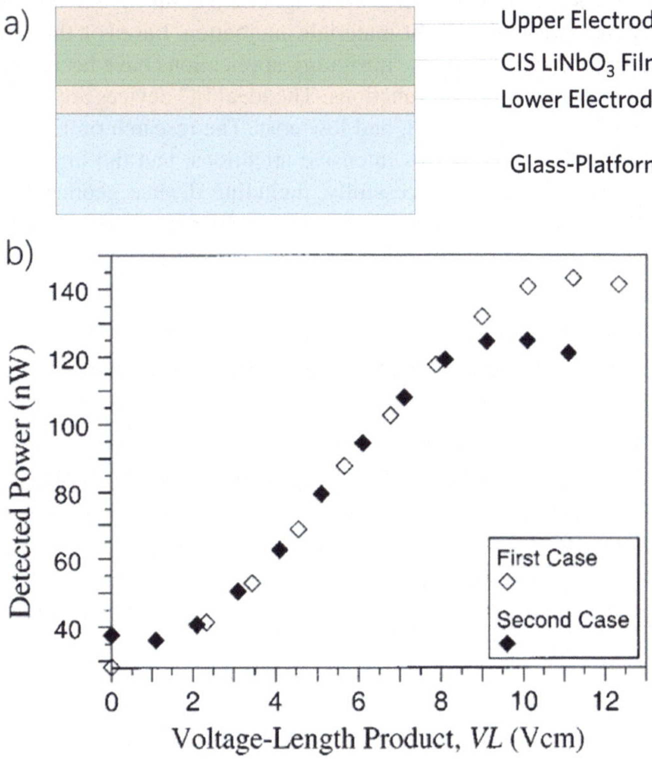

Fig. 6.4 **a** Schematic of the EO modulator based on $LiNbO_3$ freestanding membrane. **b** The variation in output power in the modulator film with $V_\pi \cdot L$. Reproduced with permission from [19], ©2000 American Institute of Physics

crystal, the value of $V_\pi \cdot L$ was ~7.4 V·cm, which indicates reasonable agreement with that from the membrane based EO modulator.

6.4 Electrooptic Modulators on LNOI Waveguides

The development of $LiNbO_3$ on insulators (LNOI) technology inspires the rapid innovation of LNOI based photonic devices [39]. Since the $LiNbO_3$ films on insulating substrates are very thin (with typical thickness of 300–900 nm), further construction of single-mode waveguides even in submicrometric scale on LNOI is realizable by using standard electron beam lithography and ion beam etching. One of the most important advantages of the LNOI-based waveguides is the very compact geometries (e.g., nanowire waveguides) that can confine light field in much tight modes, leading significant enhancement of the light-materials interaction. Based on the high-quality LNOI structures and devices, a few intriguing applications have been realized. EO modulation is one of the exciting functions. The ideal EO devices are with low loss, low drive voltage, large bandwidth, and low cost. The research on EO modulation based on LNOI platforms receives intensive attentions, and the implemented EO devices have been developed successfully, including diverse geometries, exhibiting unique performance for practical applications. In this section, several selected examples for efficient EO modulation devices are introduced.

6.4.1 EO Phase Modulator Based on Straight Waveguide (Fabry-Perot) Resonator

In straight waveguide (FP) resonators, the applied electric field induces the variation of the refractive index through the EO effect, which therefore changes the phase of the light through the waveguide. It causes the shift of the FP resonator wavelength. Cai et al. fabricated EO phase modulator on the proton exchanged in *x*-cut LNOI thin film (with SiO_2 as the insulator) [20]. The propagation loss (including the electrode absorption) of the LNOI waveguide was evaluated to be 3.5 dB/cm at ~1550 nm. Figure 6.5 shows the transmissions of the phase modulator before (black curve) and after (red curve) applying a voltage of 12 V. As it is indicated, the 12 V voltage applied to the electrodes located on the two sides of the waveguide induced a 0.097 nm shift of the Fabry-Perot resonant peak. Considering the wavelength difference of the neighboring resonant peaks (0.228 nm) and the length of the electrodes (2.3 mm), the $V_\pi \cdot L$ product was as low as 6.5 V cm, indicating the efficient EO modulation of the straight waveguide resonator.

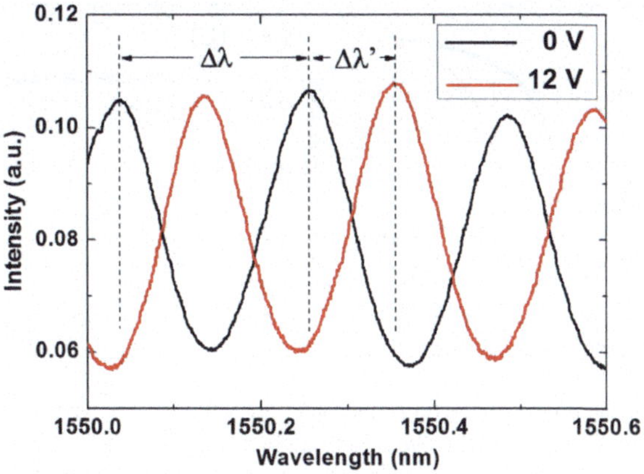

Fig. 6.5 Measured transmissions of the phase modulator before (black) and after (red) applying 12 V. A 0.097 nm shift of the resonant wavelength (Δλ') occurred. The wavelength difference of neighboring resonances (Δλ) was 0.228 nm. Reproduced with permission from [20], ©2016 Optical Society of America

6.4.2 EO Tunable Microring Resonators

Microring structures (including rings, racetracks, and disks) have been successfully produced by several groups. Benefiting from the high-Q factors up to 10^7, the high EO tunability of these resonators has been obtained.

Guarino et al. first realized the EO modulation based on LNOI microring resonator [17]. The insulator material of the LNOI was BCB. The Q-factor was 4000. The transmission and EO modulation characterization experiments were performed at 1.55 μm band. The optical resonances were with a modulation depth of 7 dB, and the FSR was 1.66 nm. The resonator finesse was determined to be ~5. For the TM mode, the wavelength shift of 105 pm was obtained under an applied voltage of 100 V (i.e., 1.05 pm/V of wavelength tunability), which corresponds to a frequency tunability of 0.14 GHz/V. About 50% EO activity was reduced in comparison to that based on bulk material, which mainly due to implantation-induced defects during the fabrication process. Since BCB is an organic material, which cannot be treated at high temperature annealing, the defects remain to a certain extent after 320 °C heating.

The shortcoming of BCB can be overcome by using SiO_2 (silica) as cladding insulators. The silica is with higher thermal stability up to 550 °C. At this temperature the annealing treatment of LNOI films for 2 h removes the defects significantly, resulting in 95% of EO coefficients of the bulk. Consequently, the EO modulators using silica-based LNOI possess better performance than BCB based devices. Wang et al. from Nankai University reported on microdisk resonators with a Q-factor up to 1.2×10^6, and with a radius of 39.6 μm [40]. At 1.55 μm wavelength band, the obtained EO modulation based on the microdisk WGM resonators was with an effective resonance-frequency tuning rate of 3.0 GHz/V. Figure 6.6 shows two typical

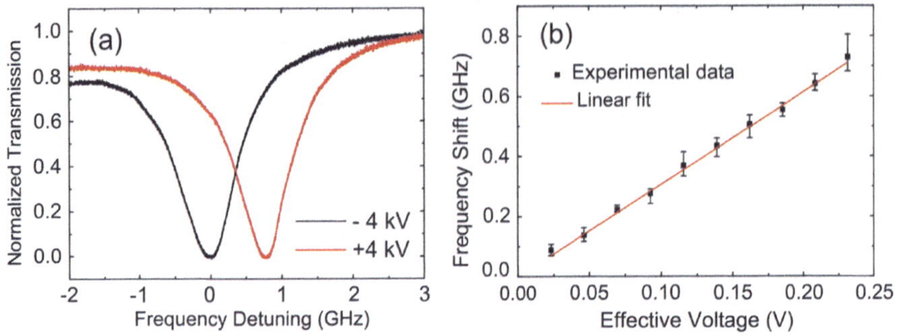

Fig. 6.6 **a** The measured normalized transmission of the LNOI microdisk resonator when −4 kV (black curve) and +4 kV (red curve) voltage were applied, and **b** the shift of resonance wavelength of the resonator with a radius 39.6 μm with respect to the applied effective voltage. Reproduced with permission from [40], ©2015 Optical Society of America

normalized transmission spectra of the LNOI resonator when the applied voltage was set to be ±4 kV, and the shift of resonance wavelength of the resonator with respect to the effective voltage.

Lončar's group from Harvard University developed high-quality microring resonators based on LNOI [41]. The fabricated WGM geometries include rings and racetracks. At wavelength of ~1.57 μm, for a racetrack resonator with the Q factor of ~50,000, the measured EO efficiency was ~7.0 pm/V with good linearity. For a resonator with Q factor of 8,000, a 3 dB electro-optic bandwidth of 30 GHz was obtained. The Q factors can be engineered by controlling the distance between the RF electrodes and the optical waveguide. The intrinsic RC bandwidth limit of the racetrack modulator was estimated to be over 100 GHz. In addition, data transmission rate as high as 40 Gbps was obtained for the racetrack resonator. At maximum frequencies, the measured extinction ratio of the racetrack modulators was 3 dB with power consumptions ($CV^2/4$) of 240 fJ/bit, which was better than silicon modulators that reach the same bandwidth (~30 pJ/bit). In addition, they also utilized specially designed EO photonic circuits containing two microrings (Q factor of 6.6×10^5) on LNOI to realize Kerr frequency comb generation and modulation. By applying a bias of 10 V to align the two resonances, an electrical tuning efficiency of 2.4 pm/V was obtained.

The EO modulation can also be used for production of optical frequency combs, which are considered as essential tools for applications in optical communication, precision metrology, timing, and spectroscopy. Zhang et al. utilized an integrated EO comb generator in a LNOI platform [25]. The core component of the device is a ring resonator (in racetrack geometry). Figure 6.7 illustrates the schematic plot of the device, the micrograph of the LNOI microring resonator, and the measured output spectrum of the EO comb generated from it. Benefiting from the large EO response ($r_{33} = 30$ pm/V), high Q-factor of 1.5×10^6, and the tight confinement of the light (the waveguide width of 1.4 μm), the comb generator operates under modulation of an external microwave synthesizer with peak voltage $V_p = 10$ V. From the measured

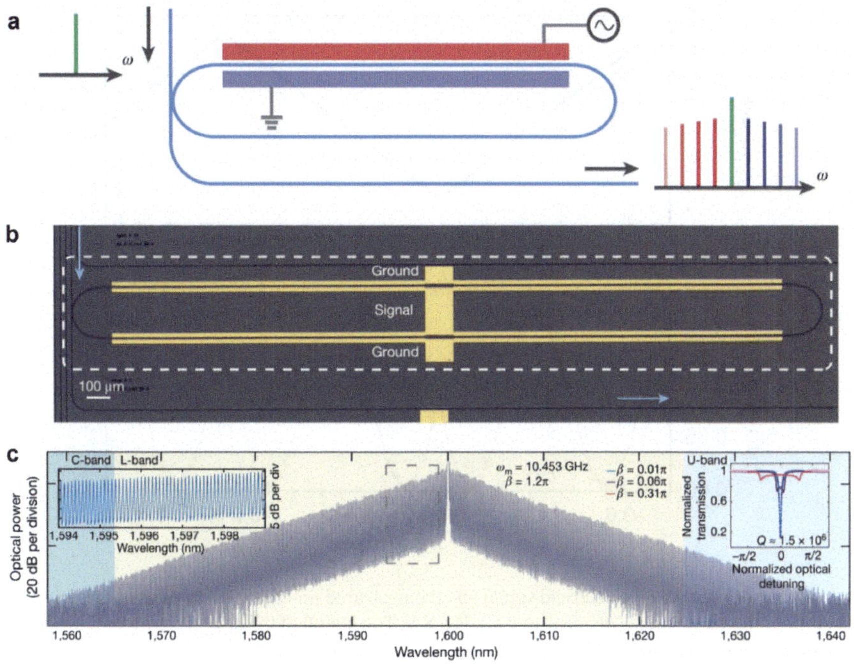

Fig. 6.7 **a** The schematic plot of the device, **b** the micrograph of the LNOI microring resonator, and **c** the measured output spectrum of the EO comb generated from it. Reproduced with permission from [25], ©2019 Nature Publishing Group

comb spectrum, a slope of about 1 dB nm^{-1} was obtained, corresponding to a power variation of less than 0.1 dB between adjacent comb lines.

6.4.3 *EO Mach-Zehnder Interferometer Resonators*

By using Mach-Zehnder geometry, high-speed EO modulations have been realized in LNOI-based EO devices. The next-generation optical communication networks require low-power and ultra-high-speed EO modulation. Rabiei et al. fabricated a 6 mm long Mach-Zehnder modulator based on LNOI on silicon substrate, and obtained a half-wave voltage of 6.8 V, equivalent to a voltage-length product of 4 V·m [42]. Wang et al. from Harvard University used a 20 mm-long Mach-Zehnder interferometer based on LNOI waveguides, and achieved a half-wave EO modulation efficiency of 1.8 V·cm [28]. The data transmission rate of the Mach-Zehnder resonator is 22 Gbps. The Mach-Zehnder interferometer modulator is of 8 dB extinction ratio with power consumptions 1.6 pJ/bit. Later the same group implemented monolithically integrated LNOI EO modulators with a switching voltage of 1.4 V while supporting very high bandwidths. This device is with 20 mm-long microwave strip

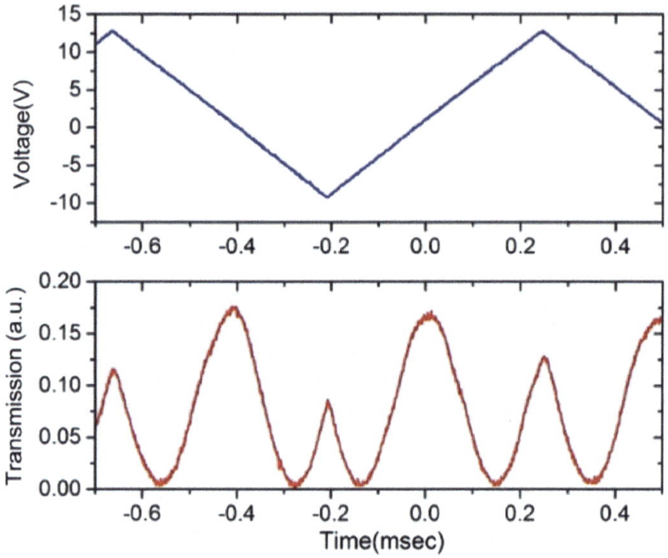

Fig. 6.8 Applied sawtooth electrical signal and the measured modulation response of a 6 mm-long Mach-Zehnder modulator. The measured V_{π} is 6.8 V. The length of the device is 6 mm. Reproduced with permission from [42], ©2013 Optical Society of America

line electrodes, enabling a very low half-wave voltage of 1.4 V that can be directly driven by a CMOS circuit. Figure 6.8 shows the normalized optical transmission as a function of the applied voltage for the monolithically integrated LNOI EO modulators. The measured small-signal EO response of the modulator shows a high 3 dB bandwidth of >45 GHz. In this case, data modulation at 70 Gbit/s has been obtained directly driven by a CMOS circuit without an electrical amplifier. By amplifying the electrical signals to 2.5 V, data transmission at a higher rate up to 210 Gbit/s was achieved.

Based on a silicon and $LiNbO_3$ (Si/LN) hybrid integration platform of LNOI, He et al. from Sun Yat-sen University realized a low voltage-length-product, high-bandwidth and high modulation-rate EO device by using Mach-Zehnder geometry [43]. To make use of the scalability of silicon with the modulation of LN, this modulator employs two layers of hybrid integrated waveguides and vertical adiabatic couplers. Figure 6.9 shows the structure of the hybrid Si/LN Mach-Zehnder modulator. The hybrid device exhibited a low insertion loss of 2.5 dB, a low voltage–length product of 2.2 V cm, a high bandwidth of >70 GHz, and a high modulation rate up to 112 Gbit/s.

The interleaver is a comb-shaped optical filter that separates a high-channel density stream into two complementary low-channel density streams, or vice versa, which is a basic element in telecommunication systems. Based on *x*-cut LNOI platform, Li et al. implemented a compact and EO tunable interleaver by using an asymmetrical Mach–Zehnder interferometer configuration [22]. The device was typically

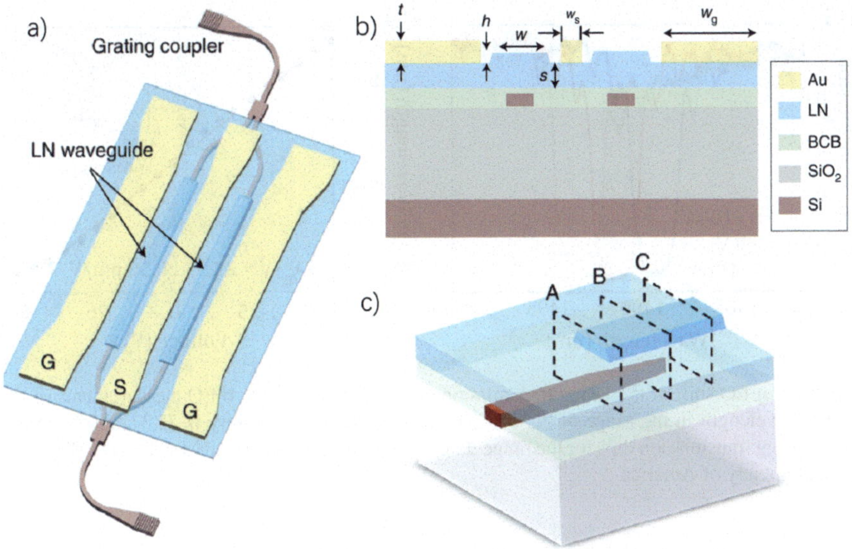

Fig. 6.9 **a** Schematic of the structure of the whole hybrid Si/LN Mach-Zehnder modulator circuit, **b** schematic of the cross-section of the hybrid waveguide, and **c** schematic of the vertical adiabatic couplers. Reproduced with permission from [43], ©2019 Nature Publishing Group

4 mm long, with EO interactive length of ~1.35 mm. It exhibited negligible polarization dependence and offers a channel spacing of ~49.7 GHz, a maximum channel isolation of 23 dB, and a maximum electrical wavelength tuning efficiency of ~18 pm/V over a wide wavelength range from 1528 to 1605 nm.

6.4.4 EO Tunable LNOI Bragg Reflectors

Bragg reflectors (BRFs) are essential elements in optical telecommunication and sensing applications. Fast EO tuning is an ideal approach to switch the BRFs. Escalé et al. demonstrated an integrated EO tunable BRF in a LNOI platform with excellent EO performance [44]. The Bragg gratings were written on an *x*-cut LNOI chip with e-beam lithography and dry-etched by an inductively coupled plasma (ICP) process, allowing the cross sections of fabricated structures to be sub-micron scale. About ~14 dB transmission dip was observed for the fabricated BRF. An EO shift of the Bragg resonance of ~23.4 pm/V was obtained, which was ~33 times larger than the Bragg tenability coefficient (0.70 pm/V) with femtosecond laser-written $LiNbO_3$ waveguides. Figure 6.10 shows the EO shift of the Bragg transmission dip (at 0 and ±100 V) and the measured Bragg wavelength (−50 to 50 V).

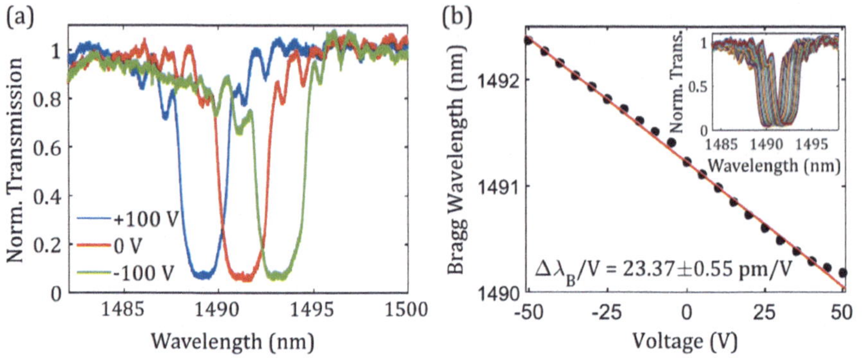

Fig. 6.10 **a** EO shift of the Bragg transmission dip for 0 and ±100 V. **b** EO measurements of the Bragg wavelength in the range of −50 to 50 V, where the EO tunability is 23.37 ± 0.55 pm/V. Inset: Bragg transmission dips for the range ±50 V. Reproduced with permission from [44], ©2018 Optical Society of America

6.4.5 Compact EO Fourier Transform Spectrometers

Fourier transform spectrometers (FTSs) are widely applied to investigate wavelength-dependent light-matter interactions and analyze the spectra of various light sources. The miniaturization of the FTSs for chip-scale integration is of particular importance for applications. Pohl et al. proposed a broadband EO FTS based on LNOI platform by using hybrid waveguides of $LiNbO_3$ and silicon nitride (LN–SiN), which were produced by electron-beam lithography (EBL) and reactive ion etching (RIE) [45]. Figure 6.11 shows the EO FTS based on LNOI. The size of the prototype device was less than 10 mm^2. The integrated single-waveguide FTS was with an operational bandwidth of 500 nm in the near-and short-wavelength infrared. The function of the device has been tested under measurement with applied voltages at monochromatic (1550 or 1040 nm), dual-wavelength (e.g., 1550 + 1480 or 1540 + 1550 nm) or a broadband (50 nm) centered at 1555 nm, the obtained data for spectra was of high-resolution in all cases.

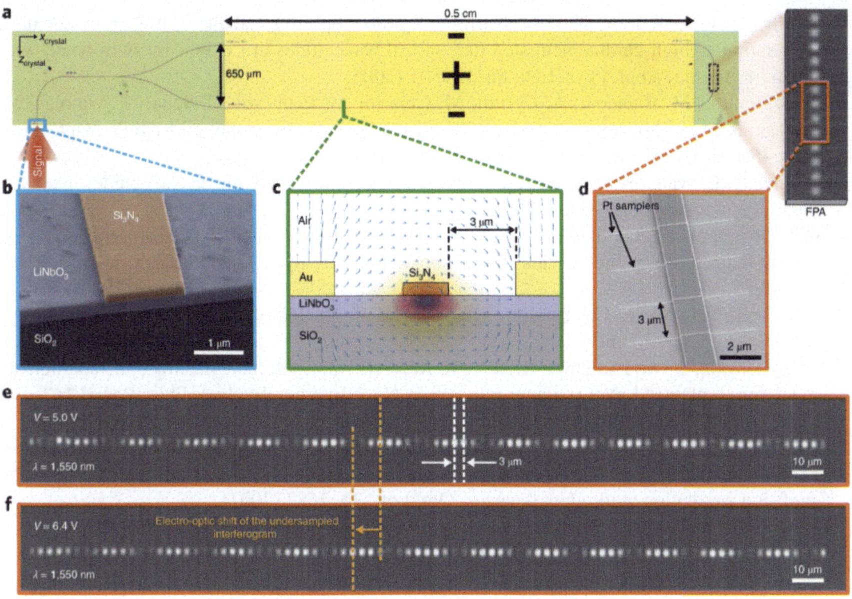

Fig. 6.11 **a** Optical microscope image of the closed-loop LN–SiN waveguide structure with gold electrodes. **b** Scanning electron microscope (SEM) image of the input of a SiN–LN ridge waveguide. **c** Cross-sectional schematic of the waveguide–electrode configuration. **d** SEM image of four platinum EFSs on top of the waveguide. **e**, **f** Images of the interferogram at 1550 nm for different applied voltages $V = 5.0$ V, 6.4 V. Reproduced with permission from [45], ©2020 Nature Publishing Group

References

1. C.C. Davis, *Laser and Electro-optics: fundamentals and Engineering* (Cambridge University Press, 1996)
2. Q. Xu, B. Schmidt, S. Pradhan, M. Lipson, Micrometre-scale silicon electro-optic modulator. Nature **435**, 325 (2005)
3. D. Janner, D. Tulli, M. García-Granda, M. Belmonte, V. Pruneri, Micro-structured integrated electro-optic $LiNbO_3$ modulators. Laser Photonics Rev. **3**, 301 (2009)
4. H. Daniel Lee, J. Cardenas, M. Lipson, Graphene electro-optic modulator with 30 GHz bandwidth. Nat. Photonics **9**, 511 (2015)
5. S. Abel, F. Eltes, J.E. Ortmann, A. Messner, P. Castera, T. Wagner, D. Urbonas, A. Rosa, A.M. Gutierrez, D. Tulli, P. Ma, B. Baeuerle, A. Josten, W. Heni, D. Caimi, L. Czornomaz, A.A. Demkov, J. Leuthold, P. Sanchis, J. Fompeyrine, Large Pockels effect in micro-and nanostructured barium titanate integrated on silicon. Nat. Mater. **18**, 42 (2019)
6. R.S. Weis, T.K. Gaylord, Lithium niobate: summary of physical properties and crystal structure. Appl. Phys. A **37**, 191 (1985)
7. F. Chen, Photonic guiding structures in lithium niobate crystals produced by energetic ion beams. J. Appl. Phys. **106**, 081101 (2009)
8. J. Olivares, A. García-Navarro, A. Méndez, F. Agulló-López, G. García, A. García-Cabañes, M. Carrascosa, Novel optical waveguides by in-depth controlled electronic damage with swift ions. Nucl. Instrum. Methods Phys. Res. B **257**, 765 (2007)

9. J. Olivares, G. García, A. García-Navarro, F. Agulló-López, O. Caballero, A. García-Cabañes, Generation of high-confinement step-like optical waveguides in $LiNbO_3$ by swift heavy ion-beam irradiation. Appl. Phys. Lett. **86**, 183501 (2005)
10. G.G. Bentini, M. Bianconi, A. Cerutti, M. Chiarini, G. Pennestrì, C. Sada, N. Argiolas, M. Bazzan, P. Mazzoldi, Integrated Mach-Zehnder micro-interferometer on $LiNbO_3$. Opt. Laser Eng. **45**, 368 (2007)
11. F. Chen, X.L. Wang, K.M. Wang, Development of ion-implanted optical waveguides in optical materials: a review. Opt. Mater. **29**, 1523 (2007)
12. G.L. Destefanis, J.P. Gailliard, E.L. Ligeon, S. Valette, B.W. Farmery, P.D. Townsend, A. Perez, The formation of waveguides and modulators in $LiNbO_3$ by ion implantation. J. Appl. Phys. **50**, 7898 (1979)
13. G.T. Reed, B.L. Weiss, Electro-optic effect in He^+-implanted optical waveguides in $LiNbO_3$. Electron. Lett. **23**, 424 (1987)
14. A. Boudrioua, S. Ould Salem, P. Moretti, R. Kremer, J.C. Loulergue, Electro-optic coefficients in H^+-ion implanted $LiNbO_3$ planar waveguide. Nucl. Instrum. Methods Phys. Res. B **147**, 393 (1999)
15. J. Olivares, A. García-Navarro, G. García, A. Méndez, F. Agulló-López, A. García-Cabañes, M. Carrascosa, O. Caballero, Nonlinear optical waveguides generated in lithium niobate by swift-ion irradiation at ultralow fluences. Opt. Lett. **32**, 2587 (2007)
16. G. Poberaj, R. Degl'Innocenti, C. Medrano, P. Günter, UV integrated optics devices based on beta-barium borate. Opt. Mater. **31**, 1049 (2009)
17. A. Guarino, G. Poberaj, D. Rezzonico, R. Degl'Innocenti, P. Günter, Electro-optically tunable microring resonators in lithium niobate. Nat. Photonics **1**, 407 (2007)
18. P. Rabiei, P. Gunter, Optical and electro-optical properties of submicrometer lithium niobate slab waveguides prepared by crystal ion slicing and wafer bonding. Appl. Phys. Lett. **85**, 4603 (2004)
19. T.A. Ramadan, M. Levy, R.M.O. Jr., Electro-optic modulation in crystal-ion-sliced z-cut $LiNbO_3$ thin films. Appl. Phys. Lett. **76**, 1407 (2000)
20. L. Cai, Y. Kang, H. Hu, Electric-optical property of the proton exchanged phase modulator in single-crystal lithium niobate thin film. Opt. Express **24**, 4640 (2016)
21. W. Qiu, A. Ndao, V.C. Vila, R. Salut, N. Courjal, F.I. Baida, M.-P. Bernal, Fano resonance-based highly sensitive, compact temperature sensor on thin film lithium niobate. Opt. Lett. **41**, 1106 (2016)
22. X.P. Li, K.X. Chen, L.F. Wang, Compact and electro-optic tunable interleaver in lithium niobate thin film. Opt. Lett. **43**, 3610 (2018)
23. L. Wang, C. Wang, J. Wang, F. Bo, M. Zhang, Q. Gong, M. Lončar, Y.F. Xiao, High-Q chaotic lithium niobate microdisk cavity. Opt. Lett. **43**, 2917 (2018)
24. C. Wang, M. Zhang, M.J. Yu, R.R. Zhu, H. Hu, M. Loncar, Monolithic lithium niobate photonic circuits for Kerr frequency comb generation and modulation. Nat. Commun. **10**, 978 (2019)
25. M. Zhang, B. Buscaino, C. Wang, A. Shams-Ansari, C. Reimer, R. Zhu, J.M. Kahn, M. Lončar, Broadband electro-optic frequency comb generation in a lithium niobate microring resonator. Nature **568**, 373 (2019)
26. Y. Enami, C.T. Derose, D. Mathine, C. Loychik, C. Greenlee, R.A. Norwood, T.D. Kim, J. Luo, Y. Tian, A.K.Y. Jen, N. Peyghambarian, Hybrid polymer/sol–gel waveguide modulators with exceptionally large electro-optic coefficients. Nat. Photonics **1**, 180 (2007)
27. P.O. Weigel, J. Zhao, K. Fang, H. Al-Rubaye, D. Trotter, D. Hood, J. Mudrick, C. Dallo, A.T. Pomerene, A.L. Starbuck, C.T. DeRose, A.L. Lentine, G. Rebeiz, S. Mookherjea, Bonded thin film lithium niobate modulator on a silicon photonics platform exceeding 100 GHz 3 dB electrical modulation bandwidth. Opt. Express **26**, 23728 (2018)
28. C. Wang, M. Zhang, B. Stern, M. Lipson, M. Lončar, Nanophotonic lithium niobate electro-optic modulators. Opt. Express **26**, 1547 (2018)
29. A.J. Mercante, S. Shi, P. Yao, L. Xie, R.M. Weikle, D.W. Prather, Thin film lithium niobate electro-optic modulator with terahertz operating bandwidth. Opt. Express **26**, 14810 (2018)

30. A.L. Chen, Q. Xu, M.G. Wood, R.M. Reano, Hybrid silicon and lithium niobate electro-optical ring modulator. Optica **1**, 112 (2014)
31. M. Zhang, C. Wang, R. Cheng, A. Shams-Ansari, M. Lončar, Monolithic ultra-high-Q lithium niobate microring resonator. Optica **4**, 1536 (2017)
32. F. Bo, J. Wang, J. Cui, S.K. Ozdemir, Y. Kong, G. Zhang, J. Xu, L. Yang, Lithium-niobate–silica hybrid whispering-gallery-mode resonators. Adv. Mater. **27**, 8075 (2015)
33. R. Degl'Innocenti, A. Majkic, P. Vorburger, G. Poberaj, P. Günter, M. Döbeli, Ultraviolet electro-optic amplitude modulation in β-BaB_2O_4 waveguides. Appl. Phys. Lett. **91**, 051105 (2007)
34. D. Fluck, T. Pliska, P. Gunter, L. Beckers, C. Buchal, Cerenkov-type second-harmonic generation in $KNbO_3$ channel waveguides. IEEE J. Quantum Electron. **32**, 905 (1996)
35. G.G. Bentini, M. Bianconi, A. Cerutti, A. Nubile, P. Cerabolini, M. Chiarini, G. Pennestrì, W. Dinicolantonio, R. Guzzi, A new miniaturised optical system for chemical species spectroscopic detection based on a scanning integrated Mach-Zehnder microinterferometer on $LiNbO_3$. Orig. Life Evol. B. **36**, 597 (2006)
36. A. Majkic, G. Poberaj, P. Gunter, Optical microring resonators in fluorine-implanted lithium niobate for electrooptical switching and filtering. IEEE Photonics Technol Lett. **21**, 639 (2009)
37. A. Majkić, M. Koechlin, G. Poberaj, P. Günter, Optical microring resonators in fluorine-implanted lithium niobate. Opt. Express **16**, 8769 (2008)
38. T. Izuhara, R. Roth, R. Osgood, S. Bakhru, H. Bakhru, Low-voltage tunable TE/TM converter on ion-sliced lithium niobate thin film. Electron. Lett. **39**, 1118 (2003)
39. A. Boes, B. Corcoran, L. Chang, J. Bowers, A. Mitchell, Status and potential of lithium niobate on insulator (LNOI) for photonic integrated circuits. Laser Photonics Rev. **12**, 1700256 (2018)
40. J. Wang, F. Bo, S. Wan, W. Li, F. Gao, J. Li, G. Zhang, J. Xu, High-Q lithium niobate microdisk resonators on a chip for efficient electro-optic modulation. Opt. Express **23**, 23072 (2015)
41. C. Wang, M. Zhang, X. Chen, M. Bertrand, A. Shams-Ansari, S. Chandrasekhar, P. Winzer, M. Loncar, Integrated lithium niobate electro-optic modulators operating at CMOS-compatible voltages. Nature **562**, 101 (2018)
42. P. Rabiei, J. Ma, S. Khan, J. Chiles, S. Fathpour, Heterogeneous lithium niobate photonics on silicon substrates. Opt. Express **21**, 25573 (2013)
43. M. He, M. Xu, Y. Ren, J. Jian, Z. Ruan, Y. Xu, S. Gao, S. Sun, X. Wen, L. Zhou, L. Liu, C. Guo, H. Chen, S. Yu, L. Liu, X. Cai, High-performance hybrid silicon and lithium niobate Mach-Zehnder modulators for 100 Gbit s −1 and beyond. Nat. Photonics **13**, 359 (2019)
44. M.R. Escalé, D. Pohl, A. Sergeyev, R. Grange, Extreme electro-optic tuning of Bragg mirrors integrated in lithium niobate nanowaveguides. Opt. Lett. **43**, 1515 (2018)
45. D. Pohl, M.R. Escalé, M. Madi, F. Kaufmann, P. Brotzer, A. Sergeyev, B. Guldimann, P. Giaccari, E. Alberti, U. Meier, R. Grange, An integrated broadband spectrometer on thin-film lithium niobate. Nat. Photonics **14**, 24–29 (2020)

Chapter 7
Photoluminescence of Dielectric Waveguides

Abstract This chapter overviews the ion-beam-implanted/irradiated dielectric waveguides in view of their applications in photoluminescence. The up-conversion and the near-infrared luminescence processes realized in ion-implanted/irradiated waveguides based on rare-earth-doped crystals and glasses are summarized in Sects. 7.1 and 7.2, respectively. Section 7.3 reviews the optical amplifier devices based on ion-implanted/irradiated waveguides.

Photoluminescence (PL) is a kind of light emission from matter after the absorption of photons, which excite electrons to a higher energy level in an atom [1]. It is photoexcitation induced luminescence. Up-conversion luminescence has been applied in a number of fields, such as three-dimensional display and bio tagging [2]. Traditional PL is the basis of lasers and amplification. Both up-conversion and traditional PL emissions require suitable energy band for electron transition. Rare-earth ion doped dielectrics, for example, Er-doped materials, are ideal platforms for both up-conversion and traditional PL [3]. The up-conversion pumping mechanism is either excited state absorption (ESA), energy transfer up-conversion (ETU) or ESA with looping, depending on the wavelength pumping conditions [4]. Traditional PL is a process in which a molecule absorbs a photon, exciting an electron to a higher excited state, and then radiates a photon as the electron returns to a lower energy state [3]. Both up-conversion and traditional PL emissions have been explored in ion beam produced waveguides, in which a compact platform is offered for more efficient PL emissions. Up-conversion of waveguides mostly focuses on Er-doped dielectrics, whilst traditional PL (particularly in near infrared wavelength region) is investigated for waveguides based on some of the favorable rare-earth ion (e.g., Nd, Yb, Er, etc.) doped materials.

F. Chen et al., *Ion Irradiation of Dielectrics for Photonic Applications*, Springer Series in Optical Sciences 231, https://doi.org/10.1007/978-981-15-4607-5_7

7.1 Up-Conversion of Waveguides

Up-conversion luminescence process refers to excitation of low-energy photons and emission of high-energy photons. It requires at least two photon absorption to achieve enough energy of excited electrons for up-conversion emissions. In addition, suitable electronic energy level systems are necessary to realize the occurrence of up-conversion process. Rare-earth (RE) ion doped dielectrics are considered as one of the most promising systems [3]. In RE-doped crystals or glasses, the RE ions are trivalent, and the $4f^n$ electronic level structures of the RE ions offer a platform of multiple long-life intermediate levels and high energy metastable levels for the infrared and visible band emissions [5]. For ion implanted waveguides, researchers mainly focus on the Er^{3+}-doped systems for green and red light up-conversion. Figure 7.1 shows the energy levels of Er^{3+} ions and an example of 800 nm light excitation for green light up-conversion luminescence emission.

A typical process of the Er-doped systems is three-step [5] for green light luminescence emission (e.g., at ~800 nm light excitation): (1) The 800 nm photon excites an electron from the fundamental level $^4I_{15/2}$ to the metastable level $^4I_{9/2}$, and the electron rapidly decay to the metastable state of $^4I_{13/2}$ without radiation; (2) The second photon is absorbed, further exciting the electron from $^4I_{13/2}$ to a higher level of $^2H_{11/2}$; and (3) transition from $^2H_{11/2}$ and $^4S_{3/2}$ to ground state of $^4I_{15/2}$ resulting in up-conversion emission in the green light band. This mechanism requires the absorption of two photons. The steps (1)–(3) are excited state absorption and energy transfer up-conversion.

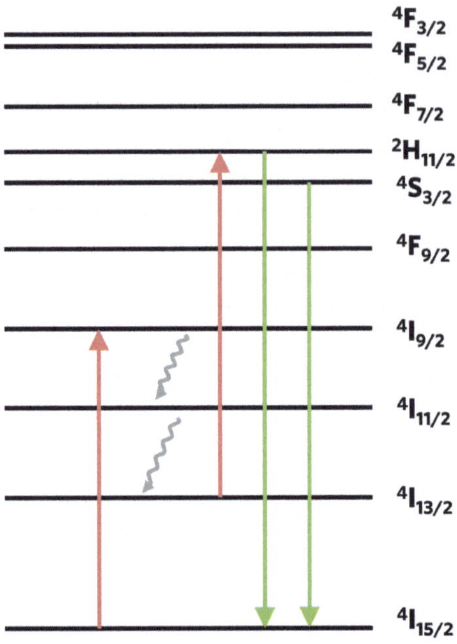

Fig. 7.1 The energy levels of Er^{3+} ions and an example of 800 nm light excitation for green light up-conversion emission

In ion implanted waveguides, the up-conversion emissions were observed in Er:$LiNbO_3$ [6] and Er:YAG [7] platforms. Herreros et al. investigated the He-ion implanted planar waveguides in Er^{3+}:$LiNbO_3$ crystal, and found two excitation regions for two-photon absorption at low temperature (liquid helium) [6]. Szachowicz et al. reported on the room-temperature green up-conversion luminescence in H^+ or He^+ ion implanted Er:YAG waveguides [7]. For as-implanted (by H^+ ions) channel waveguides, stable up-conversion emissions were observed, and the luminescence intensity increased with the ion fluence, which may be due to the better waveguiding effect in case of high-fluence H^+ ions. Figure 7.2 shows the comparison of up-conversion emission spectra (based on $^2H_{11/2}$, $^4S_{3/2}$ to $^4I_{15/2}$ transitions) from channel waveguide and bulk under excitation at 789 nm. The efficient luminescence emission was observed, and kept stable for post-implantation annealed sample. As one can also see, the relevant intensities of the up-conversion luminescence of emission lines from the waveguides are different from those from the bulk materials, which may be due to the H^+ ion modification as well as the waveguiding effects.

In ion implanted waveguide structures, the up-conversion luminescence emissions may also be affected by the implantation induced defects, which causes considerable losses of the waveguides. Liu et al. investigated the up-conversion luminescence from a 17 MeV C^{5+} ion implanted Er:MgO:SLN (stoichiometric $LiNbO_3$) planar waveguide, and observed 40% quenching of the green luminescence in comparison

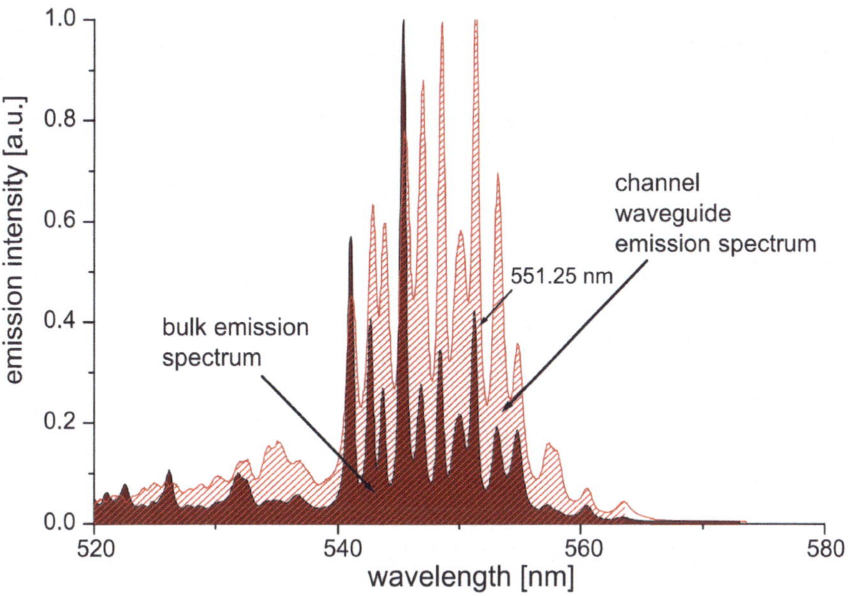

Fig. 7.2 The comparison of up-conversion emission spectra from channel waveguide and bulk under excitation at 789 nm Reproduced with permission from [7], ©2006 Elsevier

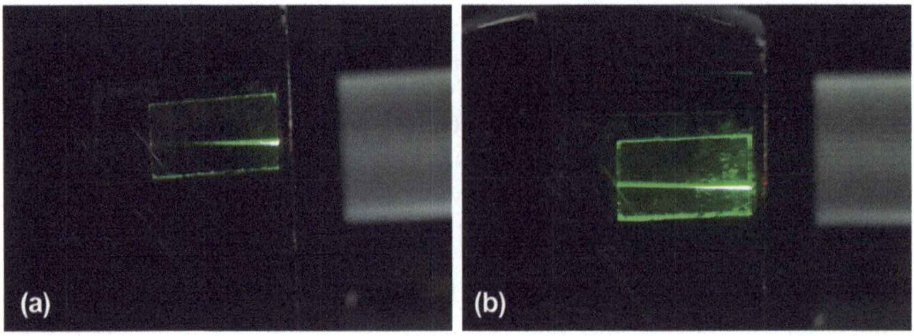

Fig. 7.3 The photographs of the green light generated in the Er^{3+}:MgO:SLN waveguide area **a** and bulk material, **b** with the pumping laser at the same power Reproduced with permission from [8], ©2014 Elsevier

to the bulk [8]. Nevertheless, one can further improve the waveguide quality by post-implantation treatment (e.g., annealing at 200–300 °C) to obtain a higher-efficiency up-conversion of green light (Fig. 7.3).

Szachowicz et al. investigated infrared-to-blue up-conversion from a H^+ ion implanted channel waveguides in Nd^{3+}, Tm^{3+}:YAG layers [9]. The system of Nd^{3+} and Tm^{3+} co-doping benefits the energy transfer between relevant levels of Nd^{3+} and Tm^{3+}. The guided blue luminescence due to the Tm^{3+} $^1G_4 \rightarrow {}^3H_6$ transition was studied under infrared excitation at 785 nm (Tm^{3+} absorption) and 808 nm (Nd^{3+} absorption), respectively. The intense Tm^{3+} emission peaked at 486 nm was observed with steady state rise time of about 4 ms. It was found that the Nd^{3+} co-doping improved the Tm^{3+} blue emission due to an efficient energy transfer up-conversion process. Figure 7.4 shows a plot of the blue fluorescence intensity versus the excitation power of 785 nm light.

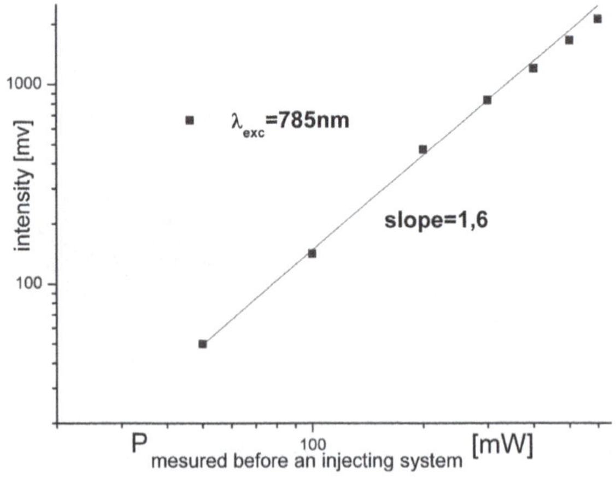

Fig. 7.4 The blue fluorescence intensity recorded at the output of the H^+ ion implanted Nd^{3+}, Tm^{3+}:YAG channel waveguide as a function of the 785 nm excitation power Reproduced with permission from [9], ©2008 American Institute of Physics

In addition to crystals, ion implanted RE-doped glass waveguides also show good up-conversion luminescence properties. Chen et al. used 15 MeV C^{5+} ion implantation to form optical waveguides in Er/Yb co-doped phosphate glass [10]. Under the optical pump laser at 980 nm, the up-conversion emission of both green and red fluorescence has been realized through the waveguide structures.

7.2 Near Infrared Luminescence in Waveguides

The RE ion doped systems offer multiple energy levels for luminescence emissions in the near infrared (NIR) bands, which is the base of solid-state lasers used in laser technology. The well-investigated systems of RE ions include Nd^{3+}, Yb^{3+}, and Er^{3+} doped crystals. The ion-beam-processed waveguides have been fabricated in many laser crystals and ceramics, including Nd:MgO:$LiNbO_3$, Nd:YAG, Nd:GGG, Er:MgO:$LiNbO_3$, Nd:YVO_4, Nd:$GdVO_4$, Nd:GdCOB, Nd:YCOB, Nd:$YLiF_4$, Cr:$LiSrAlF_6$, Ti:sapphire, Nd:KGW, Nd:LGS, Er:YAP, etc., and the NIR luminescence properties have been reported. It has also been found that the luminescence emissions related to active ions in waveguides depend on the used ion beam techniques as well as the materials (e.g., crystalline structures). By using the confocal microscope-based photoluminescence (μPL) spectroscopy, one can detect the localized modification of the luminescence signals from the waveguides and surrounding regions, revealing the physical mechanisms of the ion-induced lattice changes in specific regions. In this section, the PL properties of the waveguides will be introduced based on materials.

7.2.1 Doped $LiNbO_3$ Crystals

$LiNbO_3$ crystals doped by Nd^{3+}, Yb^{3+}, or Er^{3+} ions can serve as gain media for lasing and signal amplification [11]. The Nd^{3+} ions receive particular attentions due to the sensitive modifications of the emission lines in the NIR bands. The groups of Chen and Jaque studied the μPL properties in the NIR emission bands of 500 keV H^+, 3 MeV C^{3+}, or 3 MeV O^{3+} ion implanted Nd:MgO:$LiNbO_3$ channel waveguides [12–16]. For practical applications, the waveguiding core is of major interest. The early work by Field et al. indicated obvious quenching of photoluminescence emission (~30% reduction of intensity) in the He ion implanted Nd:MgO:$LiNbO_3$ waveguide [13]. Differently, for 500 keV H^+, 3 MeV C^{3+}, or 3 MeV O^{3+} ion implanted Nd:MgO:$LiNbO_3$ channel waveguides, no obvious quenching is detected by monitoring the full emission bands of Nd^{3+} transitions of $^4F_{3/2} \rightarrow {}^4I_{9/2}$ and $^4F_{3/2} \rightarrow {}^4I_{11/2}$. In all these three cases, the intensity of PL emissions from the waveguides is almost the same to that from the bulks, showing well-preserved PL properties of the waveguides in comparison to the bulks. Figure 7.5 shows the room-temperature

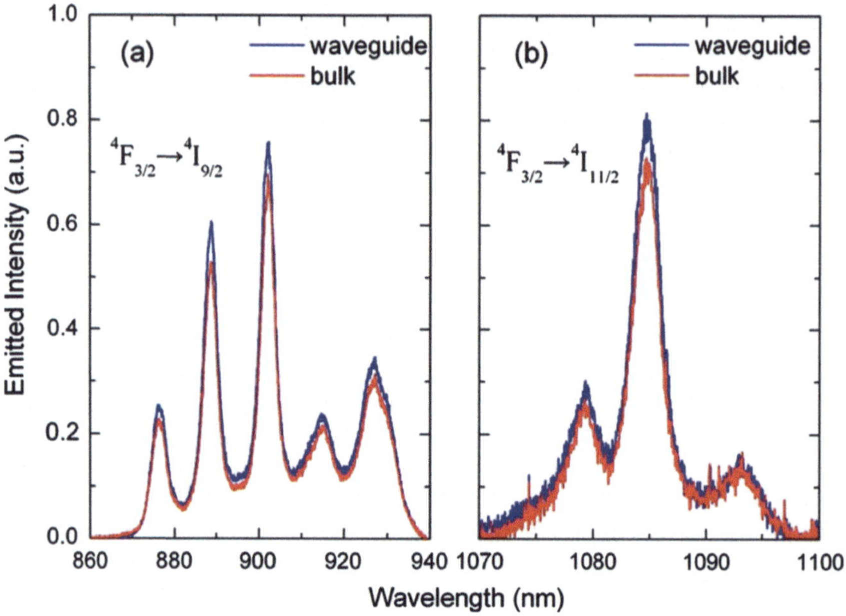

Fig. 7.5 The room-temperature μPL spectra obtained in the 3 MeV C^{3+} ion implanted Nd^{3+}:MgO:$LiNbO_3$ waveguide (after annealing at 260 °C for 2 h in air) and the bulk for **a** $^4F_{3/2} \rightarrow {}^4I_{9/2}$ and **b** $^4F_{3/2} \rightarrow {}^4I_{11/2}$ emissions Reproduced with permission from [12], ©2010 Optical Society of America

μPL spectra obtained in the 3 MeV C^{3+} ion implanted Nd^{3+}:MgO:$LiNbO_3$ waveguide (after annealing at 260 °C for 2 h in air) and the bulk for $^4F_{3/2} \rightarrow {}^4I_{9/2}$ and $^4F_{3/2} \rightarrow {}^4I_{11/2}$ emissions [12]. From the μPL spectra, one can clearly see that the PL emission bands of the waveguides does not affected significantly. This suggests that the spectroscopic properties of the $^4F_{3/2}$ metastable state of Nd^{3+} ions are almost not changed by the waveguide fabrication procedure. In particular, waveguide formation procedure does not reduce the Nd^{3+} fluorescence intensity, indicating that the quantum efficiency of the $^4F_{3/2}$ metastable state is unaffected by ion implantation.

Similar results have also been obtained in 500 keV H^+ or 3 MeV O^{3+} ion implanted Nd:MgO:$LiNbO_3$ channel waveguides [14, 15]. In addition to the PL emission intensity, Jaque et al. measured the $^4F_{3/2}$ fluorescence lifetime in the 3 MeV O^{3+} ion implanted waveguide [15]. The time decay curves obtained in the channel waveguide and that in the bulk are identical (considering the experimental uncertainty), and the $^4F_{3/2}$ fluorescence lifetime in the channel waveguide is 83 μs, which is in good agreement with that obtained from the bulk (86 μs).

Nevertheless, the ion implantation does affect the crystal lattices, and modify the PL properties of the Nd^{3+} ions in different regions. The μPL technique may detect even the slight modifications of the crystal by focusing specific emission lines. Figure 7.6 shows the typical emission spectrum related to $^4F_{3/2} \rightarrow {}^4I_{9/2}$ transition

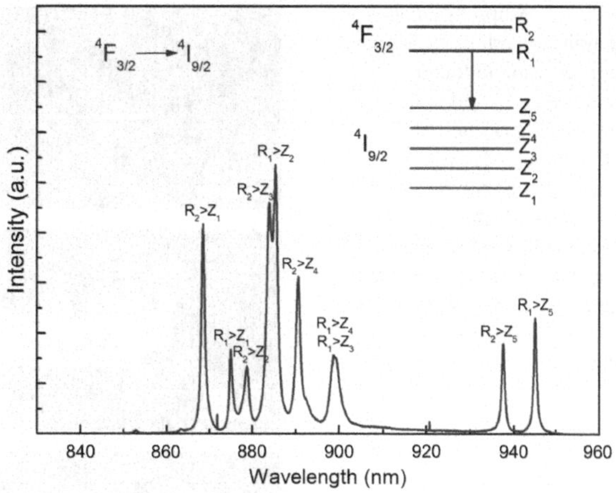

Fig. 7.6 Typical emission spectrum related to $^4F_{3/2} \rightarrow$ $^4I_{9/2}$ transition of Nd^{3+} ions in Nd:YAG crystal. The emission lines are correlated to the transition of respective sub-Stark levels

of Nd^{3+} ions in Nd:YAG crystal. The metastable state $^4F_{3/2}$ possesses two sub-Stark levels, i.e., R_1 and R_2, and state of $^4I_{9/2}$ has 5 sub-Stark levels, i.e., Z_1, Z_2, Z_3, Z_4, and Z_5. The transitions from $^4F_{3/2}$ to $^4I_{9/2}$ sub-levels construct PL emission lines in the obtained spectrum, and all the emission lines construct the full spectrum of Nd^{3+} $^4F_{3/2}$ to $^4I_{9/2}$ transition. Similar PL spectra can be obtained in Nd:MgO:$LiNbO_3$ and other Nd^{3+}-ion doped crystals.

For H^+ ion implanted Nd:MgO:$LiNbO_3$, Jaque et al. focused on the emission line centered at around 11250 cm^{-1}, corresponding to the radiative decay involving the lowest energy levels of both $^4F_{3/2}$ and $^4I_{9/2}$ states ($R_1 \rightarrow Z_1$) [14]. By using confocal μPL mapping, the modification of Nd^{3+} ion PL emission can be imaged spatially, which reveals the ion implanted effects in both waveguides and surroundings. Figure 7.7 shows the spatial distribution of the intensity, spectral shift, and width of the emission line related to $R_1 \rightarrow Z_1$ transition [14]. The spectral shift has been defined here with respect to the peak position in an untreated Nd:MgO:$LiNbO_3$ crystal. From the intensity map, the Nd^{3+} emission intensity is fairly homogeneous and free of any quenching. This is in good agreement with the full-band PL spectrum. One can also see clear spectral changes in terms of a redshift and line broadening of emission line, which is localized at the boundary of waveguide and bulk. The presence of a local spectral shift in the Nd^{3+} fluorescence lines at the waveguide/bulk boundary denotes a slight change in the energy of the involved electronic levels in this region, and it, in turn, reveals a modification in the local crystal field affecting the Nd^{3+} ions as a consequence of a local rearrangement of the lattice. At the same time, the broadening of the emission line also accounts for an increase in the local disorder, which is due to the nuclear damage induced by the ion implantation.

In summary, for Nd:MgO:$LiNbO_3$, the PL properties of the channel waveguides produced by H^+, C^{3+}, and O^{3+} ion implantation are well preserved compared with the bulks [12]. The confocal μPL spectra reveal the localized modification of the waveguides at different regions.

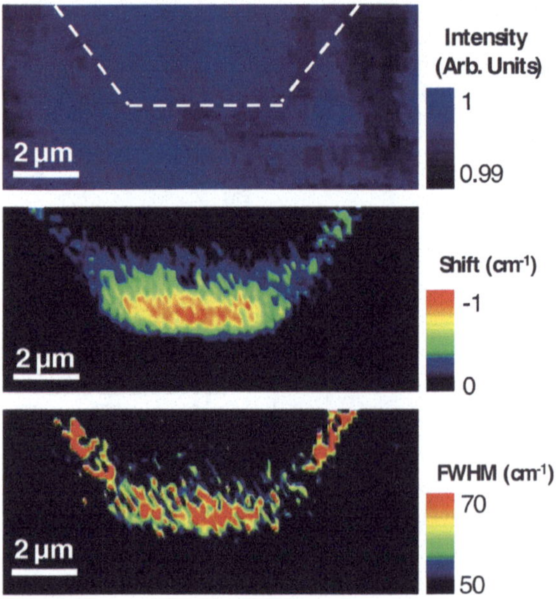

Fig. 7.7 Mapping images of **a** emission intensity, **b** ion implantation induced spectral shift, and **c**, **b** broadening of the Nd^{3+} $^4F_{3/2}(R_1) \rightarrow {}^4I_{9/2}(Z_1)$ emission line for 500 keV H^+ ion implanted Nd:MgO:$LiNbO_3$ channel waveguide Reproduced with permission from [14], ©2009 American Institute of Physics

The Er^{3+} doped optical materials are of great interest to both researchers and industries for the all-optical information systems, due to the intra-4f emissions of Er^{3+} ions at the first telecommunication wavelength of ~1.5 μm which allows to implement active photonic device. The PL properties of Er^{3+} doped $LiNbO_3$ are also investigated. Chen et al. reported on the PL properties of Er^{3+} ions in a 3 MeV O^{3+} ion implanted Er:MgO:SLN (nearly stoichiometric $LiNbO_3$) channel waveguide [17]. The *z*-cut SLN sample was doped with 0.2 mol% Er^{3+} and 1.5 mol% MgO. The transmission attenuation was determined to be ~1.7 dB/cm at the wavelength of 632.8 nm. Figure 7.8 shows the room-temperature micro-luminescence spectra of $^4I_{13/2} \rightarrow {}^4I_{15/2}$ transitions of Er^{3+} ions from the bulk and the channel waveguide.

The 1.5 μm wavelength PL properties in the waveguide were excited by using a fiber-coupled confocal microscope in back-scattering configuration. An 800 nm wavelength fiber-coupled diode at power of 40 mW is used as the excitation source, which promotes the excitation of Er^{3+} ions from their fundamental state ($^4I_{15/2}$) to the excited state ($^4I_{9/2}$). The emitted luminescence at $^4I_{13/2} \rightarrow {}^4I_{15/2}$ transition band is collected by a detector. As one can see, the emission spectra of the bulk and the channel waveguide exhibit similar characteristics (in terms of peak positions, widths and relative intensity between sub-stark transitions), and therefore it appears to be reasonable to conclude that the main emission characteristics of Er^{3+} ions are well preserved in the waveguides.

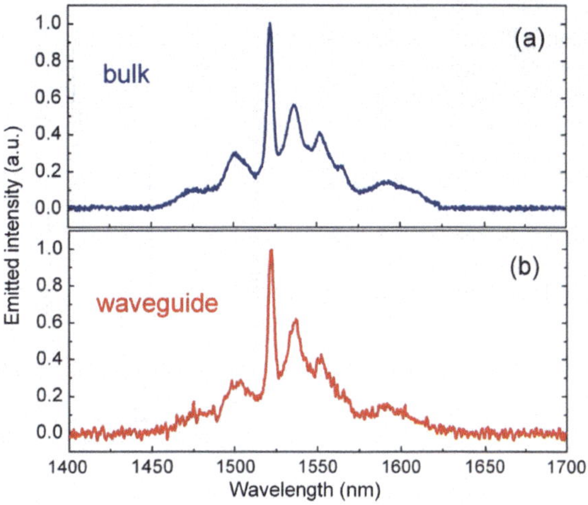

Fig. 7.8 Confocal room-temperature luminescence emission spectra from **a** the bulk and **b** the 3 MeV O^{3+} ion implanted Er:MgO:SLN channel waveguide volume, corresponding to the $^4I_{13/2} \rightarrow {}^4I_{15/2}$ transitions Reproduced with permission from [17], ©2008 Optical Society of America

7.2.2 Doped YAG Crystals

Solid-state lasers are on the basis of the gain materials with excellent PL properties. RE-ion doped YAG crystals and ceramics receive considerable attentions on their capability for diverse laser systems [18]. The PL properties have also been investigated in RE-ion doped YAG waveguides produced by ion beam irradiation. The YAG is cubic structure: the single crystalline crystal and its polycrystalline counterpart YAG ceramic are with similar behaviors under ion beam processing, which means the refractive index and PL modifications induced by ion beams follow the similar effect. So far, the PL properties of Nd^{3+}, Yb^{3+} or Er^{3+} doped YAG waveguides produced by several ion beam techniques have been investigated. The deployed techniques include light ion implantation (e.g., protons or He ions), proton beam writing, heavy ion implantation, and swift heavy irradiation.

The PL spectra of Nd:YAG under proton implantation and proton beam writing are with different properties. In case of proton implantation, the PL emissions from the waveguides are with considerable quenching in comparison to the bulk [19]. Figure 7.9 shows the comparison of emitted PL intensity obtained from the proton implanted channel waveguide (dashed line) and bulk (solid line) through the $^4F_{3/2} \rightarrow {}^4I_{9/2}$ transition of Nd^{3+} ions [19]. The waveguide was produced by multiple-energy proton implantation (at energy of (460 + 480 + 500) keV and fluence of (2 + 2 + 4) × 10^{16} ions cm^{-2}). Although thermal annealing at 300–400 °C was performed to remove the color centers, which may be responsible for PL quenching, there are ~20% reduction in the fluorescence intensity observed in the waveguide. By using the confocal μPL spectroscopy technique, one can observe the main PL modification happens in the waveguide region, from the spatial distributions of the PL mapping. Another proof for the PL quenching is the lifetime reduction from the waveguide (195 μs), compared with that from the bulk (215 μs). The reduction clearly denotes

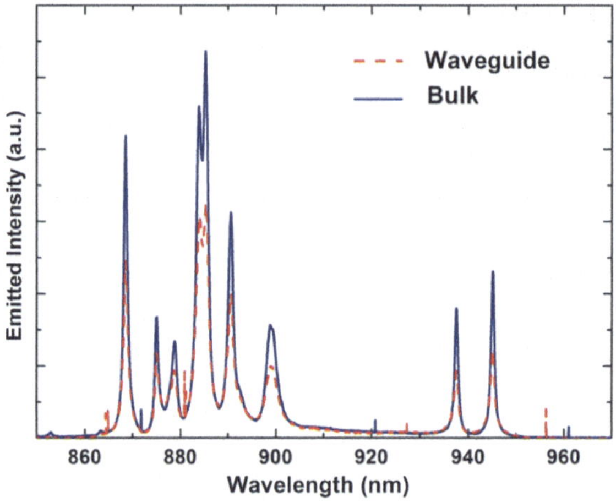

Fig. 7.9 Comparison of emitted PL intensity obtained from the proton implanted Nd:YAG waveguide (after annealing at 200 °C for 1 h + 300 °C for 30 min + 400 °C for 30 min) and bulk through the $^4F_{3/2} \rightarrow {}^4I_{9/2}$ transitions Reproduced with permission from [19], ©2010 IOP Publishing Ltd

the presence of non-radiative energy transfer from Nd^{3+} ions to proton implantation induced defects, acting as luminescence killers, in the waveguide region.

In case of proton beam writing, the waveguide cores are located at the end of the protons' range, which are buried inside the crystal at a certain depth. As a result, the PL emission intensity seems to be well preserved in comparison to that of the bulk. This is significantly different from the proton implanted Nd:YAG waveguides. Yao et al. investigated the PL emission correlated to the $^4F_{3/2} \rightarrow {}^4I_{11/2}$ transition of Nd^{3+} ions of the proton beam written Nd:YAG crystal waveguide [20]. From the obtained spectra, the outstanding luminescence features of Nd^{3+} ions are well preserved without quenching at the waveguide's volume. Further study on the $^4F_{3/2} \rightarrow {}^4I_{9/2}$ fluorescence line of the Nd ions in YAG (located at around 938 nm) shows that the main changes of PL emissions happen at the waveguide region (in this case, i.e., the end of the ion range) with slight blueshift and a line broadening close to 40% [21]. Benayas et al. explored the PL properties under thermal treatment [22]. Figure 7.10 depicts the PL images obtained in terms of the spatial variations of both intensity and spectral shift of the 938 nm hypersensitive line of Nd:YAG after 50, 700, and 900 °C annealing. The spectral blueshift still retains measurable values after 700 °C annealing, but almost disappear after the 900 °C annealing. One can also conclude that the damage-induced fluorescence reduction decreases monotonously in the whole temperature range. The complete erasing of the proton-induced damage by cumulative annealing processes suggests that the proton-induced damage consists of thermally removable defects.

The PL properties of Nd:YAG waveguides produced by heavy ion implantation or swift heavy ion irradiation were investigated by using confocal PL spectroscopy technique [23–25]. It has been found that, for both cases, the PL features are well preserved in the waveguide regions. Figure 7.11a, b shows the room-temperature μPL spectra obtained for the 3 MeV O ion implanted Nd:YAG ceramic channel waveguide (annealed) and bulk for $^4F_{3/2} \rightarrow {}^4I_{9/2}$ and $^4F_{3/2} \rightarrow {}^4I_{11/2}$ emissions, respectively

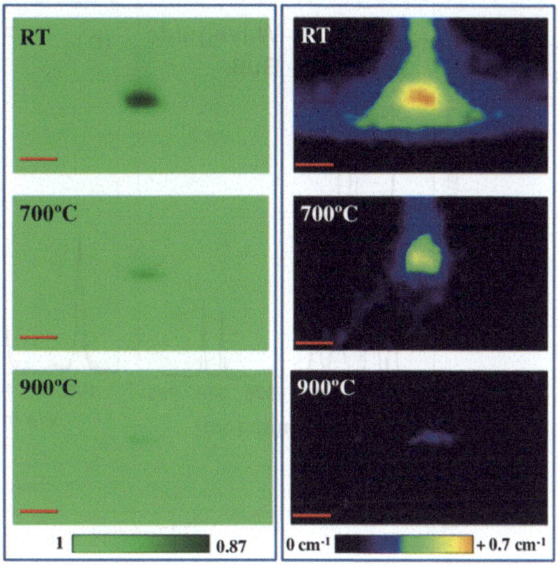

Fig. 7.10 Spatial dependence of the (left column) integrated intensity and spectral shift (right column) of the $^4F_{3/2} \rightarrow {^4I_{9/2}}$ Nd^{3+} emission line at around 938 nm measured at three different temperatures for 1 MeV proton beam written Nd:YAG waveguide. Scale bar is 5 μm in all the images Reproduced with permission from [22], ©2011 Optical Society of America

[23]. The lifetime (Fig. 7.11c) of the $^4F_{3/2}$ metastable state in the waveguide was determined to be 214 μs, slightly reduced by less than 3%, compared with that of the bulk (220 μs).

Ren et al. used 60 MeV Ar^{4+} ions (at a fluence of 2×10^{12} ions/cm^2) to form a planar waveguide in Nd:YAG crystal [25]. For as-irradiated sample (i.e., without annealing treatment), the original PL features of the Nd:YAG crystal have been well preserved in the waveguide, which is similar to heavy ion implanted waveguide. In addition, from the line-broadening spatial distribution of the $^4F_{3/2} \rightarrow {^4I_{9/2}}$ PL line around 940 nm (see Fig. 7.12), one can clearly see that the main changes happen at the core of planar waveguide layer, which is close to the peak position (at depth of 5 μm) of electronic stopping power curve. This measured slight line broadening is not accompanied by any reduction of the fluorescence intensity, indicating the absence of any relevant damage in the Nd:YAG network.

In addition to Nd^{3+} doped YAG materials, the PL properties of Yb^{3+} or Er^{3+} ion doped YAG waveguides in crystals and ceramics have been studied. The obtained results indicate that the PL properties of Yb^{3+} or Er^{3+} ions have been well-preserved in the waveguide regions [26–28].

7.2.3 Other Crystals

The PL properties of ion beam produced waveguides based on other crystals have been investigated by several groups.

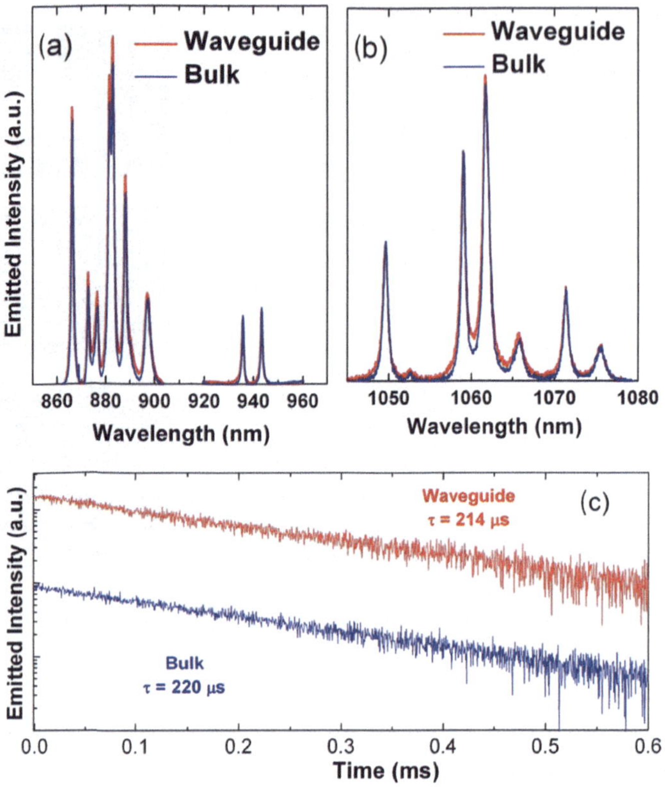

Fig. 7.11 Comparison of the μPL spectra obtained for the 3 MeV O ion implanted Nd:YAG ceramic channel waveguide and bulk for **a** $^4F_{3/2} \rightarrow {}^4I_{9/2}$ and **b** $^4F_{3/2} \rightarrow {}^4I_{11/2}$ emissions. **c** Time-decay curves obtained in the waveguide and bulk Reproduced with permission from [23], ©2009 Optical Society of America

Nd:GGG is a well-known laser material for solid state laser systems [29]. It is with cubic crystal structure, and the ion beam produced Nd:GGG waveguides possess similar properties with Nd:YAG. Field et al. utilized He ion implantation to form planar or channel waveguides in Nd:GGG crystal [30]. Ren et al. studied the room-temperature PL emissions for $^4F_{3/2} \rightarrow {}^4I_{11/2}$ transition of a proton implanted Nd:GGG waveguide [31]. The proton beams are at energy of 460, 480, and 500 keV, and with the respective fluences of 2×10^{16}, 2×10^{16}, and 4×10^{16}ions/cm^2. It has been observed that with such a high-fluence implantation, the PL quenching happens. The average PL intensity from the waveguide is about 74% of that from the bulk. This reflects the fact that the high-fluence protons create moderate modifications of the original Nd:GGG lattices. Jia et al. reported the PL properties (for $^4F_{3/2} \rightarrow {}^4I_{9/2}$ and $^4F_{3/2} \rightarrow {}^4I_{11/2}$ emissions) of swift heavy ion irradiated Nd:GGG waveguide [32]. The C^{5+} ions at an energy of 17 MeV and a fluence of 2×10^{14} ions/cm^2 were applied to form waveguide layer. The intensity of fluorescence signals decreases by about

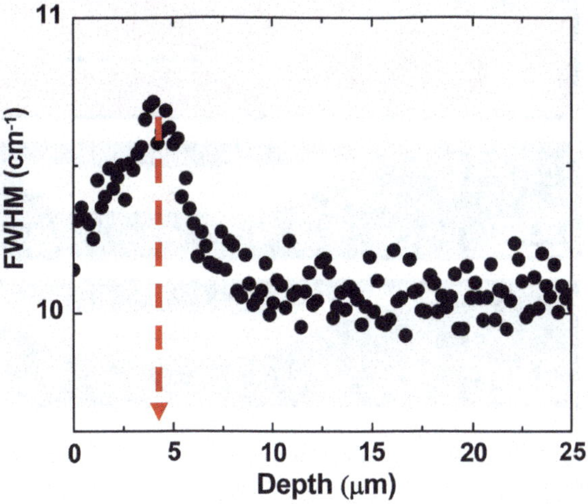

Fig. 7.12 The line-broadening spatial distribution of the $^4F_{3/2} \rightarrow {}^4I_{9/2}$ PL line around 940 nm for the 60 MeV Ar^{4+} ion irradiated Nd:YAG waveguide Reproduced with permission from [25], ©2010 Optical Society of America

10% in the waveguide with respect to the bulk, which means the majority of the PL active features has been preserved, without clear quenching.

Nd:LGS ($La_3Ga_5SiO_{14}$) crystal is a new laser gain material. LGS also possesses good electrooptic features, which could be used as Q switches for lasers [33]. For ion beam produced Nd:LGS waveguides, the PL properties are well preserved in the waveguide regions [34]. For example, Ren et al. reported the PL features of proton implanted channel waveguides in Nd:LGS [35]. Unlike the cubic crystals, the proton implantation does not affect the PL emissions of Nd^{3+} ions. The fluorescence intensity obtained from the active volume is quite close (less than 3% different) to that obtained from the original Nd:LGS network. Nevertheless, they have been strongly modified in the barrier regions (i.e., nuclear stopping region). Figure 7.13 includes the fluorescence images of the Nd:LGS waveguides obtained in terms of the spatial variation of the intensity, spectral induced shift and width of the Nd^{3+} emission line at around 9400 cm^{-1}. From these data, one can clear see the ion beam induced modification of PL emissions, which suggests that the nuclear collisions taking place at the end of proton tracks has caused the appearance of lattice defects that quenches the Nd^{3+} fluorescence.

Vanadate crystals are excellent for hosting Nd^{3+} ions for low threshold lasing [36]. The widely used vanadate laser crystals include Nd:YVO_4, Nd:$GdVO_4$, and Nd:$LuVO_4$, etc. By using MeV ion implantation of ions, e.g., H, He, C, and Si, both planar and channel waveguides were produced in Nd:YVO_4 crystals [37–40]. Sánchez-Morales et al. reported on the well-preserved PL properties in 7 MeV C^{2+} ion implanted Nd:YVO_4 waveguides [38]. Swift heavy ion irradiated Nd:YVO_4 waveguide was investigated by Yao et al. [41] It has been observed that in the 180 MeV Ar^{8+} ion irradiated Nd:YVO_4 waveguide, the obtained minimum fluorescence intensity is around 50% of the bulk, indicating clear PL quenching of Nd^{3+} ions. This reduction is mainly attributed to lattice damage induced by the electronic collision during the

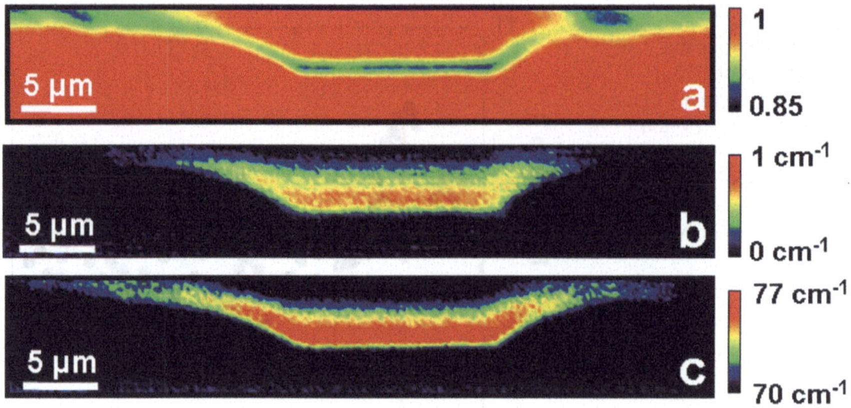

Fig. 7.13 **a** Spatial distribution of the emitted intensity, **b** spectral shift and **c** emission width at the cross section of the proton implanted Nd:LGS channel waveguide Reproduced with permission from [35], ©2010 Optical Society of America

irradiation process. Similar results have been obtained for 20 MeV N^{3+} ion irradiated Nd:$GdVO_4$ waveguide [42]. Figure 7.14 shows the comparison of the spectra of the emitted luminescence intensity at different points obtained from the waveguide region and bulk related to the $^4F_{3/2} \rightarrow {}^4I_{9/2}$ transition of Nd^{3+} ions. It is clear that the emission spectra are quite different in both the shape and the intensity, which means that the original fluorescence property of the Nd^{3+} ions has been considerable modified in the waveguide volume after 20 MeV N^{3+} ions irradiation.

Fig. 7.14 Nd^{3+} ion PL emission spectra as a function of layer depth for 20 MeV N^{3+} ion irradiated Nd:$GdVO_4$ planar waveguide Reproduced with permission from [42], ©2012 Elsevier

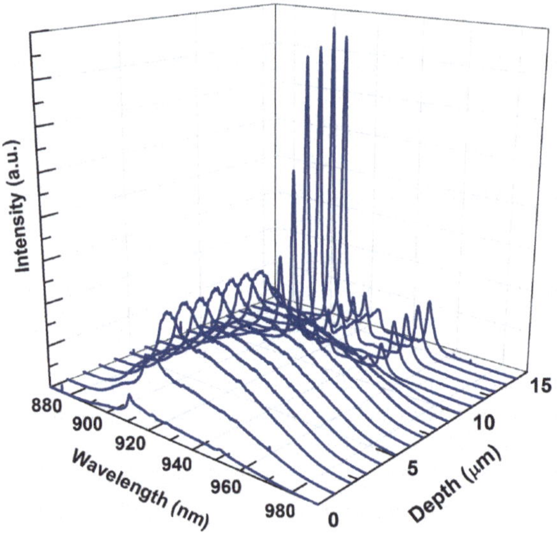

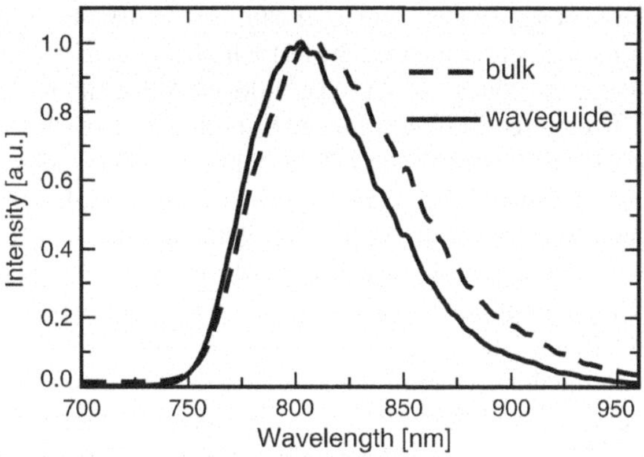

Fig. 7.15 Measured spectrum of the fluorescence light emitted out of the He^+ ion implanted Cr:LiSAF waveguide (solid line) and the bulk material (dashed line) Reproduced with permission from [46], ©2007 Springer-Verlag

Trivalent titanium-doped sapphire (Al_2O_3:Ti^{3+}, Ti:Sapphire) [43] is one of the most commercially successful crystals for vibronic lasers, allowing tunability from ~670 to 1100 nm. Cr:$LiSrAlF_6$ (Cr:LiSAF) is an interesting fluorescent material characterized by a wide emission bandwidth in a wavelength range of 750–930 nm, enabling potential applications for tunable laser action [44]. So far, only light ion implantation has been applied to produce waveguides in these two crystals. For proton implanted Ti:Sapphire channel waveguides, the room-temperature PL emission is very similar to that of the un-implanted bulk crystal [45]. Majkic et al. investigated the He^+ ion implanted Cr:LiSAF channel waveguide as broadband fluorescence source [46]. Figure 7.15 shows the measured spectra of the fluorescence light emitted from the waveguide and from the bulk. The spectra differ only slightly, which suggests that the local environment of the active Cr^{3+} ions remained mostly unaltered after the implantation.

The PL properties of ion beam produced waveguides in other crystals such as Nd:YAP, Nd:YCOB, Nd:GdCOB, Yb:CaF_2, Yb:SBN, and Nd:CBN [47–52], have also been investigated. Similar to the above-mentioned crystals, the modification on PL emissions depends on the ion beam conditions as well as the material properties. Nevertheless, the preserved PL features in these waveguides still enable further lasing on compact geometries.

7.3 Waveguide Amplifiers

The erbium-doped waveguide amplifiers (EDWA) attracts much attention because of the extensive applications in the 1.55 μm telecommunication field. The operation principle of EDWA is identical to that of erbium-doped fiber amplifiers (EDFA), which has been widely used in the telecommunication systems. EDWAs offer low-cost solution in compact geometries to realize signal amplification. By using direct Er ion implantation to achieve doping, the Er^{3+}-implanted Al_2O_3 waveguide amplifiers have been produced with a net gain of 0.58 dB/cm with 9 mW pump at 1.48 μm [53]. Some dielectrics are ideal hosts for highly doped Er^{3+} ions, such as phosphate glasses. One can produce EDWA in Er^{3+}-doped dielectrics by using ion beam techniques. Liu et al. reported EDWA produced by proton beam writing at fluences of 10^{14}–10^{15} cm^{-2} at 2.0 MeV in an Er^{3+}/Yb^{3+} co-doped phosphate glass [54]. The co-doping of Yb^{3+} ions is to avoid PL quenching of Er^{3+} ion at high concentration. Their results have indicated that the fluorescence related to telecommunication bands of the channel waveguides was slightly different from that of the related bulk materials. The result of light amplification was obtained with a pump power of 100 mW at a wavelength of 975 nm. The net gain of the formed EDWA was found to have a close relationship with the fluences of protons. In case of low-fluence writing, e.g., of 4×10^{14} cm^{-2}, the net gain of 0.52 dB/cm was obtained. As higher fluence implantation up to 8×10^{14} cm^{-2} was used, the gain was further increased up to 1.72 dB/cm.

Other Er^{3+} ion doped dielectrics may be also used as substrates for amplification. Tan et al. reported on the amplification at C band (1530–1565 nm) and L band 1585 nm (1565–1625 nm) in a carbon ion implanted Er:YAG ceramic waveguide [55]. The authors utilized a double-energy implantation to construct a cladding waveguide: by 6 MeV C^{3+} ions at fluence 5×10^{14} ions/cm^2 and by 15 MeV C^{5+} ions at 2×10^{14} ions/cm^2, respectively. The waveguide has a multilayer structure that assures good overlap between the pump beam and the input signal. Under the 200 mW pump at 980 nm with a fiber-coupled diode laser, the Er:YAG cladding waveguide possess a peak internal gain of 2.6 dB/cm at 1550 nm and of 4.0 dB/cm at 1585 nm. Figure 7.16 shows the performance of Er:YAG ceramic waveguide in the optical signal amplification.

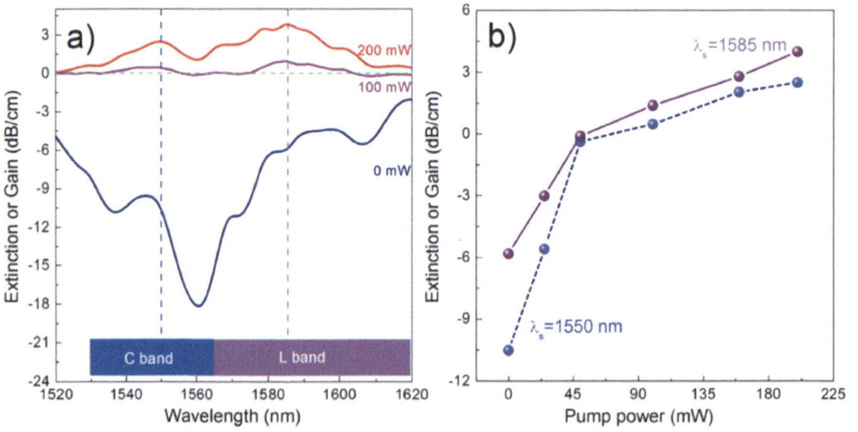

Fig. 7.16 **a** Measured absorption and gain spectra in the wavelength range of 1520–1620 nm. **b** Measured extinction/gain as a function of the pump power at the wavelength of 1550 nm and 1585 nm, respectively Reproduced with permission from [55], ©2016 Optical Society of America

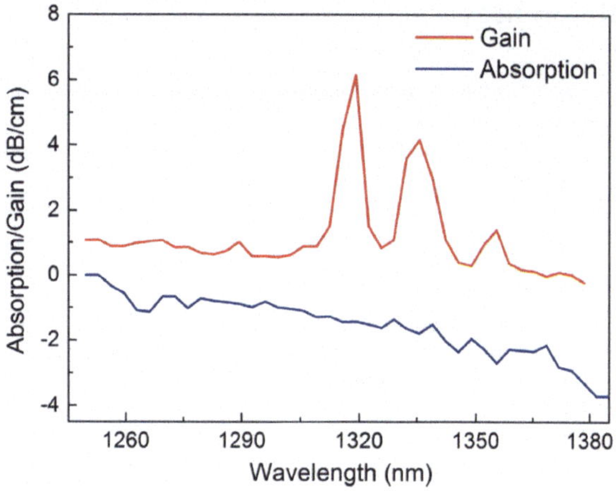

Fig. 7.17 Measured absorption and gain spectrum in the wavelength range of 1260–1380 nm at the pump wavelength of 808 nm Reproduced with permission from [56], ©2015 Optical Society of America

Although Er-doped waveguides receive major attention of researchers, the Nd^{3+} doped waveguide amplifiers have been realized operating at the second telecommunication wavelength band (~1.3 μm, correlated to Nd^{3+} transition $^4F_{3/2} \rightarrow {}^4I_{13/2}$) and 1064 nm ($^4F_{3/2} \rightarrow {}^4I_{11/2}$ transition of Nd^{3+} ions), respectively. Tan et al. reported on a high-gain optical waveguide amplifier based on proton beam writing of Nd:YAG channel waveguides [56]. Under the pumping of a continuous wave laser, the high-gain optical amplification at single wavelength of 1064 nm and wavelength band of 1300–1360 nm were obtained. Figure 7.17 shows the observed optical amplification of the second telecommunication band when the 808 nm pump power was increased to 80 mW. At the signal wavelength of 1319 nm, 1336 nm and 1356 nm, the gain was 6 dB/cm, 4 dB/cm and 1.4 dB/cm, respectively.

The 1064 nm optical amplification is much stronger than that of the second telecommunication band because the $^4F_{3/2} \rightarrow {}^4I_{11/2}$ transition of Nd^{3+} ions is dominating over the $^4F_{3/2} \rightarrow {}^4I_{13/2}$. The maximum gain was 24 dB/cm at 1064 nm (a net gain of 9.6 dB for a 4 mm long sample) for the proton beam written Nd:YAG waveguide amplifier under the 100 mW continuous wave pump laser. The value of gain was around 23 dB/cm within the pump laser of 803–817 nm, which indicates broad absorption band for stable optical signal amplification [56]. Similar results have been obtained for Nd:YAG ceramic waveguide amplifier, which is produced by the swift carbon ion irradiation [57]. The 1064 nm pulse laser amplification ($g = 26.3$ dB/cm) has been realized by the enhanced florescence in conjunction with the excellent guiding properties in the waveguide. The net small signal gain value of the 7 mm long waveguide is determined to be ~18.4 dB at 1064 nm [57].

References

1. E. Marsden, *Photoluminescence: advances in Research and Applications* (Nova Science Publishers, 2018)
2. F.E. Auzel, Materials and devices using double-pumped-phosphors with energy transfer. Proc. IEEE **6**, 758 (1973)
3. G.K. Liu, B. Jacquier, *Spectroscopic Properties of Rare Earths in Optical Materials* (Springer, 2005)
4. F. Wang, X. Liu, Rare-earth doped upconversion nanophosphors. Compr. Nanosci. Technol. **1**, 607 (2011)
5. J.A. Capobianco, G. Prevost, P.P. Proulx, P. Kabro, M. Bettinelli, Upconversion properties of Er^{3+} doped lead silicate glasses. Opt. Mater. **6**, 175 (1996)
6. B. Herreros, G. Lifante, F. Cussó, P.D. Townsend, P.J. Chandler, Green up-conversion emission in Er^{3+}-doped $LiNbO_3$ ion-implanted planar waveguides. J. Lumin. **72**, 198 (1997)
7. M. Szachowicz, S. Tascu, M.F. Joubert, P. Moretti, M. Nikl, Realization and infrared to green upconversion luminescence in Er^{3+}:$YAlO_3$ ion-implanted optical waveguides. Opt. Mater. **28**, 162 (2006)
8. H. Liu, Y. Jia, Y. Ren, S. Akhmadaliev, S. Zhou, F. Chen, Green up-conversion of swift C^{5+} ion irradiated planar waveguide in Er^{3+}, MgO codoped nearly stoichiometric $LiNbO_3$ crystal. Nucl. Instrum. Meth. Phys. Res. B **320**, 22 (2014)
9. M. Szachowicz, M. Joubert, P. Moretti, M. Couchaud, B. Ferrand, C. Borca, A. Boudrioua, H + implanted channel waveguides in buried epitaxial crystalline YAG:Nd, Tm layers and infrared-to-blue upconversion characterization. J. Appl. Phys. **104**, 113104 (2008)
10. C. Chen, R.Y. He, Y. Tan, B. Wang, S. Akhrnadaliev, S.Q. Zhou, J.R.V. de Aldana, L.L. Hu, F. Chen, Optical ridge waveguides in Er^{3+}/Yb^{3+} co-doped phosphate glass produced by ion irradiation combined with femtosecond laser ablation for guided-wave green and red upconversion emissions. Opt. Mater. **51**, 185 (2016)
11. J. Amin, B. Dussardier, T. Schweizer, M. Hempstead, Spectroscopic analysis of Er^{3+} transitions in lithium niobite. J. Lumin. **69**, 17 (1996)
12. N. Dong, F. Chen, D. Jaque, Carbon ion implanted Nd:MgO:$LiNbO_3$ optical channel waveguides: an intermediate step between light and heavy ion implanted waveguides. Opt. Express **18**, 5951 (2010)
13. S.J. Field, D.C. Hanna, D.P. Shepherd, A.C. Tropper, P.J. Chandler, P.D. Townsend, L. Zhang, Ion-implanted Nd:MgO:$LiNbO_3$ planar waveguide laser. Opt. Lett. **16**, 481 (1991)
14. D. Jaque, F. Chen, High resolution fluorescence imaging of damage regions in H^+ ion implanted Nd:MgO:$LiNbO_3$ channel waveguides. Appl. Phys. Lett. **94**, 011109 (2009)
15. D. Jaque, F. Chen, Y. Tan, Scanning confocal fluorescence imaging and micro-Raman investigations of oxygen implanted channel waveguides in Nd:MgO:$LiNbO_3$. Appl. Phys. Lett. **92**, 161908 (2008)
16. F. Chen, Y. Tan, Active waveguide in Nd^{3+}:MgO:$LiNbO_3$ crystal produced by low-dose carbon ion implantation. Appl. Phys. Lett. **92**, 021110 (2008)
17. F. Chen, Y. Tan, A. Ródenas, Ion implanted optical channel waveguides in Er^{3+}/MgO co-doped near stoichiometric $LiNbO_3$: a new candidate for active integrated photonic devices operating at 1.5 μm. Opt. Express **16**, 16209 (2008)
18. T.Y. Fan, Heat generation in Nd:YAG and Yb:YAG. IEEE J. Sel. Top. Quantum Electron. **29**, 1457 (1993)
19. Y. Tan, F. Chen, Proton-implanted optical channel waveguides in Nd:YAG laser ceramics. J. Phys. D Appl. Phys. **43**, 075105 (2010)
20. Y. Yao, Y. Tan, N. Dong, F. Chen, A.A. Bettiol, Continuous wave Nd:YAG channel waveguide laser produced by focused proton beam writing. Opt. Express **18**, 24516 (2010)
21. A. Benayas, D. Jaque, Y. Yao, F. Chen, A.A. Bettiol, A. Rodenas, A.K. Kar, Microstructuring of Nd:YAG crystals by proton-beam writing. Opt. Lett. **35**, 3898 (2010)
22. A. Benayas, N. Dong, Y. Yao, F. Chen, A.A. Bettiol, D. Jaque, Thermal optimization and erasing of Nd:YAG proton beam written waveguides. Opt. Lett. **36**, 3278 (2011)

23. F. Chen, Y. Tan, D. Jaque, Ion-implanted optical channel waveguides in neodymium-doped yttrium aluminum garnet transparent ceramics for integrated laser generation. Opt. Lett. **34**, 28 (2009)
24. Y. Kong, F. Chen, D. Jaque, Y. Tan, N. Dong, Q. Lu, H. Ma, Low-dose O^{3+} ion-implanted active optical planar waveguides in Nd:YAG crystals: guiding properties and micro-luminescence characterization. J. Phys. D Appl. Phys. **41**, 175112 (2008)
25. Y. Ren, N. Dong, F. Chen, A. Benayas, D. Jaque, F. Qiu, T. Narusawa, Swift heavy-ion irradiated active waveguides in Nd:YAG crystals: fabrication and laser generation. Opt. Lett. **35**, 3276 (2010)
26. Y. Cheng, J. Lv, S. Akhmadaliev, I. Hernández-Palmero, C. Romero, J.R.V. de Aldana, S. Zhou, F. Chen, Optical ridge waveguides in Yb:YAG laser crystal produced by combination of swift carbon ion irradiation and femtosecond laser ablation. Opt. Laser Technol. **72**, 100 (2015)
27. N. Dong, A. Benayas, D. Jaque, Y. Tan, F. Chen, Fluorescence-quenching free channel waveguides in Yb:YAG ceramics by carbon ion implantation. J. Lightwave Technol. **29**, 1460 (2011)
28. Y. Tan, L. Ma, S. Akhmadaliev, S. Zhou, F. Chen, Ion irradiated Er:YAG ceramic cladding waveguide amplifier in C and L bands. Opt. Mater. Express **6**, 711 (2016)
29. L.J. Qin, D.Y. Tang, G.Q. Xie, C.M. Dong, Z.T. Jia, X.T. Tao, High-power continuous wave and passively -switched laser operations of a Nd:GGG crystal. Laser Phys. Lett. **5**, 100 (2008)
30. S.J. Field, D.C. Hanna, A.C. Tropper, D.P. Shepherd, P.J. Chandler, P.D. Townsend, L. Zhang, Ion-implanted Nd:GGG channel waveguide laser. Opt. Lett. **17**, 52 (1992)
31. Y. Ren, N. Dong, Y. Tan, J. Guan, F. Chen, Q. Lu, Continuous wave laser generation in proton implanted Nd:GGG planar waveguides. J. Lightwave Technol. **28**, 3578 (2010)
32. Y. Jia, N. Dong, F. Chen, J.R.V. de Aldana, S. Akhmadaliev, S. Zhou, Ridge waveguide lasers in Nd:GGG crystals produced by swift carbon ion irradiation and femtosecond laser ablation. Opt. Express **20**, 9763 (2012)
33. Y. Yu, J. Wang, H. Zhang, Z. Wang, H. Yu, M. Jiang, Continuous wave and Q-switched laser output of laser-diode-end-pumped disordered Nd:LGS laser. Opt. Lett. **34**, 467 (2009)
34. Y. Cheng, J. Lv, S. Akhmadaliev, S. Zhou, F. Chen, Optical ridge waveguides in Nd:LGS crystal produced by combination of swift C^{5+} ion irradiation and precise diamond blade dicing. Opt. Laser Technol. **81**, 122 (2016)
35. Y. Ren, Y. Tan, F. Chen, D. Jaque, H. Zhang, J. Wang, Q. Lu, Optical channel waveguides in Nd:LGS laser crystals produced by proton implantation. Opt. Express **18**, 16258 (2010)
36. J.E. Bernard, A.J. Alcock, High-efficiency diode-pumped Nd:YVO_4 slab laser. Opt. Lett. **18**, 968 (1993)
37. G.V. Vázqueza, M.E. Sánchez-Moralesa, H. Márquezb, J. Rickardsc, R. Trejo-Lunac, Analysis of ion implanted waveguides formed on Nd:YVO_4 crystals. Opt. Commun. **240**, 351 (2004)
38. M.E. Sánchez-Morales, G.V. Vázquez, H. Márquez, J. Rickards, R. Trejo-Luna, Optical waveguides formed in Nd:YVO_4 crystals by C^{2+} implantation. J. Mod. Opt. **53**, 539 (2006)
39. F. Chen, X. Wang, K. Wang, Optical waveguides formed in Nd:YVO_4 by MeV Si^+ implantation. Appl. Phys. Lett. **80**, 3473 (2002)
40. F. Chen, L. Wang, Y. Jiang, X. Wang, K. Wang, Optical channel waveguides in Nd:YVO_4 crystal produced by O^+ ion implantation. Appl. Phys. Lett. **88**, 071123 (2016)
41. Y. Yao, N. Dong, F. Chen, L. Pang, Z. Wang, Q. Lu, Continuous wave waveguide lasers of swift argon ion irradiated Nd:YVO_4 waveguides. Opt. Express **19**, 24252 (2011)
42. N. Dong, Y. Yao, F. Chen, Optical waveguides in Nd:$GdVO_4$ crystals fabricated by swift N^{3+} ion irradiation. Opt. Mater. **35**, 310 (2012)
43. A.M. Kowalevicz, T. Ko, I. Hartl, J.G. Fujimoto, M. Pollnau, R.P. Salathe, Ultrahigh resolution optical coherence tomography using a superluminescent light source. Opt. Express **10**, 349 (2002)
44. S.A. Payne, L.L. Chase, L.K. Smith, W.L. Kway, H.W. Newkirk, Laser performance of $LiSrAlF_6{:}Cr^{3+}$. J. Appl. Phys. **66**, 1051 (1989)
45. C. Grivas, D.P. Shepherd, R.W. Eason, L. Laversenne, P. Moretti, C.N. Borca, M. Pollnau, Room-temperature continuous-wave operation of Ti: sapphire buried channel-waveguide lasers fabricated via proton implantation. Opt. Lett. **31**, 3450 (2006)

46. A. Majkić, G. Poberaj, R. Degl'Innocenti, M. Döbeli, P. Günter, Cr:$LiSrAlF_6$ channel waveguides as broadband fluorescence sources. Appl. Phys. B **88**, 205 (2007)
47. S.J. Field, D.C. Hanna, D.P. Shepherd, A.C. Tropper, P.J. Chandler, P.D. Townsend, L. Zhang, Ion-implanted Nd:YAP planar waveguide laser. Electron. Lett. **26**, 1826 (1990)
48. Y. Ren, N. Dong, Y. Jia, L. Pang, Z. Wang, Q. Lu, F. Chen, Efficient laser emissions at 1.06 μm of swift heavy ion irradiated Nd:YCOB waveguides, Opt. Lett. **36**, 4521 (2011)
49. Y. Ren, Y. Jia, F. Chen, Q. Lu, S. Akhmadaliev, S. Zhou, Second harmonic generation of swift carbon ion irradiated Nd:GdCOB waveguides. Opt. Express **19**, 12490 (2011)
50. V. Petit, P. Moretti, P. Camy, J.L. Doualan, R. Moncorgé, Active waveguides produced in Yb^{3+}:CaF_2 by H^+ implantation for laser applications. J. Alloy. Compd. **451**, 68 (2008)
51. G. Liu, N. Dong, J. Wang, S. Akhmadaliev, S. Zhou, F. Chen, Opt. Mater. **51**, 31 (2016)
52. Y. Tan, F. Chen, D. Jaque, W. Gao, H. Zhang, J.G. Solé, H. Ma, Ion-implanted optical-stripe waveguides in neodymium-doped calcium barium niobate crystals. Opt. Lett. **34**, 1438 (2009)
53. G. N. van den Hoven, R. J. I. M. Koper, A. Polman, C. van Dam, J. W. M. van Uffelen, M. K. Smit, Net optical gain at 1.53 μm in Er-doped Al_2O_3 waveguides on silicon. Appl. Phys. Lett. **68**, 1886 (1996)
54. K. Liu, E.Y.B. Pun, T.C. Sum, A.A. Bettiol, J.A. van Kan, F. Watt, Erbium-doped waveguide amplifiers fabricated using focused proton beam writing. Appl. Phys. Lett. **84**, 684 (2004)
55. Y. Tan, L. Ma, S. Akhmadaliev, S. Zhou, F. Chen, Ion irradiated Er:YAG ceramic cladding waveguide amplifier in C and L bands. Opt. Mater. Express **3**, 711 (2016)
56. Y. Tan, Z. Shang, S.K. Vanga, A.A. Bettiol, F. Chen, High-gain optical waveguide amplifier based on proton beam writing of Nd:YAG crystal. Opt. Express **23**, 14612 (2015)
57. Y. Tan, Q. Luan, F. Liu, S. Akhmadaliev, S. Zhou, F. Chen, Swift carbon ion irradiated Nd:YAG ceramic optical waveguide amplifier. Opt. Express **21**, 13992 (2013)

Chapter 8
Nonlinear Optical Dielectric Waveguides

Abstract Nonlinear optical effects in waveguides are of potential applications in the development of various integrated photonic devices. Ion beam produced nonlinear waveguides possess well preserved properties of the bulks and could be utilized for nonlinear optical applications. In this chapter, the ion beam modification of nonlinear optical crystals and the applications of the nonlinear waveguides are introduced. In Sect. 8.1, the nonlinear properties of the waveguides, including the modified nonlinear coefficients in the guiding regions, are presented. In Sect. 8.2, one of the major applications of the nonlinear waveguides, the frequency/wavelength conversion, is overviewed. Section 8.3 reviews the research results on the photorefractive waveguides, which are based on a "weak" light induced effect. Section 8.4 is devoted to the nonlinear propagation of light (spatial solitons) in waveguides and waveguide arrays.

Nonlinear Optics became an intriguing branch of *Optics* since the invention of laser in 1960s. In linear case, the media in which light propagates do not affect the property of light itself because the intensity of light is low enough to avoid obvious nonlinear phenomena. The invention of laser offers high intensity of optical fields, enabling effective modification of media (dielectrics) properties, which in turn change the light behaviors [1–4]. Nonlinear optical phenomena, such as harmonic generation, optical parametric amplification or oscillation, sum or difference frequency generation, are necessary to realize light sources in broadband wavelength regimes. Efficient nonlinear optical effects depend on properties of light as well as nonlinear media. Second harmonic generation (SHG) is widely used to realize wavelength conversion in bulks and waveguides. SHG has been achieved in ion-implanted waveguides based on phase-matching (PM), quasi phase-matching (QPM), and Cerenkov mechanisms. In addition, as a weak light induced effect, photorefractive property also brings out some potential applications.

F. Chen et al., *Ion Irradiation of Dielectrics for Photonic Applications*, Springer Series in Optical Sciences 231, https://doi.org/10.1007/978-981-15-4607-5_8

8.1 Nonlinear Optical Properties of Waveguides

The nonlinear waveguides are constructed on the basis of nonlinear media. Nonlinear optical crystals are major materials for light-field induced nonlinear effects. These crystals, for frequency conversion like SHG, must offer platforms to achieve both energy and momentum conservation of specific photons of fundamental and second harmonic (SH) frequencies. In bulk materials, these conditions are satisfied based on the original birefringence of the crystals (PM) by angle or temperature tuning, enabling the refractive indices of the fundamental and SH light with the same value, or on the periodically engineered domains of materials (QPM). In waveguides, the situation becomes more complicated due to the possible anisotropic behaviors of the guided wave with different polarizations. Another possible reason is that the influence of waveguide fabrication technology on the optical nonlinearity of the media, which may induce modification of the properties of bulk materials.

For ion beam produced waveguides, the degradation of nonlinearity to a certain extent in most cases is inevitable due to the interaction of incident ions with the atoms of the crystals. However, the positive point is that the ion induced distortion or damage of the original lattices may not happen at the waveguide core regions, or the lattice damage could be removed after some treatment, e.g., post-implantation thermal annealing at moderate temperatures. Typically, thermal annealing at 200–350 °C for a half to a few hours is considerably effective to remove the point defects in the waveguiding cores induced by the ion bombardment [5]. The severe damage induced by the nuclear energy deposition remains after moderate annealing, which happens at the boundary region between the waveguide cores and surrounding [6]. For nonlinear optical crystals, the modification of SH properties at different positions reflects the level of the localized lattice damage induced by ion beams. By using micro-spectroscopy technique (see Chap. 1 for details), one can clearly see the modification of SH signals in the waveguide region, which can be considered as an effective method to tailor the optical nonlinearity of the waveguide [7].

Light-ion implantation is a typical solution to produce barrier-type waveguide, in which the waveguide cores are surrounded by low-index optical barriers (refractive index cladding layer). Figure 8.1 shows the optical transmission photographs and the corresponding confocal micro-SH (μ-SH) cross sectional images of the planar and channel waveguides in $KTiOPO_4$ (KTP) crystal by multiple He^+ ion implantation at energies of $(1.9 + 2.0 + 2.1)$ MeV and fluencies of $(2.7 + 2.7 + 4.5) \times 10^{15}$ ions/cm^2, respectively [8]. One can see that the nonlinear responses (the back-scattered SH signal intensity) at the waveguide active volumes are not obviously deteriorated with respect to the bulk, neither in the planar nor the channel waveguides. The obtained SH intensities are ~80% in the waveguide regions with respect to that from the bulk. And there is a strong reduction (by 60% in comparison with the bulk) of the SH efficiency in the boundary of the waveguide and bulk (i.e., in the barrier region), which is located at the end of ion range and leads to the largest refractive index change due to the nuclear collisions of the incident ions and target nuclei. This indicates that the optical nonlinearities are well preserved in the waveguide, which could be used for

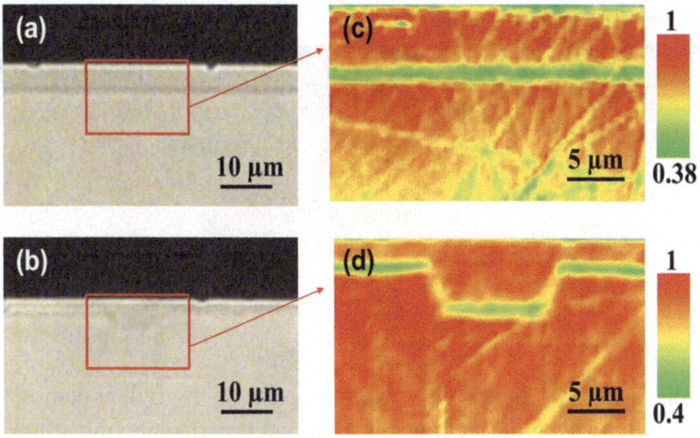

Fig. 8.1 Optical transmission photographs obtained from the cross sections of **a** planar and **b** channel waveguides in $KTiOPO_4$. **c** and **d** are the corresponding confocal μ-SH cross sectional images of the planar and channel waveguides after annealing 200 °C for 30 min in the air, respectively. Reproduced with permission [8], ©2011 Optical Society of America

the frequency conversion. Further investigation shows that the SH intensity profile matches the refractive index profile [8]: the nonlinear features are well preserved in the waveguide's volumes, and then decreased to the minimum value at the barrier regions (see Fig. 8.2).

Focused ion beam writing (protons or He ions) could be used to form buried waveguides, and the waveguide cores are located at the ion irradiated regions (i.e., at

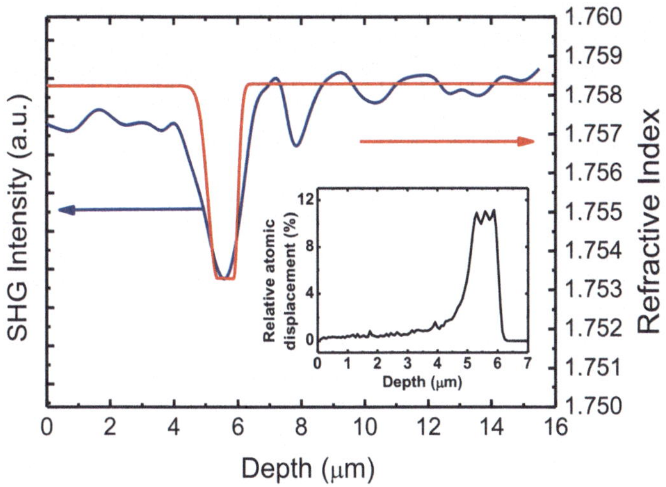

Fig. 8.2 1-D spatial scan of the SHG intensity (blue color) and refractive index profile (red color) of the planar $KTiOPO_4$ waveguide as a function of the depth below the surface. Reproduced with permission [8], ©2011 Optical Society of America

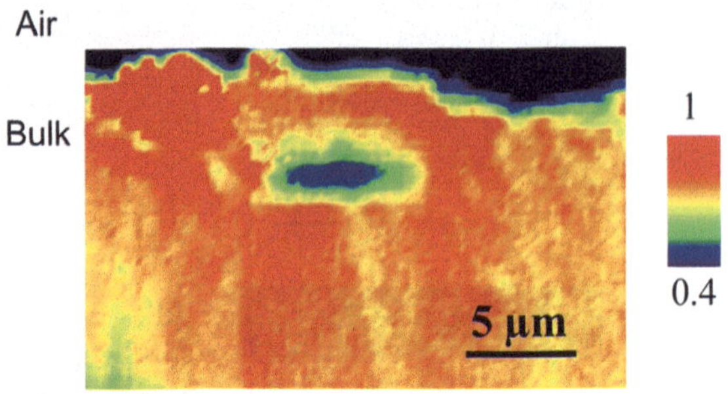

Fig. 8.3 The confocal μ-SH cross sectional image of He^+ ion written waveguide in $KTiOPO_4$. The He^+ ion beam is at energy of 2 MeV and fluence of 2×10^{15} ions/cm^2

the end of ion range) [9]. In this case, there is relatively high degradation of optical nonlinearity in the waveguide region since the nuclear damage most happen at the end of the ion range. Figure 8.3 indicates the confocal μ-SH cross sectional image of 2 MeV He^+ ion written waveguide in $KTiOPO_4$. There is about 60% reduction of SH intensity from the waveguide core compared with the bulk.

As a widely applied nonlinear crystal, $LiNbO_3$ receives considerable attentions from the scientific community. For $LiNbO_3$, the modification of the ion beams on the second-order nonlinear optical coefficient d_{33} has been investigated by a few groups. It has been found that the ion-sliced $LiNbO_3$ or PPLN membrane [10, 11] and H-ion implanted waveguides possess almost the same d_{33} values as the bulk [12]. The ion sliced membrane benefits from the high-temperature annealing, which can remove the ion-induced damage, leading to almost perfect recovery of the lattice. For He-ion-implanted $LiNbO_3$ samples, the d_{33} has been reduced to 50–80% of that of the bulk, depending on the fluences [13]. The d_{33} of the swift heavy ion (e.g., 20 MeV F, 46 MeV Cl, 809 MeV Kr or 1.4 GeV Xe) irradiated $LiNbO_3$ waveguides was reduced to 60–90% compared with the bulks [14–18]. Figure 8.4 shows the d_{33} values of the 46 MeV Cl ion irradiated $LiNbO_3$ waveguide as function of ion fluence [17]. For a fluence of 2×10^{12} ions/cm^2, the d_{33} was decreased by ~50%.

The C ion implantation seems to be an efficient method to obtain $LiNbO_3$ waveguides with well-preserved nonlinearity. For instance, in a 15 MeV C^{5+} ion irradiated

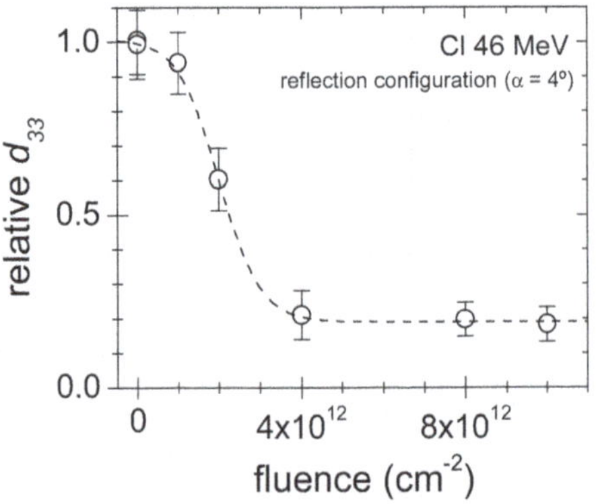

Fig. 8.4 Evolution of the relative d_{33} nonlinear coefficient of $LiNbO_3$ crystal with the irradiation fluence of Cl 46 MeV ions. Reproduced with permission [17], ©2007 Optical Society of America

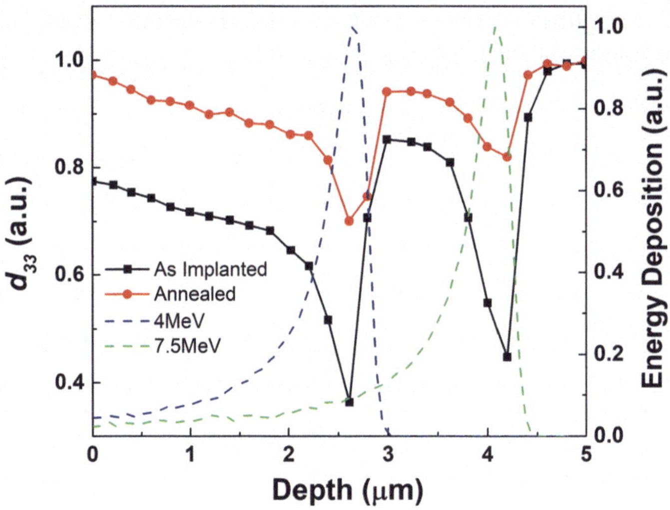

Fig. 8.5 The d_{33} profile (normalized to that of bulk $LiNbO_3$) of the double energy (4 + 7.5 MeV) carbon ion implanted PPLN waveguide before and after the thermal annealing. The nuclear energy deposition of 4 and 7.5 MeV carbon ions into $LiNbO_3$ crystal as function of depth based on SRIM simulation is also depicted (dashed lines)

PPLN waveguide, the effective coefficient d_{eff} was determined to be 9.43 pm/V, which means ~73% of the second-order nonlinearity has been preserved [19]. Nevertheless, thermal annealing up to 350 °C was performed to recover the original lattices, which at same time reduced the propagation loss down to 0.5 dB/cm. In addition, for a PPLN waveguide produced by double-energy C ion implantation at 4 and 7.5 MeV, the d_{33} in the waveguide core reached 10.3 pm/V, about 80% of the bulk, after the annealing at 300 °C for 30 min [20], which suggested the partial recovery of the damaged lattice induced by ion bombardment. The effective coefficient d_{eff} was 80% of the bulk, considering the mode overlap effect. It has been also found that the modification of d_{33} is strongly dependent on the nuclear energy deposition (see Fig. 8.5 for the d_{33} profile).

8.2 Nonlinear Frequency Conversion Based on Waveguides

Based on nonlinear waveguides, frequency/wavelength conversion has been realized. The investigated crystals include $KNbO_3$, GdCOB, Nd:GdCOB, β-BaB_2O_4 (β-BBO), LiB_3O_5 (LBO), $LiNbO_3$, and $LiTaO_3$. The applied mechanisms are PM, QPM, and Cerenkov configurations. Since Cerenkov SHG is relatively weak for waveguides, and basically cannot be used for integrated optical devices, we will only focus on PM and QPM frequency conversion in waveguides.

An important figure of merit to characterize a nonlinear waveguide is the normalized conversion efficiency η (η') given by [3].

$$\eta = \frac{P_{2\omega}}{L^2 P_\omega^2} \tag{8.1}$$

or

$$\eta' = \frac{P_{2\omega}}{P_\omega^2} \tag{8.2}$$

where L denotes the waveguide (sample) length, P_ω and $P_{2\omega}$ are the incident powers of the fundamental and SH light. η includes the device length, whilst η' is independent of the device length. At pump of pulsed lasers, the conversion efficiency may also be written as the ratio of SH and fundamental light powers:

$$\eta'' = \frac{P_{2\omega}}{P_\omega} \tag{8.3}$$

In case of low propagation loss waveguides, the longer the sample is, the higher the conversion efficiency is.

In practice, the nonlinear optical properties of ion implanted waveguides are preserved to some extent in comparison to respective bulk crystals, which offers possibility to achieve guided-wave SHG. Table 8.1 summarizes the obtained data on the frequency conversion (SHG) of the ion beam processed waveguides in some nonlinear crystals.

Table 8.1 The obtained data on the frequency doubling of the ion beam processed waveguides in some nonlinear crystals. The mechanisms for SHG include either Type I and Type II configurations, depending on the orientation of the nonlinear crystals.

As one can see, the $KNbO_3$ waveguides [21] seem to be more efficient for blue light SHG owing to both the large nonlinear coefficients and good guiding quality. The obtained maximum conversion efficiency was 12.1%W^{-1} for a ridge waveguide, and the maximum SH power was 320 mW for a planar waveguide.

The blue light SHG in the He ion implanted GdCOB waveguides was investigated by a French group [22, 23]. The obtained conversion efficiency for planar and channel waveguides was 0.4 and 0.32%W^{-1} at cw 822 nm pump laser power of 250 mW, respectively. Ren et al. reported the SHG of green light (at 532 nm) for the swift heavy ion irradiated Nd:GdCOB planar waveguide (produced by irradiation of 17 MeV C ions at fluence of 2×10^{14} cm^{-2}) based on Type I PM [24]. The maximum output powers of SH signals were ~0.53 mW (at pump of 334.6 mW) and ~0.72 mW (at pump of ~102.7 mW) for the cw and pulsed light beams, resulting in the conversion efficiencies of 0.48%W^{-1} and 6.8%W^{-1}, respectively. Jia et al. further realized SHG conversion efficiency of 11.4% (corresponding to 43.4%W^{-1}) of a Nd:GdCOB ridge waveguide produced by 17 MeV C ion irradiation combined with femtosecond laser ablation [25]. Figure 8.6 shows the SHG at 532 nm from the waveguide under pump

Table 8.1 The obtained data on the frequency doubling of the ion beam processed waveguides in some nonlinear crystals

Materials	Fabrication technique	Waveguide configuration	Regime	SHG configuration	λ_{SH} (nm)	η	$P_{2\omega}$ (mW)	Refs.
$KNbO_3$	He	Planar	Pulsed		434	8%	4×10^5 (peak)	Fluck and Gunter [21]
	He ion implantation	Planar	cw	PM	433	$1.2\%W^{-1}$	2.4	Fluck and Gunter [21]
	He ion implantation	Channel	cw	PM	441	$3.3\%W^{-1}$	2.6	Fluck and Gunter [21]
	He ion implantation	ridge	cw	PM	438	$12.1\%W^{-1}$	14	Fluck and Gunter [21]
GdCOB	He ion implantation	planar	cw	PM	411.5	$0.4\%W^{-1}$	–	Boudrioua et al. [22]
	He ion implantation	Planar	cw	PM	480.2	$0.5\%W^{-1}$	–	Boudrioua et al. [22]
	He ion implantation	Channel	cw	PM	405.5	$0.3\%W^{-1}$	–	Vincent et al. [23]
Nd:GdCOB	C ion irradiation	Planar	cw	PM	532	$0.48\%W^{-1}$	0.53	Ren et al. [24]
	C ion irradiation	Planar	pulsed (11 ns)	PM	532	4.6%	1.8×10^4 (peak)	Ren et al. [24]
	C ion irradiation	Ridge	Pulsed (11 ns)	PM	532	11.4%	1.1×10^5 (peak)	Jia et al. [25]
β-BBO	He ion implantation	Ridge	cw	PM	278	$0.1\%W^{-1}$	0.024	Poberaj et al. [26]
	He ion implantation	Ridge	cw	PM	266	$0.07\%W^{-1}$	0.3	Degl'Innocenti et al. [27]

(continued)

Table 8.1 (continued)

Materials	Fabrication technique	Waveguide configuration	Regime	SHG configuration	λ_{SH} (nm)	η	$P_{2\omega}$ (mW)	Refs.
$LiNbO_3$	He ion implantation	Planar	pulsed (10 ns)	PM	567	$0.018\% W^{-1}$	–	Hamelin et al. [28]
	Xe ion irradiation	Planar	Pulsed (11 ns)	PM	532	6.9%	2.95×10^4 (peak)	Chen et al. [29]
	He ion slicing	Membrane	pulsed (200 fs)	PM	775	$8 \times 105\% W^{-1}$	2×10^{-4} (average)	Gaathon et al. [30]
$MgO:LiNbO_3$	He ion implantation	Planar	pulsed (10 ns)	PM	567	$0.54\% W^{-1}$	–	Hamelin and Chow [31]
LBO	He ion implantation	Channel	cw	PM	469	$6 \times 10^{-4}\% W^{-1}$	1.5×10^{-6}	Davis et al. [32]
Nd:YCOB	Ar ion irradiation	Planar	cw	PM	530.89	$0.35\% W^{-1}$	0.9	Ren et al. [33]
	C ion irradiation	Planar	cw	PM	532	$0.5\% W^{-1}$	0.62	Li et al. [34]
KTP	O ion irradiation	Planar	cw	PM	532	$0.2\% W^{-1}$	0.0746	Cheng et al. [35]
	O ion irradiation	Planar	Pulsed	PM	532	11.5%	8.2×10^3 (peak)	Cheng et al. [35]
	C ion irradiation	Ridge	cw	PM	532	$1.12\% W^{-1}$	1.08	Chen et al. [36]
	C ion irradiation	Ridge	Pulsed	PM	532	12.4%	1.1×10^5 (peak)	Chen et al. [36]
PPLN	C ion irradiation	Ridge	cw	QPM	807.5	$77.9\% W^{-1} cm^{-2}$	0.0021	Ai et al. [19]

(continued)

Table 8.1 (continued)

Materials	Fabrication technique	Waveguide configuration	Regime	SHG configuration	λ_{SH} (nm)	η	$P_{2\omega}$ (mW)	Refs.
	He ion implantation	Planar	cw	QPM	490.5	$0.0065\% W^{-1}$	–	Vincent et al. [37]
	O ion irradiation	Channel	cw	QPM	492.5	$34.5\% W^{-1} cm^{-2}$	1.11	Huang et al. [38]
	He ion implantation	Planar	cw	QPM	532	$1.23\% W^{-1}$	2.9	Ripault et al. [39]
Zn:PPLN	He ion implantation	Planar	cw	QPM	518.5	$0.01\% W^{-1}$	0.1	Vincent et al. [40]
PPMgO:$LiNbO_3$	C ion irradiation	Ridge	cw	QPM	806	$20.3\% W^{-1} cm^{-2}$	4.15×10^{-4}	Wang et al. [41]

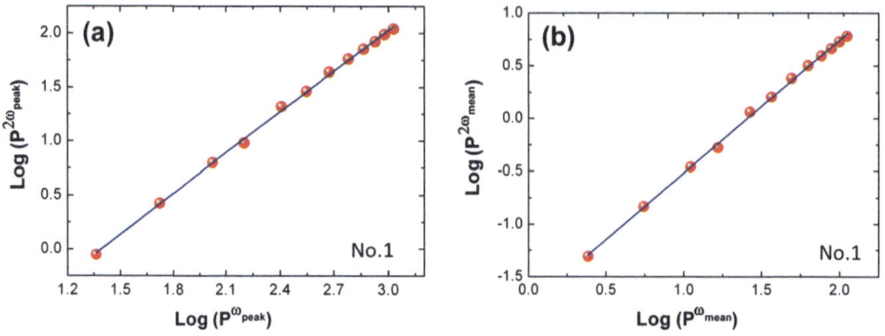

Fig. 8.6 The SH powers at 532 nm as a function of the fundamental pump powers for the 40 μm wide Nd:GdCOB ridge waveguide: **a** in peak power and **b** in mean power. Reproduced with permission [25], ©2012 Elsevier

of pulsed laser at fundamental wavelength of 1064 nm. The conversion efficiency of the cw SHG in swift heavy ion irradiated Nd:GdCOB waveguide was comparable to that of the He ion implanted ones; however, the SHG in pulsed regime was of conversion efficiency one or two orders higher of magnitude.

$KTiOPO_4$ crystals are widely used in green light SHG based on Type II PM. The ion implanted $KTiOPO_4$ waveguides are used for 1064 → 532 nm SHG under both the cw and pulsed laser pump. In the cw case, for the 17 MeV O ion implanted planar waveguide, the SH efficiency was ~0.2%W^{-1}, whilst for the ridge waveguide, the efficiency reaches 1.12%W^{-1}, which was about 6 times of that of the planar waveguide [35]. In the pulsed regime, the maximum SHG efficiency has reached 12.4% for $KTiOPO_4$ ridge waveguide with loss of ~1 dB/cm, and 11.5% for planar waveguide [36].

The β-BBO waveguides have been recently applied for SHG at UV wavelength based on Type I PM configuration [26, 27]. A maximum conversion efficiency of 0.1% W^{-1} was measured at 278 nm for cw SHG in an 8 mm long ridge waveguide corresponding to 24 μW with input power of 153 mW [26]. In addition, the SH power of about 300 μW at 266 nm was obtained at 532 nm cw fundamental laser (with input power of 650 mW), corresponding to a conversion efficiency of 0.07%W^{-1} [27].

For $LiNbO_3$ crystals with PM configuration, the He ion implanted and MgO doped planar waveguides reach 0.02%W^{-1} and 0.54%W^{-1} conversion efficiency at 567 nm SHG with pump of pulsed laser [28, 31]. Chen et al. report on SHG efficiency of 6–7% for 120 or/and 240 MeV Xe ion irradiation under pulsed 1064 nm fundamental light pump via non-critical phase matching configuration [29]. In addition, the ion sliced $LiNbO_3$ thin film was found to be with better SHG performance than the bulk [30]. PPLN waveguides are more efficient for SHG operating in diverse wavelength bands [19, 37–41]. QPM SHG of blue (490.5 nm) and green (518.5 nm) light was realized in He^+ ion implanted periodically poled $LiNbO_3$ (PPLN) planar waveguides, reaching up to 0.01%W^{-1}, which was considerably higher than that of the bulk SHG [37, 40]. Moreover, by using the C ion irradiated Mg-doped PPLN diamond saw diced ridge waveguide, the normalized SHG conversion efficiency from ~1.6 μm fundamental

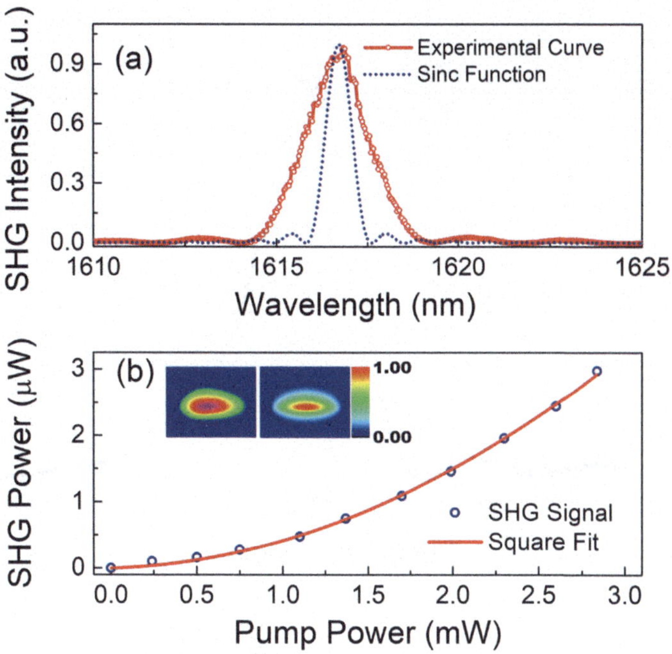

Fig. 8.7 The SHG signal versus the wavelength of the pump (**a**) and the power dependence between the SHG signal and the pump light (**b**). The mode patterns are also included in the figure. The scale bar in the inset is 2 μm. Reproduced with permission [19], ©2017 IEEE

wavelength is up to 78%W^{-1} cm^{-2}, which benefits from the low propagation loss (0.5 dB/cm) and the well-preserved nonlinearity (73% of the bulk) [19]. In addition, for O ion implanted PPLN waveguide, the SH (at 492.5 nm) efficiency reaches 34.5%W^{-1} cm^{-2} [38]. Figure 8.7 presents the SHG power versus the wavelength of the pump light in a ridge waveguide with a width of 7.38 μm after annealing at 280 °C for 30 min.

Other ion implanted nonlinear waveguides includes LBO, YCOB, etc., and guided-wave SHG in respective regimes has been realized [32–34].

Best results of efficient SHG are obtained for LNOI based waveguides. By using ridge waveguides of SiN optical loading on LNOI wafer, and the periodic domain patterns achieved by using digitated electrodes, conversion efficiency of up to 160% W^{-1} cm^{-2} have been achieved for QPM SHG from fundamental wavelength at 1.53 μm by Chang et al. [42], as indicated in Fig. 8.8.

Wang et al. from Harvard University further realized an extremely high conversion efficiency of 2600%W^{-1} cm^{-2}, in a ridge waveguide LNOI platform produced by inductively coupled plasma reactive ion etching of LNOI wafer [43]. The electron beam lithography was used to pattern the stripes for ridges. Such a high conversion efficiency benefits from the very low propagation loss of 0.03 dB/cm.

By using high-Q microring structures based on PPLNOI (periodically poled LNOI) thin-film platform, the ultrahigh conversion efficiency can be achieved. Chen

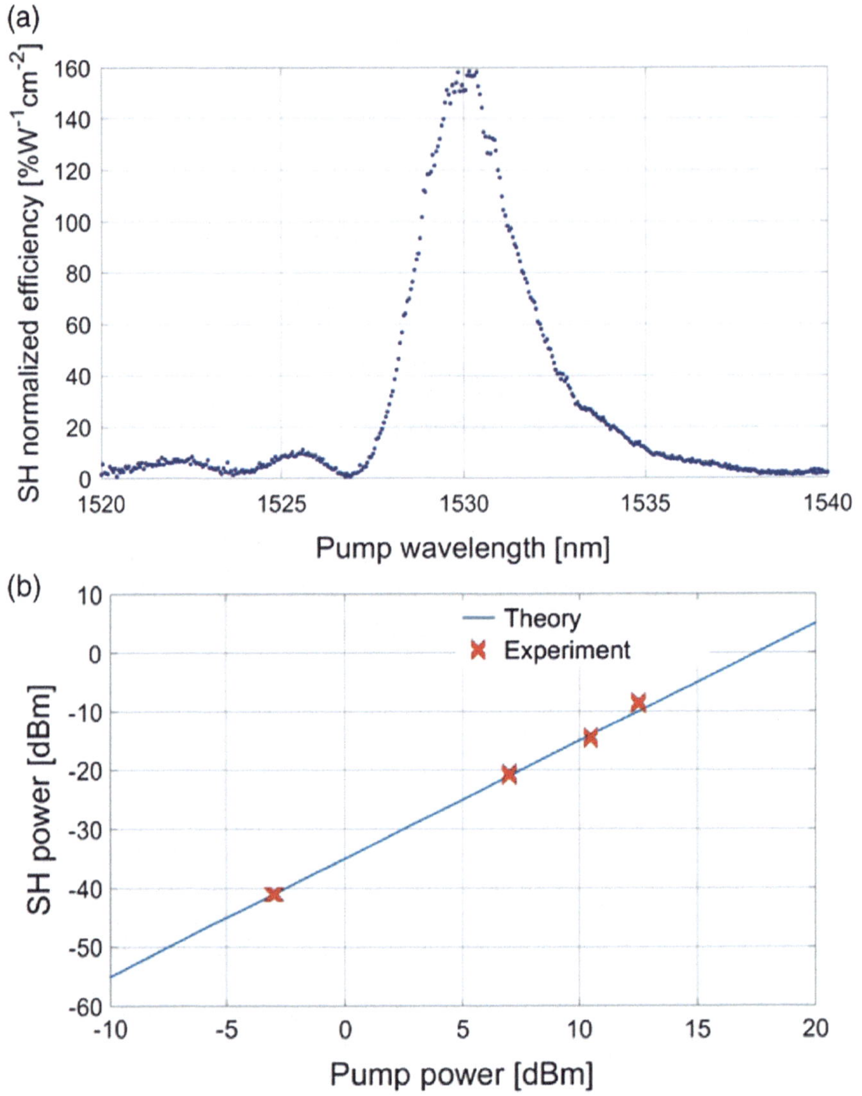

Fig. 8.8 **a** Normalized efficiency as a function of the pump wavelength for a periodically poled SiN loaded ridge waveguide in LNOI. **b** SH peak power under different pump powers (red crosses). Reproduced with permission [42], ©2016 Optical Society of America

et al. demonstrated QPM frequency conversion in a PPLNOI microring resonator, and achieved a normalized efficiency of 230,000%W^{-1} [44]. Similar results have been obtained by Lu et al., in which an on-chip SHG efficiency of up to 250,000%W^{-1} was achieved under low pump power regime at ~1617 nm [45].

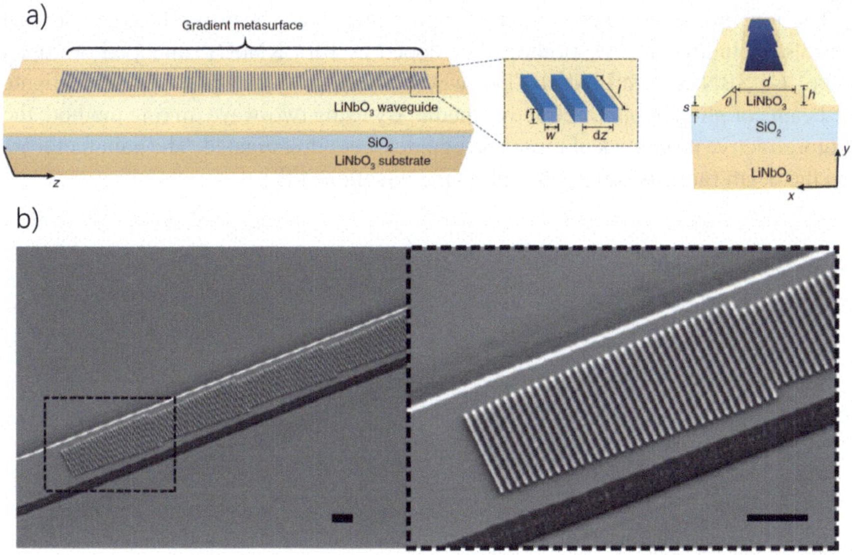

Fig. 8.9 **a** The schematic plots of the LNOI ridge waveguide with metasurfaces and **b** the SEM of the fabricated device with phased antenna arrays. Reproduced by permission from [46], ©2019 Springer-Nature

In addition to the PM that is used for frequency conversion, recently, Wang et al. realized SHG in LNOI waveguides by metasurfaces consisting of densely packed arrays of optical antennas [46]. Such gradient metasurface structures are used to circumvent the PM requirement in on-chip wavelength conversion. Under pump of light wavelength of 1580–1680 nm, the efficient SHG has been realized, and the measured conversion efficiency was up to 6×10^{-5}. Taking the short metasurface-covered device length of 19 μm into account, the normalized conversion efficiency was up to $1660\%\mathrm{W}^{-1}\ \mathrm{cm}^{-2}$. Figure 8.9 shows the schematic plots of the LNOI ridge waveguide with metasurfaces and the SEM of the fabricated device with phased antenna arrays.

8.3 Photorefractive Waveguides

In a photorefractive material, refractive index changes can be induced at relatively low optical power levels, which is of particular interest for applications in holographic storage and optical communications [47]. When using waveguide geometries, a better performance may be expected when compared to the bulk, e.g., higher gain factors in wave mixing and shorter response time caused by the high photoconductivity, which is proportional to light intensity [48].

The ion implanted photorefractive waveguides have been produced in several crystals, including $BaTiO_3$ (undoped or doped by Rh), $KNbO_3$ (undoped or doped by Fe), Ce:SBN, and Fe:$LiNbO_3$. Two wave mixing is a typical process to investigate the beam coupling in photorefractive media. By using two-wave mixing method, the photorefractive properties of these waveguides were investigated. The gain factor γ_0 and the beam ratio β_0 satisfy the following equations [49].

$$\beta_0 = \frac{P_s}{P_p} \tag{8.4}$$

$$\gamma_0 = \frac{I_s(\text{with pump})}{I_s(\text{without pump})} \tag{8.5}$$

and

$$\gamma_0 = \frac{(1+\beta_0)e^{\Gamma L}}{1+\beta_0 e^{\Gamma L}} \tag{8.6}$$

where L is the interaction length of the waves in the crystal. The exponential gain factor Γ can be calculated by reasonable fit to the experimental data.

The inverse of the (Maxwell) time constant τ is proportional to the product of specific photoconductivity $\sigma_{\text{ph},0}$ and intensity I [49].

$$\tau^{-1} = \frac{\sigma_{\text{ph},0}}{\varepsilon_0 \varepsilon_r} I \tag{8.7}$$

where ε_0 is the dielectric constant in vacuum, and ε_r is the static dielectric constant along the particular orientation of the crystals.

Figure 8.10 shows the layout of a typical two-wave mixing experimental setup [50]. Both pump and signal light beams are focused by one microscope objective and then are coupled into the waveguide, and out-coupled by another microscope objective for periodical measurement on the output powers of the two beams through a chopper.

Table 8.2 summarizes the results of two-wave mixing in ion implanted waveguides.

Benefiting from the very compact waveguide structures, higher gain coefficients as well as shorter response time than those of bulk crystals may be obtained. The light-ion implanted $BaTiO_3$ waveguides show excellent photorefractive properties [51–54]. For example, for the He ion implanted $BaTiO_3$ waveguides, the gain coefficient was as high as 58 cm^{-1} and 24 cm^{-1} at 514.5 and 854 nm, respectively [51, 52]. The response time for the grating built-up was reduced by one to three orders of magnitude. In addition, the Fe:$KNbO_3$ waveguides produced by H ion implantation also possess high gain of 40 cm^{-1} at 514.5 nm and low response time of 0.12 ms at 632.8 nm [56, 57]. Moreover, the investigation of the He ion implanted SBN waveguides at both visible and wavelength of 1.55 μm has shown that the high gain

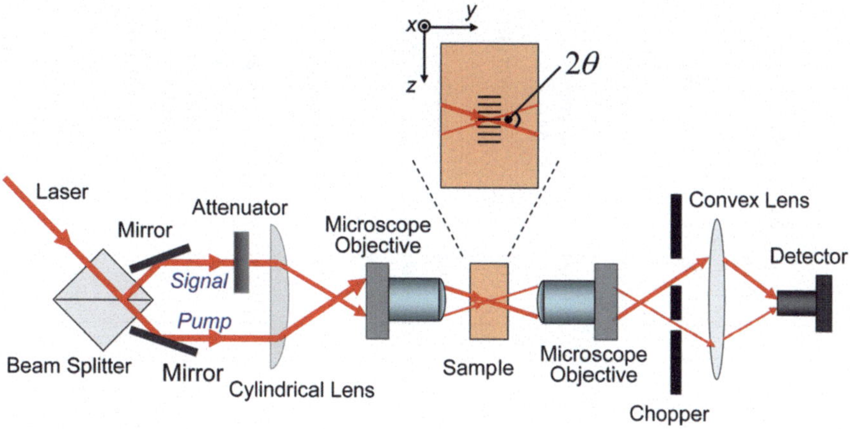

Fig. 8.10 The layout of a typical two-wave mixing experimental setup. Reproduced by permission from [50], ©2009 Springer

and short response time were obtained in the waveguides, compared with those of the bulks [58].

The two-wave mixing of the ion implanted $LiNbO_3$ waveguides has also been investigated [50, 59]. For O^{3+} ion implanted Fe:$LiNbO_3$ (congruent) waveguide, the gain coefficient of 18 cm^{-1} and response time of 5 s are obtained at 632.8 nm [50]. By using a near-stoichiometric Fe:$LiNbO_3$ waveguide produced by H^+ ion implantation, Peng et al. obtained a gain coefficient of 15 cm^{-1} at 632.8 nm (see Fig. 8.11) and response time of the gratings are almost same for waveguides and bulk [59]. This unaltered feature also suggested the photorefractive properties were well-preserved in the waveguides. Taking this advantage, some photonic structures could be fabricated. For example, by using selective illumination of either coherent (green laser) [60] or incoherent (white) light [61], one can produce reconfigurable channel waveguides and Y-branch beam splitters in ion implanted $LiNbO_3$ planar waveguide platforms (see Fig. 8.12). The refractive index changes for the waveguides are due to the spatial dark soliton-like structures: the photovoltaic effect in $LiNbO_3$ results in negative index changes in illuminated regions. The channel waveguide cores are therefore formed in the non-illuminated regions of the planar waveguide layer. These structures could work well at the infrared wavelength regimes, in which the photorefractive effects of the materials are much weak or even negligible at low power level. The homogeneous white-light illumination (e.g. in a cold white-light lamp at 100 W to illuminate the sample surface for ~30 min) will erase the formed structures completely so that the channel waveguides can be reconfigurable by new selective light illumination [60]. In addition, Villarroel et al. investigated the photorefractive properties of the swift heavy ion irradiated $LiNbO_3$ waveguides (by 20 MeV F ion irradiation), and obtained the maximum refractive index change in the waveguide was about $\sim 5 \times 10^{-5}$ under the 632.8 nm light illumination of intensity of ~1 W/cm^2 [62].

Table 8.2 Summary of two-wave mixing in ion irradiated photorefractive waveguides

Materials	Implanted ions	Energies (MeV)	Doses ($\times 10^{15}$ cm^{-2})	Gain coefficients (cm^{-1})	Response time	Pump wavelength (nm)	Pump intensity (W cm^{-2})	Refs.
$BaTiO_3$	He^+	1.8 + 1.9 + 2	2 + 2 + 2	Up to 58	10^{-3} of bulk	514.5	~20	Dazzi et al. [51]
Rh:$BaTiO_3$	He^+	1.8 + 1.9 + 2	4 + 5 + 5	Up to 24	10^{-1}–10^{-2} of bulk	854	~1000	Mathey et al. [52, 53]
$BaTiO_3$	H^+	1.5	10	–	280 μs	488	~100	Youden et al. [54]
$BaTiO_3$	H^+	1.5	10	16.5	10^{-2} of bulk	568	~100	Youden et al. [54]
$KNbO_3$	He^+	2.6	0.75	7.8	–	514.5	~100	Zha et al. [55]
$KNbO_3$	He^+	2.6	0.75	1.5	120 μs	925	~100	Zha et al. [55]
Fe:$KNbO_3$	H^+	0.6	10	Up to 40	3.3 ms	514.5	~200	Brülisauer et al. [56]
Fe:$KNbO_3$	H^+	3.0	30	2.2	60 ms	1313	~200	Brülisauer et al. [57]
Fe:$KNbO_3$	H^+	3.0	30	11	0.12 ms	632.8	~200	Brülisauer et al. [57]
Fe:$KNbO_3$	H^+	3.0	30	0.9	130 ms	1550	~200	Brülisauer et al. [57]
Ce:SBN61	He^+	2.0	0.5–5	45	2.2–10.3 ms	514.5	~200	Kip et al. [58]
Fe:SLN	H^+	0.48	40	15	5 s	632.8	~2	Peng et al. [59]
Fe:CLN	O^{3+}	6.0	0.2	18	120 s	632.8	~2	Tan et al. [50]

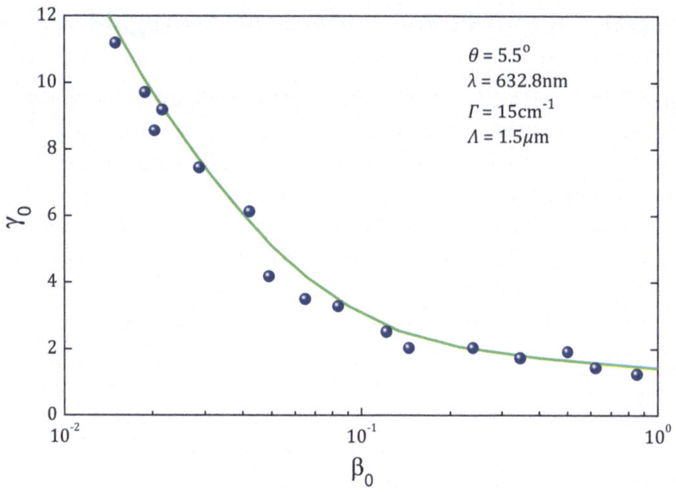

Fig. 8.11 Two wave mixing gain factor γ_0 versus incident beam power ratio β_0 at wavelength of 632.8 nm, and the intersection angle inside the simple is $2\theta = 11°$. The green line is a numerical fit to the complete set of experimental data at the gain coefficient of $\Gamma \approx 15\ \text{cm}^{-1}$. Reproduced by permission from [59], ©2011 Elsevier

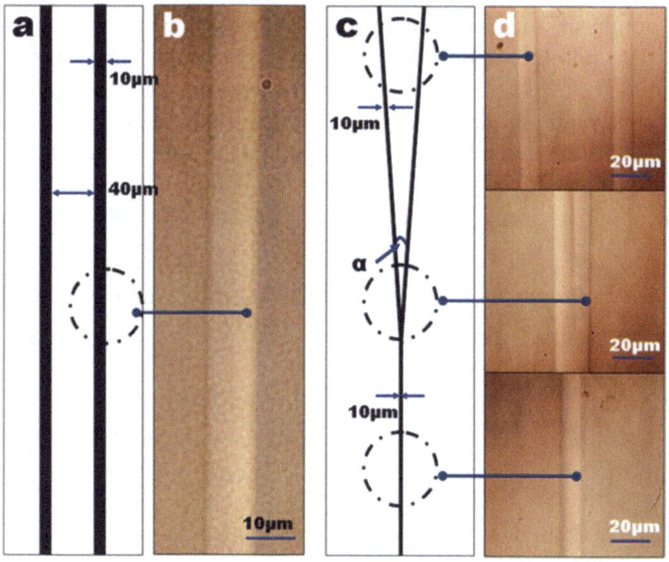

Fig. 8.12 Reconfigurable channel waveguides and Y-branch beam splitters in ion implanted $Fe:LiNbO_3$ planar waveguide platforms by selective green laser (at 532 nm) illumination. Reproduced by permission from [60], ©2008 Institute of Physics

8.4 Discrete Solitons in Waveguides and Waveguide Arrays

Another attractive application for the ion processed photorefractive waveguides is the investigation of optical spatial solitons. In the early works, Kip and colleagues observed the spatial solitons in H or He ion implanted SBN waveguides [63, 64], and realized the all-optical switches [65] based on SBN planar waveguides. Figure 8.13 shows the all optical switches achieved in a He ion implanted planar SBN waveguide by using either a chopped pump light beam at 514.5 nm or a cw light at 633 nm. At suitable optical pump power (in nonlinear case), the pump light was self-focused to a narrow beam, and, due to the diffusion of excited charge carriers, it was also self-bent, which enabled the trapped signal beam to move slowly along this bent channel with a large deflection angle of 0.23 rad [65]. In addition, Shandarov and colleagues realized the photovoltaic dark spatial soliton generation in a proton implanted planar waveguide of Cu:$LiNbO_3$ crystal. The stationary soliton regimes were achieved at powers 90 and 30 μW at wavelengths 633 and 532 nm, respectively [66].

Wave propagation in periodic media is characterized by a dispersion relation showing both pass bands and stop bands [67]. Within a linear theory, only a wave whose frequency lies outside the stop band may propagate through the system. It is shown that an intense (nonlinear) wave can change the refractive index of the periodic structure in a way to shift itself (i.e., the propagation constant) out of the stop band [68]. Such nonlinear structures reside within the band-gap as a defect and are widely known as gap solitons. Nonlinear waveguide arrays (WAs) represent one possible realization of such media possessing a periodic refractive index potential. Tan et al. observed the gap solitons (see Fig. 8.14) in the Fe:$LiNbO_3$ binary waveguide array (sometimes called photonic lattices) produced by H ion implantation and selective green laser illumination [69]. In addition, by using similar technique, the nonlinear light localization has been observed in the ion implanted $LiNbO_3$ waveguide array

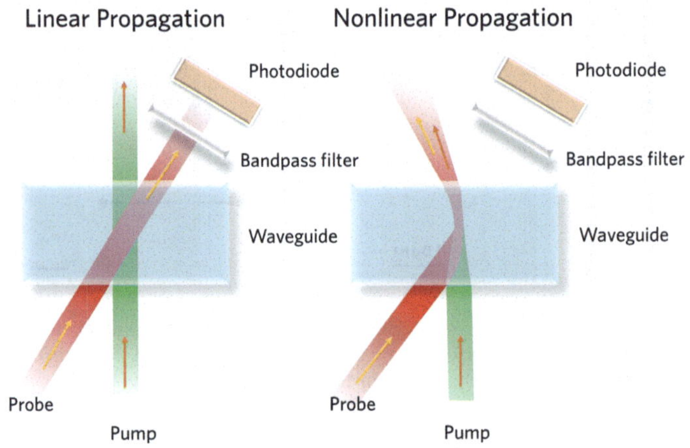

Fig. 8.13 Interaction scheme of pump (green light from an Ar^+ ion laser) and probe beam (red light from a He–Ne laser) inside the SBN waveguide

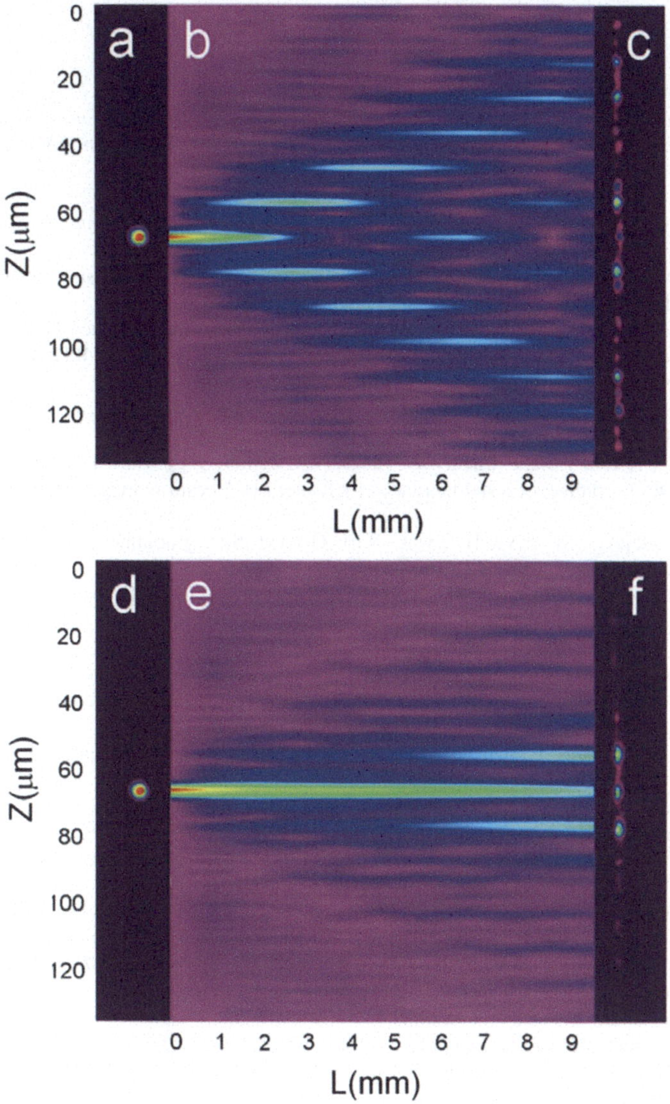

Fig. 8.14 Light propagation and modal distributions of the linear discrete diffraction (top) and the nonlinear gap soliton (bottom) from the ion-implanted Fe:$LiNbO_3$ binary waveguide array. **a** and **d** show the profile of input beams, **b** and **e** show the light propagation in the waveguide array for linear and nonlinear case, and **c** and **f** illustrate the output beam profiles for the discrete diffraction and gap solitons. Reproduced by permission from [69], ©2009 Springer

containing phase-slip defects [70]. The obtained results demonstrated the existence of various robust linear and nonlinear localized modes in $LiNbO_3$ waveguide arrays exhibiting saturable defocusing nonlinearity.

References

1. R.W. Boyd, *Nonlinear Optics* (Academic Press, 2008)
2. Z. Chen, R. Morandotti, *Nonlinear Photonics and Novel Optical Phenomena* (Springer, 2012)
3. P.F. Liao, P.L. Kelley, *Quantum Electronics: principles and Applications* (Academic Press, 1992)
4. J. Yao, Y. Wang, *Nonlinear Optics and Solid-State Lasers* (Springer, 2012)
5. P.D. Townsend, P. J. Chandler, L. Zhang, *Optical Effects of Ion Implantation* (Cambridge University Press, 1994)
6. F. Chen, Micro-and submicrometric waveguiding structures in optical crystals produced by ion beams for photonic applications. Laser Photonics Rev. **6**, 622 (2012)
7. N. Dong, Y. Tan, A. Benayas, J.R.V. de Aldana, D. Jaque, C. Romero, F. Chen, Q. Lu, Femtosecond laser writing of multifunctional optical waveguides in a Nd:YVO_4+KTP hybrid system. Opt. Lett. **36**, 975 (2011)
8. N. Dong, F. Chen, D. Jaque, Q. Lu, Micro second harmonic and Raman spectra of He^+ implanted $KTiOPO_4$ waveguides. Opt. Express **19**, 13934 (2011)
9. N. Dong, Y. Yao, Y. Jia, F. Chen, S.K. Vanga, A.A. Bettiol, Q. Lu, Buried channel waveguides in $KTiOPO_4$ nonlinear crystal fabricated by focused He^+ beam writing. Opt. Mater. **35**, 184 (2012)
10. A.M. Radojevic, M. Levy, H. Kwak, R.M. Osgood, Strong nonlinear optical response in epitaxial liftoff single-crystal $LiNbO_3$ films. Appl. Phys. Lett. **75**, 2888 (1999)
11. A.M. Radojevic, M. Levy, R.M. Osgood, D.H. Jundt, A. Kumar, H. Bakhru, Second-order optical nonlinearity of 10 μm-thick periodically poled $LiNbO_3$ films. Opt. Lett. **25**, 1034 (2000)
12. R. Kremer, A. Boudrioua, P. Moretti, J.C. Loulergue, Measurements of the non-linear d_{33} coefficients of light-ion implanted lithium niobate by second harmonic generation in total reflection geometry. Opt. Commun. **219**, 389 (2003)
13. J. Rams, J. Olivares, P.J. Chandler, P.D. Townsend, Second harmonic generation capabilities of ion implanted $LiNbO_3$ waveguides. J. Appl. Phys. **84**, 5180 (1998)
14. J. Olivares, A. García -Navarro, A. Méndez, F. Agulló-López, G. García, A. Garcıa-Cabanes, M. Carrascosa, Nucl. Instrum. Meth. Phys. Res. B **257**, 765 (2007)
15. F. Chen, Photonic guiding structures in lithium niobate crystals produced by energetic ion beams. J. Appl. Phys. **106**, 081101 (2009)
16. J. Olivares, M.L. Crespillo, O. Caballero-Calero, M.D. Ynsa, A. García-Cabañes, M. Toulemonde, C. Trautmann, F. Agulló-López, Thick optical waveguides in lithium niobate induced by swift heavy ions (~10 MeV/amu) at ultralow fluences. Opt. Express **17**, 24175 (2009)
17. J. Olivares, A. García-Navarro, G. García, A. Méndez, F. Agulló-López, A. García-Cabañes, M. Carrascosa, O. Caballero, Nonlinear optical waveguides generated in lithium niobate by swift-ion irradiation at ultralow fluences. Opt. Lett. **32**, 2587 (2007)
18. J. Olivares, G. García, A. Garcıa, F. Agulló-López, O. Caballero, A. García-Cabanes, Appl. Phys. Lett. **86**, 183501 (2005)
19. L. Ai, L. Wang, Y. Tan, S. Akhmadaliev, S. Zhou, F. Chen, Efficient second harmonic generation of diced ridge waveguides based on carbon ion-irradiated periodically poled $LiNbO_3$. J. Lightwave Technol. **35**, 2476 (2017)
20. L. Ai, L. Wang, X. Zhang, C. Chen, F. Chen, Depth profile of the nonlinear susceptibility of $LiNbO_3$ ridge waveguides fabricated by ion implantation and dicing. Opt. Mater. Express **7**, 3836 (2017)
21. D. Fluck, P. Gunter, Second-harmonic generation in potassium niobate waveguides. IEEE J. Sel. Top. Quant. Electron. **6**, 122 (2000)
22. A. Boudrioua, J.C. Loulergue, P. Moretti, B. Jacquier, G. Aka, D. Vivien, Second-harmonic generation in He^+-implanted gadolinium calcium oxoborate planar waveguides. Opt. Lett. **24**, 1299 (1999)

23. B. Vincent, A. Boudrioua, J.C. Loulergue, P. Moretti, S. Tascu, B. Jacquier, G. Aka, D. Vivien, Channel waveguides in $Ca_4GdO(BO_3)_3$ fabricated by He^+ implantation for blue-light generation. Opt. Lett. **28**, 1025 (2003)
24. Y. Ren, Y. Jia, F. Chen, Q. Lu, S. Akhmadaliev, S. Zhou, Second harmonic generation of swift carbon ion irradiated Nd:GdCOB waveguides. Opt. Express **19**, 12490 (2011)
25. Y. Jia, F. Chen, J.R.V. de Aldana, S. Akhmadaliev, S. Zhou, Femtosecond laser micromachining of Nd:GdCOB ridge waveguides for second harmonic generation. Opt. Mater. **34**, 1913 (2012)
26. G. Poberaj, R. Degl'Innocenti, C. Medrano, P. Günter, UV integrated optics devices based on beta-barium borate, Opt. Mater. **31**, 1049 (2009)
27. R. Degl'Innocenti, A. Majkic, F. Sulser, L. Mutter, G. Poberaj, P. Günter, UV second harmonic generation at 266 nm in He^+ implanted β-BaB_2O_4 optical waveguides. Opt. Express **16**, 11660 (2008)
28. N. Hamelin, G. Lifante, P.J. Chandler, P.D. Townsend, S. Pityana, A.J. Mccaffery, Second harmonic generation in ion implanted lithium niobate planar waveguides. J. Mod. Opt. **41**, 1339 (1994)
29. C. Chen, L. Pang, Q. Lu, L. Wang, Y. Tan, Z. Wang, F. Chen, Refractive index engineering through swift heavy ion irradiation of $LiNbO_3$ crystal towards improved light guidance. Sci. Rep. **7**, 10805 (2017)
30. O. Gaathon, A. Ofan, D. Djukic, J. Dadap, R. M. Osgood, S. Bakhru, H. Bakhru, *Proceedings of the Conference on Lasers and Electro-Optics/Quantum Electronics and Laser Science Conference and Photonic Applications Systems Technologies, OSA Technical Digest (CD)* (Optical Society of America, San Jose, 2008), paper CWG6
31. N. Hamelin, Y.T. Chow, Guided-type second harmonic generation in ion implanted $MgO:LiNbO_3$. J. Mod. Opt. **45**, 2125 (1998)
32. G.M. Davis, L. Zhang, P.J. Chandler, P.D. Townsend, Planar and channel waveguide fabrication in LiB_3O_5 using MeV He^+ ion implantation. J. Appl. Phys. **79**, 2863 (1996)
33. Y. Ren, Y. Jia, N. Dong, L. Pang, Z. Wang, Q. Lu, F. Chen, Guided-wave second harmonics in Nd:YCOB optical waveguides for integrated green lasers. Opt. Lett. **37**, 244 (2012)
34. R. Li, W. Nie, Z. Shang, C. Cheng, S. Akhmadaliev, S. Zhou, Q. Lu, F. Chen, Guided-wave second harmonics in Nd:YCOB ridge waveguides produced by combination of carbon ion irradiation and precise diamond blade dicing. Opt. Mater. **57**, 153 (2016)
35. Y. Cheng, Y. Jia, S. Akhmadaliev, S. Zhou, F. Chen, Guided-wave phase-matched second-harmonic generation in $KTiOPO_4$ waveguide produced by swift heavy-ion irradiation. Opt. Eng. **53**, 117102 (2014)
36. C. Chen, C.E. Ruter, M.F. Volk, C. Cheng, Z. Shang, Q. Lu, S. Akhmadaliev, S. Zhou, F. Chen, D. Kip, Second harmonic generation of diamond-blade diced $KTiOPO_4$ ridge waveguides. Opt. Express **24**, 16434 (2016)
37. B. Vincent, A. Boudrioua, R. Kremer, P. Moretti, Second harmonic generation in helium-implanted periodically poled lithium niobate planar waveguides. Opt. Commun. **247**, 461 (2005)
38. Q. Huang, P. Liu, T. Liu, L. Zhang, Y. Zhou, X. Wang, Second harmonic generation in periodically poled $LiNbO_3$ waveguides formed by oxygen-ion implantation. Phys. Status Solidi-R. **6**, 205 (2012)
39. Q. Ripault, M.W. Lee, F. Mériche, T. Touam, B. Courtois, E. Ntsoenzok, L. Peng, A. Fischer, A. Boudrioua, Investigation of a planar optical waveguide in 2D PPLN using Helium implantation technique. Opt. Express **21**, 7202 (2013)
40. B. Vincent, R. Kremer, A. Boudrioua, P. Moretti, Y.C. Zhang, C.C. Hsu, L.H. Peng, Green light generation in a periodically poled Zn-doped $LiNbO_3$ planar waveguide fabricated by He^+ implantation. Appl. Phys. B **89**, 235 (2007)
41. L. Wang, C.E. Haunhorst, M.F. Volk, F. Chen, D. Kip, Quasi-phase-matched frequency conversion in ridge waveguides fabricated by ion implantation and diamond dicing of $MgO:LiNbO_3$ crystals. Opt. Express **23**, 30188 (2015)
42. L. Chang, Y. Li, N. Volet, L. Wang, J. Peters, J.E. Bowers, Thin film wavelength converters for photonic integrated circuits. Optica **3**, 531 (2016)

43. Cheng Wang, Carsten Langrock, Alireza Marandi, Marc Jankowski, Mian Zhang, Boris Desiatov, Martin M. Fejer, Marko Lončar, Ultrahigh-efficiency wavelength conversion in nanophotonic periodically poled lithium niobate waveguides. Optica **5**, 1438 (2018)
44. J. Chen, Z. Ma, Y. Sua, Z. Li, C. Tang, Y. Huang, Ultra-efficient frequency conversion in quasi-phase-matched lithium niobate microrings. Optica **6**, 1244 (2019)
45. J. Lu, J.B. Surya, X. Liu, A.W. Bruch, Z. Gong, Y. Xu, H.X. Tang, Periodically poled thin-film lithium niobate microring resonators with a second-harmonic generation efficiency of 250,000%/W. Optica **6**, 1455 (2019)
46. C. Wang, Z. Li, M.H. Kim, X. Xiong, X. Ren, G. Guo, N. Yu, M. Lončar, Metasurface-assisted phase-matching-free second harmonic generation in lithium niobate waveguides. Nat. Commun. **8**, 2098 (2017)
47. P. Günter, J.P. Huignard, *Photorefractive Materials and Their Applications 1: basic Effects* (Springer, 2006)
48. D. Kip, Photorefractive waveguides in oxide crystals: fabrication, properties, and applications. Appl. Phys. B **67**, 131 (1998)
49. P. Günter, J.P. Huignard, *Photorefractive Materials and Their Applications*, vol. 2 (Springer 2007)
50. Y. Tan, F. Chen, D. Kip, Photorefractive properties of optical waveguides in Fe:$LiNbO_3$ crystals produced by O^{3+} ion implantation. Appl. Phys. B **94**, 467 (2009)
51. A. Dazzi, P. Mathey, P. Lompré, P. Jullien, P. Moretti, D. Rytz, High performance of two-wave mixing in a $BaTiO_3$ waveguide realized by He^+ implantation. J. Opt. Soc. Am. B **16**, 1915 (1999)
52. P. Mathey, A. Dazzi, P. Jullien, D. Rytz, P. Moretti, Guiding properties and nonlinear wave mixing at 854 nm in a rhodium-doped $BaTiO_3$ waveguide implanted with He^+ ions. J. Opt. Soc. Am. B **18**, 344 (2001)
53. P. Mathey, A. Dazzi, P. Jullien, D. Rytz, P. Moretti, Two-wave mixing at 854 nm in $BaTiO_3$: Rh planar waveguide implanted with He^+. Opt. Mater. **18**, 69 (2001)
54. K.E. Youden, S.W. James, R.W. Eason, P.J. Chandler, L. Zhang, P.D. Townsend, Photorefractive planar waveguides in $BaTiO_3$ fabricated by ion-beam implantation. Opt. Lett. **17**, 1509 (1992)
55. M. Zha, D. Fluck, P. Günter, M. Fleuster, C. Buchal, Two-wave mixing in photorefractive ion-implanted $KNbO_3$ planar waveguides at visible and near-infrared wavelengths. Opt. Lett. **18**, 577 (1993)
56. S. Brülisauer, D. Fluck, P. Günter, High gain two-wave mixing in H^+ implanted photorefractive Fe:$KNbO_3$ planar waveguides. Electron. Lett. **31**, 312 (1995)
57. S. Brülisauer, D. Fluck, P. Günter, L. Beckers, C. Buchal, Photorefractive effect in proton-implanted Fe-doped $KNbO_3$ waveguides at telecommunication wavelengths. J. Opt. Soc. Am. B **13**, 2544 (1996)
58. D. Kip, B. Kemper, I. Nee, R. Pankrath, P. Moretti, Photorefractive properties of ion-implanted waveguides in strontium barium niobate crystals. Appl. Phys. B **65**, 511 (1997)
59. B. Peng, Y. Tan, F. Chen, D. Kip, Two-wave mixing of ion-implanted photorefractive waveguides in near-stoichiometric Fe:$LiNbO_3$ crystals. Opt. Mater. **33**, 773 (2011)
60. Y. Tan, F. Chen, X.L. Wang, L. Wang, V. Shandarov, D. Kip, Formation of reconfigurable optical channel waveguides and beam splitters on top of proton-implanted lithium niobate crystals using spatial dark soliton-like structures. J. Phys. D **41**, 102001 (2008)
61. Y. Tan, F. Chen, M. Stepic, V. Shandarov, D. Kip, Reconfigurable optical channel waveguides in lithium niobate crystals produced by combination of low-dose O^{3+} ion implantation and selective white light illumination. Opt. Express **16**, 10465 (2008)
62. J. Villarroel, M. Carrascosa, A. García-Cabañes, O. Caballero-Calero, M. Crespillo, J. Olivares, Photorefractive response and optical damage of $LiNbO_3$ optical waveguides produced by swift heavy ion irradiation. Appl. Phys. B **95**, 429 (2009)
63. D. Kip, M. Wesner, C. Herden, V. Shandarov, Interaction of spatial photorefractive solitons in a planar waveguide. Appl. Phys. B **68**, 971 (1999)
64. D. Kip, M. Wesner, V. Shandarov, P. Moretti, Observation of bright spatial photorefractive solitons in a planar strontium-barium niobate waveguide. Opt. Lett. **23**, 921 (1998)

65. D. Kip, M. Wesner, E. Krätzig, V. Shandarov, P. Moretti, All-optical beam deflection and switching in strontium-barium-niobate waveguides. Appl. Phys. Lett. **72**, 1960 (1998)
66. V.G. Kruglov, V.M. Shandarov, Y. Tan, F. Chen, D. Kip, Dark photovoltaic spatial solitons in a planar waveguide obtained by proton implantation in lithium niobate. Quant. Electron. **38**, 1045 (2008)
67. D.N. Christodoulides, F. Lederer, Y. Silberberg, Discretizing light behavior in linear and nonlinear waveguide lattices. Nature **424**, 817 (2003)
68. Y. Eisenberg, Y. Silberberg, R. Morandotti, A.R. Boyd, J.S. Aitchison, Discrete spatial optical solitons in waveguide arrays, Phys. Rev. Lett. **81**, 3383 (1998)
69. Y. Tan, F. Chen, P.P.P.P. Beličev, M. Stepić, A. Maluckov, C.E. Rüter, D. Kip, Gap solitons in defocusing lithium niobate binary waveguide arrays fabricated by proton implantation and selective light illumination. Appl. Phys. B **95**, 531 (2009)
70. P.P. Beličev, I. Ilić, M. Stepić, A. Maluckov, Y. Tan, F. Chen, Observation of linear and nonlinear strongly localized modes at phase-slip defects in one-dimensional photonic lattices. Opt. Lett. **35**, 3099 (2010)

Chapter 9
Lasing Based on Dielectric Waveguides

Abstract Waveguide lasers are miniature light sources that possess high efficiency and low lasing thresholds. In this chapter, we will demonstrate the techniques applied to achieve solid-state waveguide lasers in ion implanted active waveguides, and introduce the major progress in this area. Section 9.1 presents a description of waveguide lasers and reviews the progress on waveguide lasing at near infrared wavelength bands. Section 9.2 is devoted to the pulsed waveguide lasers based on 2D materials that is growing rapidly. In Sect. 9.3, a specific branch of self-frequency doubled waveguide lasing is introduced, which combines the lasers and nonlinear optics.

The generic nature of solid-state waveguide laser was recognized in the early 1960s, soon after the first laser was built [1]. At the time when considerable efforts were put on the laser performance optimization by strengthening the optical confinement and hence upgrading the intracavity light intensity, the first waveguide laser timely came into being. Originating from the design philosophy in pursue of integration and miniaturization, waveguide lasers based on solid-state gain media, generally referring to the rare-earth (RE) and transition-metal ion doped glassy and crystalline host materials, are nowadays becoming a very active field as implied by the increase in the quantity and diversity of the emerging applications of such lasers over the last years [2–4]. The greatest interest in such sources largely stems from the compact geometry of waveguide structures, which allows for optical energy storage within extremely compressed volumes with dimensions of micrometric or sub-micrometric scales, reaching considerably high optical intracavity intensities with respect to that in bulks materials. Benefiting from this, small and robust laser sources with enhanced optical gain, reduced lasing threshold, and miniature footprint can therefore be implemented. Unquestionably, being provided with miniaturization, integration, and multi-functionality, solid-state waveguide laser is opening exciting possibilities for multifunctional photonic integrated circuits and lab-on-a-chip platforms.

F. Chen et al., *Ion Irradiation of Dielectrics for Photonic Applications*, Springer Series in Optical Sciences 231, https://doi.org/10.1007/978-981-15-4607-5_9

Ion implanted waveguide lasers were firstly realized in 1990s based on He or H ion implanted Nd:YAG, Nd:GGG, and Nd:MgO:$LiNbO_3$ waveguides [5–7]. Recently, the laser technology offers new opportunities to construct much higher performance waveguide laser systems that operate in both continuous wave (cw) and pulsed regimes. And the lasing wavelength can be tunable based on the feature of bulk systems.

9.1 Waveguide Lasing at Near Infrared Wavelength Regimes

The waveguide lasers operating at near infrared wavelength bands receive most attentions due to their applications in various areas. The construction of waveguide laser systems in ion implanted waveguides is mainly for lasing at NIR band. In this subsection, we focus on the cw waveguide lasing at 1 μm wavelength.

The waveguide lasers are typically realized in end-pump geometry. Figure 9.1 illustrates the schematic of a typical waveguide laser system, which mainly consists of a pump source, an active waveguide cavity, and a few optical elements to adjust the light behaviors. The pump source for the 1 μm waveguide lasers is usually a diode laser (e.g., at 808 or 980 nm) or a wavelength-tunable Ti:sapphire laser. Other

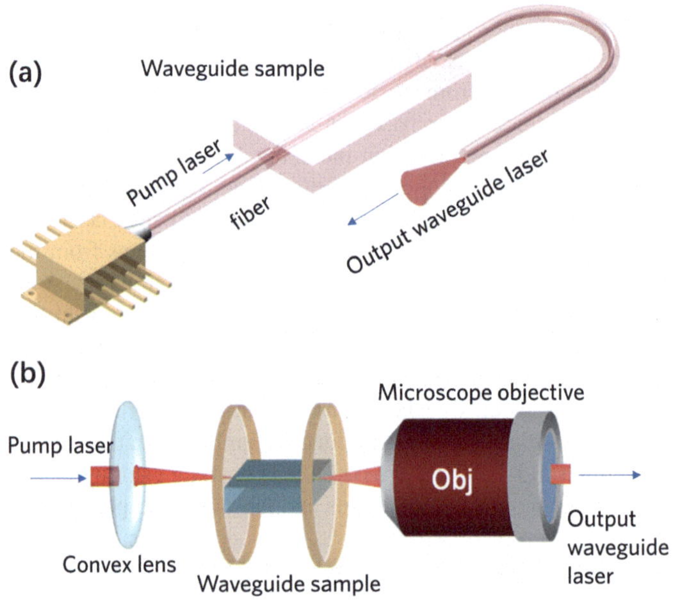

Fig. 9.1 The schematic plot of waveguide laser geometry: input and output coupling **a** by optical fiber and **b** by convex lens

solid-state lasers with single waveguide lasing may be also suitable due to the main absorption wavelength band of the gain media. For low-cost waveguide laser systems, the diode lasers are more favorite, whilst for the first-step realization of waveguide lasers, solid-state lasers with high-quality beam profiles are able to provide more efficient coupling of pump beam into waveguides.

The key parameters for a waveguide laser mainly include lasing wavelength, threshold, slope efficiency, maximum output power, and beam quality. For a four-level system, the lasing threshold and slope efficiency can be expressed as [3].

$$P_{\text{th}} = \frac{hc}{\lambda_P} \frac{1}{\eta \sigma_e \tau} \frac{\delta}{2} A_{\text{eff}} \tag{9.1}$$

$$\Phi = \eta \frac{(T_1 + T_2)}{\delta} \frac{\lambda_P}{\lambda_L} \tag{9.2}$$

where h is the Planck's constant, c is the light velocity in the vacuum, λ_L and λ_P are the wavelengths of the laser and pump beams, respectively, σ_e is the stimulated emission cross section, τ is the fluorescence lifetime, A_{eff} is the effective pump area, η is the fraction of absorbed photons, and T_1, T_2 are the transmittance of the end-faces, δ is the round-trip cavity loss exponential factor.

Table 9.1 summarizes the obtained results for solid-state waveguide lasers from the ion-beam processed active waveguides. The applied laser media for waveguide lasers include Nd:YAG, Nd:GGG, Nd:YVO_4, etc. For an efficient waveguide laser system, one of the advantages is the low pump threshold. For Nd:YAG waveguide system [5, 8–24], the lowest threshold for Nd:YAG single-crystal waveguide system was only 1.6 mW [5]. However, the maximum output of the generated laser at 1064 nm was also as low as 4.6 mW. This is because the coupling efficiency of the pump beam with the waveguide mode is relatively low. The coupling efficiency η_c can be determined by calculating the overlap integral between pump and waveguide mode [34].

$$\eta_c = \frac{\left| \int_{-\infty}^{+\infty} \mathrm{d}x\mathrm{d}y \psi_1(x, y) \psi_2^*(x, y) \right|^2}{\int_{-\infty}^{+\infty} \mathrm{d}x\mathrm{d}y |\psi_1(x, y)|^2 \cdot \int_{-\infty}^{+\infty} \mathrm{d}x\mathrm{d}y |\psi_2(x, y)|^2} \tag{9.3}$$

where $\psi_1(x, y)$ and $\psi_2(x, y)$ refer to the electric field of pump beam and waveguide mode. The pump beam can be in fiber-coupled or free space light focused by convex lens. In this case, the z-axis is assumed to be the laser propagation orientation.

There is a balance for all the parameters to achieve optimized waveguide laser system. The most remarkable Nd:YAG waveguide laser was produced by proton beam writing, in which an output power of 60 mW at 1064 nm was realized [14]. Also, in this system the guided modes are highly symmetric (nearly circular shapes), which enables it easily connected with the optical fibers (see Fig. 9.2 for modal profile of the laser beam, the laser emission spectrum, and the measured output laser powers data). In addition, for the first ion implanted Nd:YAG ceramic planar waveguide, the 1064 nm laser was generated with a 19.5 mW threshold and 11% slope efficiency

Table 9.1 The obtained results for waveguide lasers from the ion-processed laser crystal waveguides

Materials	Fabrication technique	Waveguide configuration	Working wavelength (nm)	Pump threshold (mW)	Slope efficiency (%)	Max. output power (mW)	Refs.
Nd:YAG	H ion implantation	Planar	1064	50	21%	16.1	Flores-Romero, et al. [8]
	He ion implantation	Channel	1064	1.6	29%	4.6	Field et al. [5]
	H ion implantation	Planar	1064	11	8%	5	Domenech et al. [9]
	C ion implantation	Planar	1064	22	8%	4	Domenech et al. [9]
	H ion implantation	Channel	1064	6.9	16%	7	Flores-Romero et al. [10]
	H ion implantation	Planar	1338	62	13%	3.5	Domenech et al. [11]
	Ar ion irradiation	Planar	1064	26	5.9%	2.9	Ren et al. [12]
	N ion irradiation	Planar	1064	79	16%	23	Ren et al. [13]
	H beam writing	Buried channel	1064	94	40%	63	Yao et al. [14]
	C ion implantation	Planar	1064	19.5	11%	2.5	Tan et al. [15]
	O ion irradiation and femtosecond laser ablation	Ridge	1064	39	35%	21	Jia et al. [16]
	C ion irradiation and femtosecond laser ablation	Ridge	1064	79	43%	84	Jia et al. [17]

(continued)

Table 9.1 (continued)

Materials	Fabrication technique	Waveguide configuration	Working wavelength (nm)	Pump threshold (mW)	Slope efficiency (%)	Max. output power (mW)	Refs.
	Kr ion irradiation and femtosecond laser ablation	Ridge	1064	215	25%	182	Jia et al. [18]
	C ion irradiation and femtosecond laser ablation	Ridge	1064	65	56%	72	Tan et al. [19]
	C ion irradiation and femtosecond laser ablation	Ridge	1064	110	38%	36	Yao et al. [20]
	C ion irradiation and femtosecond laser writing	Cladding	1064	24.4	27.8%	42	Shang et al. [21]
	C ion irradiation	Planar	1064	43	10%	10	Shang et al. [22]
	C ion irradiation and diamond blade dicing	Ridge	1064	60	46.1%	30	Shang et al. [22]
Nd:YAG ceramic	He ion implantation	Planar	1064	133.4	17.4%	13	Jia et al. [23]
	He ion implantation	Ridge	1064	64.9	42.5%	46	Jia et al. [23]
	C ion irradiation	Planar	1064	50	36%	300	Tan et al. [24]

(continued)

Table 9.1 (continued)

Materials	Fabrication technique	Waveguide configuration	Working wavelength (nm)	Pump threshold (mW)	Slope efficiency (%)	Max. output power (mW)	Refs.
Nd:GGG	H ion implantation	Planar	1062	49.3	30%	21.9	Ren et al. [25]
	He ion implantation	Channel	1062	6.6	27%	4	Field et al. [7]
	H beam writing	Buried channel	1064	60.8	66%	25	Yao et al. [26]
	C ion irradiation and femtosecond laser ablation	Ridge	1063	71.6	41.8%	25.6	Jia et al. [27]
Nd:YVO_4	C ion implantation	Channel	1064	45	30%	9	Sánchez-Morales et al. [28]
	Ar ion irradiation	Planar	1067	9.3	8.5%	3.1	Yao et al. [29]
Ti:Sapphire	H ion implantation	Channel	780	230	3%	17	Grivas et al. [30]
Nd:YCOB	Ar ion irradiation	Planar	1061	31.5	68%	35	Ren et al. [31]
	Ar ion irradiation	Planar	531	17	–	0.037	Ren et al. [31]
Nd:YAP	He ion implantation	Planar	1079	50	–	–	Field et al. [32]
Nd:MgO:$LiNbO_3$	He ion implantation	Planar	1085	8.3	–	–	Field et al. [6]
Yb:YSGG	C ion irradiation and diamond blade dicing	Ridge	1023.6	130.5	46%	50	Ma et al. [33]

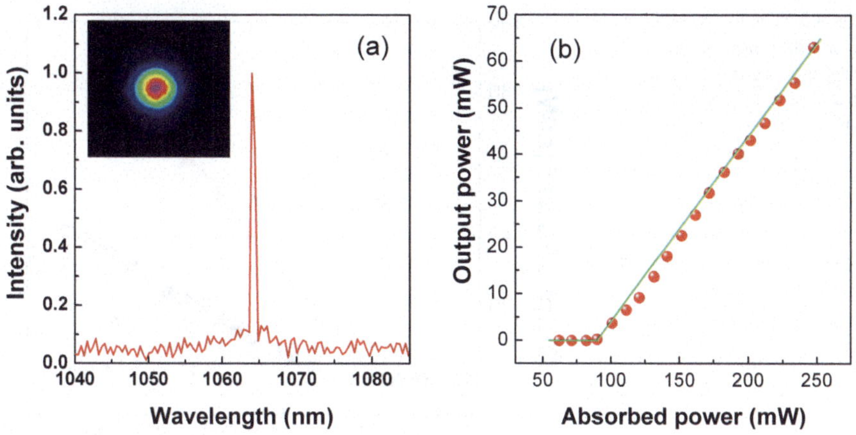

Fig. 9.2 **a** Modal profile of the laser beam and the laser emission spectrum of the proton-beam-written Nd:YAG channel waveguide. **b** The measured output laser power as a function of the absorbed pump power (balls) from the waveguide. The green solid line shows the linear fit of the experimental data. Reproduced with permission from [14], ©2010 Optical Society of America

[15]. It should be pointed out that the proton beam written Nd:YAG waveguide laser possesses the smallest beam scale (FWHM width of ~2 μm) for the solid-state laser systems.

Swift ion irradiated Nd:YAG waveguide laser systems in early works have similar performances with the ion implanted samples [12, 13]. Ren et al. reported on the first cw laser in Nd:YAG planar waveguide produced by swift heavy ion irradiation (60 MeV Ar^{4+} at 2×10^{12} ions/cm^2) [13]. Under the optical pump at 808 nm, the power threshold for the laser oscillation is ~26 mW, and the slope efficiency of ~6% can be achieved. The achieved maximum laser power at 1064 nm is ~2.9 mW for the absorbed pump power of ~77 mW, leading to an optical conversion efficiency of ~3.8%. In addition, the optimization of waveguide system significantly improves the laser performance. By using geometries of ridge waveguides, i.e., the surface patterned planar waveguides, the laser performance and beam quality could be further improved due to the more compact waveguide volume. Jia et al. used a diamond saw diced Nd:YAG ridge waveguide as gain medium, and achieve slope efficiency of 43% and lasing thresholds for the 1064 nm oscillations of 79 mW, respectively [17]. Figure 9.3 shows the cw waveguide laser output powers at 1064 nm as a function of the absorbed light power at 808 nm for the Nd:YAG ridge waveguides. These values are superior to those of the related planar waveguide partner (20% and 111 mW) in Nd:YAG produced by 15 MeV C^{5+} ion irradiation at a fluence of 3×10^{14} ions/cm^2. With a high optical conversion efficiency of 32.8%, the maximum output power of the Nd:YAG ridge waveguide laser reaches 84 mW, which is much higher than that in early works. It should be also noted that the roughness of sidewalls of the ridges is closely correlated to the propagation losses of the waveguides. The diamond saw dicing creates much smooth sidewalls (roughness of ~10 nm), whilst

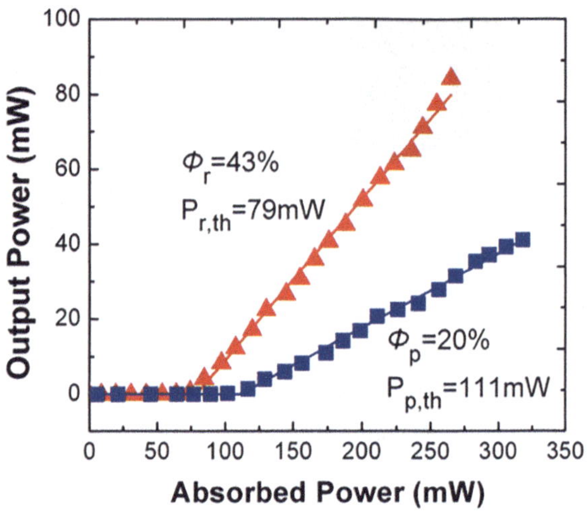

Fig. 9.3 The cw waveguide laser output powers at 1064 nm as a function of the absorbed light power at 808 nm. The triangular and rectangular symbols stand for the data of ridge and channel waveguides, respectively. The solid lines represent the linear fit of the experimental data. Reproduced with permission from [17], ©2013 Optical Society of America

the femtosecond laser ablation causes 0.5–4 μm roughness of sidewalls. This also reflects on the waveguide laser performance.

The parameters of swift heavy ions are crucial factors for the waveguide properties. By using 670 MeV Kr^{8+} ion irradiation, a 50 μm thick planar waveguide can be produced in Nd:YAG crystal [18]. Combined by the following femtosecond laser ablation, a 50 × 50 μm^2 cross section of ridge waveguide can be finally formed. The 182 mW maximum output power at 1064 nm is achieved in this system, whilst the lasing threshold is as high as 215 mW. The improvement of output power and drawback of threshold is the result of the dimension effect of waveguide geometry. Larger area of the cross section improves the coupling efficiency of pump beam and waveguide modes, but the optical intensity decreases considerably in the larger-volume waveguides.

The recently reported H implanted Nd:GGG planar waveguide laser system reached higher slope efficiency (30%) and output power (22 mW) [25] than the early demonstrated He ion implanted Nd:GGG channel waveguide system [7]. The most attractive system was the proton beam written Nd:GGG buried channel waveguide, in which a very high slope efficiency of 66% and a maximum output power of 25 mW were obtained [26]. The significant improvement may be owing to the low prorogation loss and the well-matched modal profile with the pump beam field.

Jia et al. investigated on the laser performance in a Nd:GGG ridge waveguide produced by 17 MeV C^{5+} ion irradiation and subsequent femtosecond laser ablation [27]. For comparison, the planar and ridge waveguide laser s are developed in the same crystal. The lasing thresholds are 102.7 mW and 71.6 mW for Nd:GGG planar and ridge waveguides, respectively. And the slope efficiencies are 11.5% and 41.8%, respectively. The maximum output laser power of these two waveguides is 5.9 and 25.6 mW at absorbed pump powers of 151 and 113 mW, corresponding to optical conversion efficiency of 4% and 23%, respectively. One can clearly conclude that

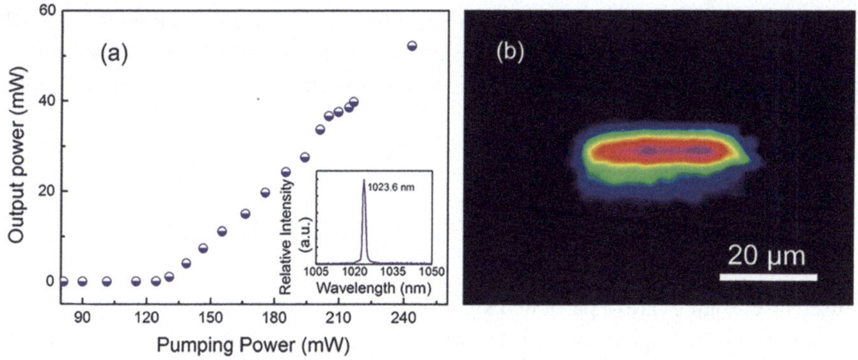

Fig. 9.4 **a** The output power of the waveguide laser at 1023.6 nm as a function of the pump power at 946 nm. The inset shows the spectrum of the output laser. **b** The recorded near-field modal profile of the output laser. Reproduced with permission from [33], ©2017 IEEE

the ridge waveguide is with much better performance than the planar one, resulting in reduced lasing threshold, higher slope efficiency and output laser power.

The 7 MeV C ion implanted Nd:YVO_4 channel waveguide lasers at 1064 nm could be generated at pump threshold power of 45 mW with a slope efficiency of 30% [28]. In addition, the 100 MeV Ar ion irradiated Nd:YVO_4 planar waveguide system generates laser oscillations at 1067 nm with a slope efficiency of 8.5% [29].

Ma et al. realized cw waveguide laser at ~1024 nm in a Yb:YSGG ridge waveguide, which is produced by the swift ion irradiation and the precise diamond blade dicing [33]. Under the optical pump of a 946 nm laser, the obtained maximum output power is 52.3 mW, corresponding to the pump power of 243 mW. The slope efficiency and the lasing threshold are 46% and 130.5 mW, respectively. Figure 9.4 shows the output power of the waveguide laser at 1024 nm as a function of the pump power at 946 nm.

Grivas et al. realized the cw waveguide lasers based on Ti:sapphire crystal, which is usually used for wavelength-tunable solid-state laser systems [30]. The waveguide is with channel geometry, which is produced by proton implantation with an implantation mask. The implantation is a three-energy process. The energy of the protons utilized in the experiment is 0.5, 0.95, and 1 MeV, by which the refractive index barriers could be broadened to obtain better confinement of light. The channel waveguide cross section is 5 μm (height) × 10 or 15 μm (width). Under optical pump at 501 nm, the laser emission at 780 nm is with maximum output power of 12.4 and 17.5 mW for the 10 or 15 μm wide channel waveguides, respectively. It should be noted that the Ti:sapphire is an ideal gain medium for tunable lasers, so the Ti:sapphire waveguides may be optimized for potential tunable waveguide lasers operation from 700 to 900 nm wavelength band.

Other ion implanted waveguide laser systems include the early reported Nd:YAP and Nd:MgO:$LiNbO_3$ produced by He ion implantation, in which cw waveguide laser oscillations could be generated with pump threshold power of 50 and 8.3 mW, respectively [6, 32].

9.2 Pulsed Waveguide Lasers Based on 2D Materials

Pulsed laser sources are of particular significance for a broad spectrum of applications ranging from the high-peak-power output for sensitive high-resolution nonlinear microscopy to the generation of minimum collateral thermal and photomechanical damage for high-precision medical diagnostics [35–38]. The passive modulation, including passively Q-switched and mode-locked operation regimes, using special saturable absorber (SA) elements (i.e., materials with intensity-dependent absorption features) is more predominant due to their ease of intracavity integration and production of shorter pulses. The simplest and most commonly used short cavity for generation of laser pulses is the plane-parallel (or Fabry-Pérot) laser resonator consisting of two facing mirrors or reflective optical films coated on end faces of gain media. Since the first demonstration in 1994, there has been a tremendous development in the research area of waveguide-based compact pulsed laser sources, covering various host laser materials, e.g. single crystals, polycrystalline ceramics, and amorphous glasses.

SA elements play a significant role in initializing and stabilizing the delivered laser pulses during both passive Q-switching and mode-locking operations. Thanks to the astonishing development in material science and microfabrication technology, an increasing number of novel material families have been employed as efficient SAs in various laser systems [39–53]. Particularly, in ion implanted waveguide laser systems, SESAMs, graphene and other two-dimensional (2D) layered materials, nanoparticle embedded crystal waveguides have been applied to achieve pulses through the Q-switching and mode-locking process.

The SAs for pulsed waveguide lasers work in two configurations, i.e., direct-field interaction or evanescent field interaction schemes (see Fig. 9.5 for schematic plots). In the direct-field interaction scheme, the interaction occurs when the central peak of the propagating optical mode passes through the nonlinear media (Fig. 9.5a). This method is the most commonly used scheme for passively Q-switching and mode-locking of solid-state waveguide lasers and it has been applied in almost all types of waveguide structures. Since the SA elements employed in this configuration usually act as one cavity mirror or output coupler, it is therefore convenient to tune the distance between the SA and the waveguide end facet, offering short laser cavities with flexible lengths for efficient dispersion management. Alternatively, SA elements that are wrapped around active waveguide cores could also interact with the laser modes via the evanescent field (Fig. 9.5b). Practically, this method is usually realized by transferring the SA materials onto the top surface of the target waveguides, using polymer composites as a supporting layer. In contrast to the direct-field interaction,

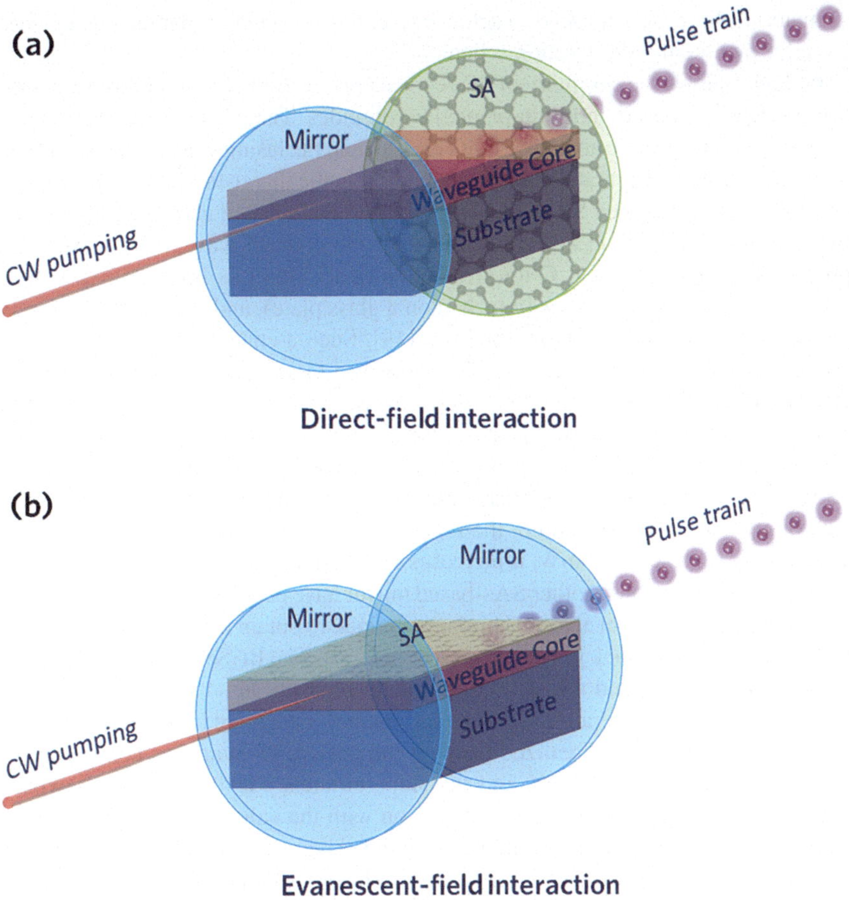

Fig. 9.5 The schematic plots for pulsed waveguide lasers in ion implanted waveguides through **a** direct-field interaction and **b** evanescent field interaction with SAs

evanescent-field interaction involves only a portion of the original optical power, thus naturally raising the optical damage threshold of the SA material. The increase in interaction length (by several to tens orders of magnitudes) enhances the overall nonlinear interaction and allows for simpler SA transfer, easier alignment to the guiding core, and less sensitive operation conditions compared to the direct-interaction method, thus resulting in less complicated passive Q-switching and mode-locking operations for pulsed laser generation. Furthermore, since the SAs in evanescent-field-interaction scheme are not in the path of the laser beam, mechanical damages to the SA elements are much less critical. It is worthy to note that one of the most fundamental requirements, and the most significant intrinsic limitations, of such scheme for practical application in solid-state waveguide lasers is whether the SA elements have the access to interact with the evanescent field of the guiding mode. As yet, a few

2D materials have been applied to achieve laser pulses in ion implanted waveguides through evanescent field iteration scheme.

Table 9.2 shows the summary of the reported results for demonstrations of pulsed waveguide lasers based on ion implanted waveguides. The used gain media include Nd:YAG crystal/ceramic and Yb:YSGG crystal, and the lasing wavelength is ~1 μm. The mostly used configuration for SA operation is the direct-field interaction. Usually, a glass mirror coated SA materials is utilized and located at the end face of the waveguide sample. Tan et al. used a 20 μm wide channel waveguide in Nd:YAG (produced by mask-assisted C ion irradiation at 15 MeV) as laser cavity, with 6-layer CVD graphene as the SA (attached on a glass plate), and realize the passively Q-switched waveguide lasers at 1064 nm [19]. Such a channel waveguide is with low propagation loss of 0.7 dB/cm. The observed Q-switched pulses are with the pulse duration of 57 ns. More interestingly, the modulation depth of the Q-switched pulses is ~ as high as 50%. In this case, the average output power of the laser is 24 mW, the corresponding Q-switched pulse energy was ~11 nJ with the repetition of 2.2 MHz. As the pump power increases, the repetition rate increases linearly as well. For example, at ~108 mW pump, the repetition rate is around 1.5 MHz, whilst the pump power is set at ~192 mW, it increases to 4.1 MHz.

In addition to graphene, other SAs-based mirror have been applied on the Nd:YAG waveguide system. Tan et al. used similar technique to obtain high-quality SAs, such as Bi_2Se_3, WS_2, and black phosphorus (BP) [24, 56]. The Bi_2Se_3 film is synthesized via a polyol method, and the saturation optical intensity is ~4 GW/cm^2, allowing highly intense pulse generation. Based on a channel waveguide in Nd:YAG ceramic, pulsed waveguide laser is realized through the passive Q-switching configuration. The lasing threshold is 72 mW, and the slope efficiency is 40.4%. The drawback on the threshold and efficiency in comparison with the value of cw laser system is due to the additional cavity loss induced by the absorption of Bi_2Se_3 SA, which is expectable for any Q-switched laser systems. The measured output power of the Q-switched waveguide laser is 168.6 mW. Figure 9.6 shows the average output power as a function of the launched pump power. The pulsed waveguide laser has a tunable repetition from 2.7 to 4.7 MHz, and a minimum pulse duration of 46 ns, corresponding to the modulation depth of 0.9%.

For WS_2 based Q-switched waveguide laser, the maximum output power was 144 mW corresponding to the slope efficiency of 14.4% [22]. The threshold of the pumping power was 90 mW. For the BP based laser, the maximum output power was 126 mW, and threshold was 100 mW corresponding to the slope efficiency of 12.9%. In all these systems, the repetition rate reaches a few MHz, and the pulse duration is with tens of ns.

For the scheme based on the modulation of evanescent field interaction, Tan et al. demonstrated the Q-switched pulse laser operation the Nd:YAG planar waveguide [55]. The planar waveguide is produced by 15 MeV C ion irradiation. The graphene with ~16 layers manufactured by the chemical vapor deposition (CVD) on the copper and nickel disks is transferred to the surface of the planar waveguide. The absorption of graphene depends on the overlap between the evanescent-field from the waveguide

Table 9.2 The summary of the reported results for demonstrations of pulsed waveguide lasers based on ion implanted waveguides

Materials	Fabrication technique	Waveguide configuration	Operation regime	Saturable absorber	Modulation type	Working wavelength (nm)	Pump threshold (mW)	Slope efficiency	Max. output power (mW)	Pulse duration	Repetation rate	Refs.
Yb:YSGG	C ion implantation	Planar	Q-switched	WS_2	Evanescent field	1024	130.5	46%	52.3	125 ns	360 kHz	Ma et al. [33]
Nd:GGG	C ion irradiation and femtosecond laser ablation	Ridge	mode-locked	SnS_2	Attached directly	1064	60	12.5%	115	37	17.9 GHz	Dong et al. [54]
Nd:YAG	Kr ion irradiation and femtosecond laser ablation	Ridge	Q-switched	graphene	Attached directly	1064	215	19%	110	90 ns	4.2 MHz	Jia et al. [18]
	C ion implantation	Planar	Q-switched	graphene	Attached directly	1064	45	56%	72	60 ns	4.1 MHz	Jia et al. [19]
	C ion implantation	Planar	Q-switched	graphene	Evanescent field	1064	72	7.4%	14	9.8 μs	26 kHz	Tan et al. [55]
	C ion implantation	Planar	Q-switched	Bi_2Se_3	Attached directly	1064	53	40%	168.6	46 ns	4.7 MHz	Tan et al. [56]

(continued)

Table 9.2 (continued)

Materials	Fabrication technique	Waveguide configuration	Operation regime	Saturable absorber	Modulation type	Working wavelength (nm)	Pump threshold (mW)	Slope efficiency	Max. output power (mW)	Pulse duration	Repetation rate	Refs.
	O ion irradiation	Planar	mode-locked	Ag nanoparticles	Evanescent field	1064	–	–	168	29.5 ps	10.53 GHz	Li et al. [57]
	C ion irradiation and femtosecond laser writing	Cladding	Q-switched	WS_2	Attached directly	1064	120	–	–	45 ns	4.6 MHz	Lv et al. [21]
	Kr ion irradiation and femtosecond laser ablation	Ridge	Q-switched	WS_2	Attached directly	1064	65	34.5%	118.5	209 ns	870 kHz	Jiang et al. [58]
	C ion irradiation	Planar	Q-switched	Graphene	Evanescent field	1064	126	4%	15	40 ns	2 MHz	Ma et al. [59]
Nd:YAG ceramic	C ion irradiation	Planar	Q-switched	WS_2	Attached directly	1064	90	14.4%	144	24 ns	6.1 MHz	Tan et al. [24]
	C ion irradiation	Planar	Q-switched	BP	Attached directly	1064	100	12.9%	126	55 ns	5.6 MHz	Tan et al. [24]

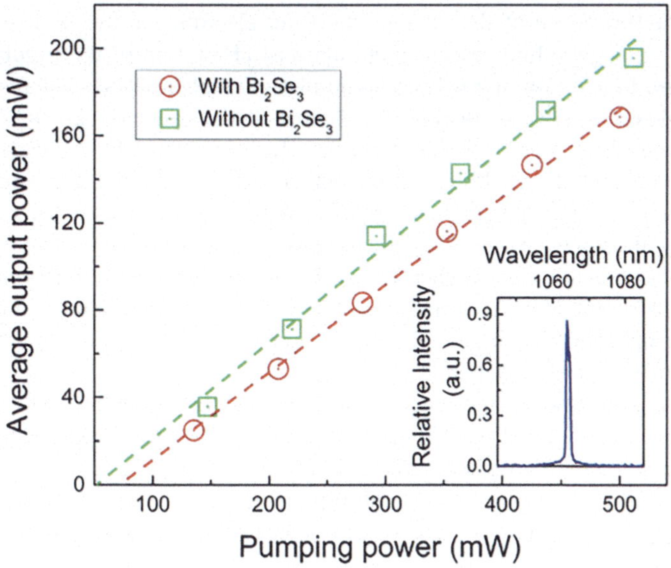

Fig. 9.6 The average output power as a function of the launched pump power for the Bi_2Se_3 based Nd:YAG waveguide laser. Reproduced with permission from [56], ©2015 Optical Society of America

and the graphene layers. To enhance the evanescent-field near the surface, the refractive index matching liquid was dripped onto the surface of the graphene, which was with refractive index of ~1.52 and thickness of ~200 μm. Owing to the coupling of the evanescent-field, the passive Q switching was realized with the 10 mm long waveguide laser cavity. The maximum output power of 10 mW was obtained with the pulse duration ~9.8 μs, and the maximum repetition rate of 29 kHz. The lasing threshold is 99 mW corresponding to the slope efficiency of ~6.6% for the Q-switched waveguide laser. Here the problem is the low-efficiency interaction between the waveguide modes and graphene, which affects the lasing performance of the evanescent field interaction.

Shang et al. applied 6 MeV C ion irradiation at a fluence of 1×10^{15} ions/cm^2 on a surface coated few-layer graphene Nd:YAG crystal [60]. This processing results in two major effects: (a) to construct a planar waveguide layer on surface region of Nd:YAG crystal, and (b) to enhance the coupling between graphene and waveguide mode by reducing the graphene layer total thickness and the space between graphene and Nd:YAG surface. Through this one-step processing, stable Q-switched waveguide lasers based on evanescent field interaction configuration are realized, reaching reduced pulse duration to ~100 ns, and the higher repetition rate of 2.3 MHz.

The ion implantation at ~MeV energy levels could enhance the layer-to-layer coupling effect by compress the few-layer 2D materials. Another successful example is the graphene/WSe_2 heterostructure [61]. During the ion irradiation process, the morphology of monolayer graphene is modified, promoting the contact with WSe_2.

Experimental evidence of the tunable interlayer electron transfer is displayed by investigation of photoluminescence and ultrafast absorption of the irradiated heterostructure. In addition, it has been found that in graphene/WSe_2 heterostructure, graphene serves as a fast channel for the photoexcited carriers to relax in WSe_2, and the nonlinear absorption of WSe_2 could be effectively tuned by the carrier transfer process in graphene, enabling specific optical absorption of the heterostructure in comparison with separated graphene or WSe_2. With these ion beam induced features, by applying the ion beam modified graphene/WSe_2 heterostructure as a SA through direct-field interaction, Q-switched pulsed lasing with optimized performance has been realized in the Nd:YAG waveguide cavity. For example, the pulse duration of modified graphene/WSe_2 based waveguide laser is 43.4 ns, which is much shorter than that of WSe_2 based system (117 ns) [61].

The waveguide laser performance could also be efficiently controlled by electrical signals imposed on SAs. Ma et al. proposed an electrically controlled waveguide laser by applying a voltage to graphene as capacitance and heater [59]. Through modifying the Fermi level of graphene, the linear absorption has increased by 2.9% at −8 V and decreased by 11.3% at 15 V. Figure 9.7 shows the evolution of laser parameters with

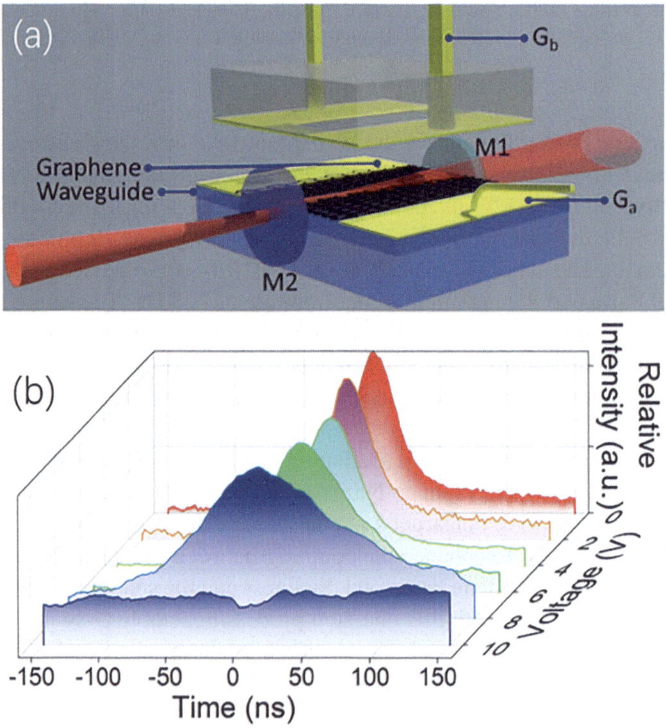

Fig. 9.7 **a** Schematic diagram of electrically controlled hybrid graphene-Nd:YAG waveguide. **b** Variations in the pulse profile corresponding to the electric signal from 0 to 10 V [59]

different applied voltage. As the increase of the voltage, the output power increases gradually and the pulse duration of the waveguide laser can be manipulated from 32 ns to 160 ns, corresponding to the decrease of modulation depth from 1.2 to 0.2%.

In addition to the 2D materials, one can also use phase changing materials as SAs. Tan et al. investigated the temperature dependent waveguide laser performance based on phase transition of VO_2, which is a well-known phase changing nanomaterial [62]. The structural phase transition from the monoclinic phase (insulating) to the rutile phase (metallic) has been observed at ~330 K with a sudden decrease of reflectivity. The nonlinear absorption coefficient of VO_2 SA is dramatically changed from 9.9 $\times$ 10^{-7} to 1.1 $\times$ 10^{-7} m/W after transition to the metallic phase. Based on the insulator-to-metal phase transition (IMPT) of VO_2, Q-switched pulsed waveguide laser with tunable repetition rate is achieved with pulse duration as short as 700 ps while maintaining high peak power. Figure 9.8a shows the recorded pulse trains at different temperatures. As shown in Fig. 9.8b, the repetition of the output laser

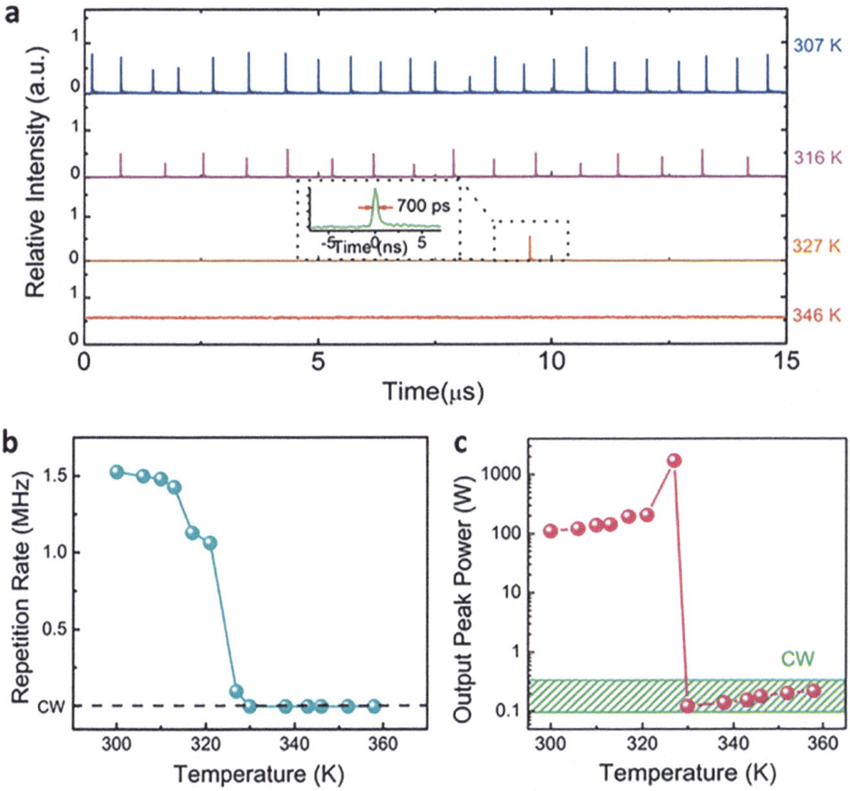

Fig. 9.8 **a** Pulse trains of Q-switched waveguide laser at different temperatures. Inset: pulse duration at 327 K. **b** Repetition rate and **c** peak power at different temperatures [62]

gradually decreases as the temperature rises and experiences a sudden transition to continuous laser when exceeding the MIPT threshold. The dependence of output peak power on temperature is also shown in Fig. 9.8c, in which the operation of the laser could be tuned between pulsed and cw regime according to the practical applications.

Mode-locking is the necessary process to achieve ultrashort pulses in laser systems. The ion beam processed optical waveguides have a lot of potential in mode-locked lasing. Recently, Q-switched mode-locked pulses have been achieved at 1 μm based on Nd:GGG ridge waveguide and tin disulfide (SnS_2) saturable absorber (SA) [54]. The ridge waveguide was fabricated by combined techniques of swift carbon ion irradiation and femtosecond laser ablation. The ultrafast saturable absorption of SnS_2 was confirmed by femtosecond Z-scan technique. The mode-locked pulse performances were recorded by an InGaAs photodetector and real-time oscilloscope. The pulse duration of individual mode-locked pulses was measured to be as short as 30 ps. By measuring the radio frequency (RF) spectrum, the generated laser was operated at a fundamental repetition rate up to multi-gigahertz level as high as 17.9 GHz and high stability with a signal-to-noise ratio (SNR) up to 53 dB. Figure 9.9a–d show the characterization of Q-switched mode-locked lasers in the Nd:GGG ridge waveguide.

9.3 Self-Frequency Doubled Waveguide Lasing

The self-frequency doubled (SFD) waveguide laser depends on a crystal that possesses both nonlinear optical and PL properties. Nd:YCOB is one of the promising SFD crystals for green laser emission under ~800 nm pump. The YCOB crystal is doped by Nd^{3+} ions for 1 μm lasing and the wafer is cut for 1 μm to ~500 nm SHG. In this configuration, the SFD laser can be generated on a monolithic chip, instead of the traditional laser plus nonlinear optical hybrid element (e.g., the typical Nd:YVO_4+KTP green laser unit). The advantage of such an intracavity nonlinear process is that the generated infrared lasers from the activations of the Nd^{3+} ions in the YCOB host matrix produces the green laser oscillations without requirement of additional nonlinear optical components.

For the realization of the intracavity SFD waveguide lasers, Ren et al. used an 810 nm cw linearly polarized laser as the optical pump [63]. The fundamental wave (i.e., the generated infrared light at 1.06 μm) is with same polarization of the pump beam. The 1061 nm waveguide laser was generated at a pump threshold of 31.5 mW, reaching a slope efficiency of 68% and a maximum output power of 35 mW. The SFD green laser is with TM polarization, which is correlated to the $TE_{0\omega}$ to $TM_{02\omega}$ conversion. Figure 9.10 shows the generated SFD green laser spectrum of the 170 MeV Ar^{8+} ion irradiated Nd:YCOB planar waveguide. The output laser spectrum is peaked at ~530.8 nm, which is in good agreement with the SFD green lasers from the bulk gain medium. The modal profile of the output SFD green waveguide laser is shown in Fig. 9.10 as an inset, which confirms that the generated laser emission is with

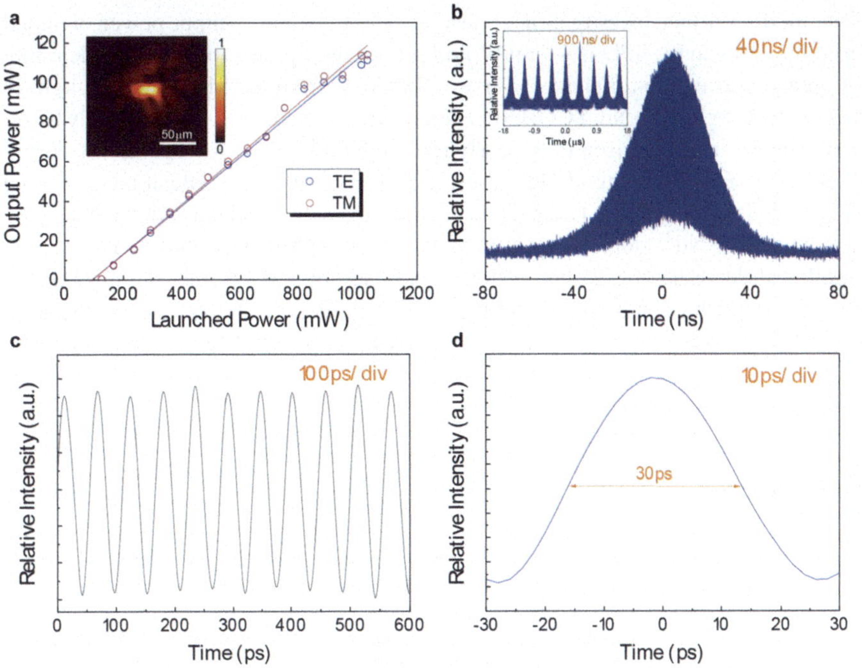

Fig. 9.9 **a** The output power as a function of launched power. The inset shows the laser modal profile. **b** Mode-locked pulse trains on nanosecond time scale. **c** The pulse trains of the waveguide laser on picosecond time scale. **d** Single mode-locked pulse profile. Reproduced with permission from [54], ©2020 Elsevier

Fig. 9.10 Emission spectrum of the green laser generated from the Nd:YCOB waveguide in the SFD experiment. The inset shows the near field distribution of the output SH laser centered at a wavelength of ~530.8 nm. Reproduced with permission from [63], ©2012 Optical Society of America

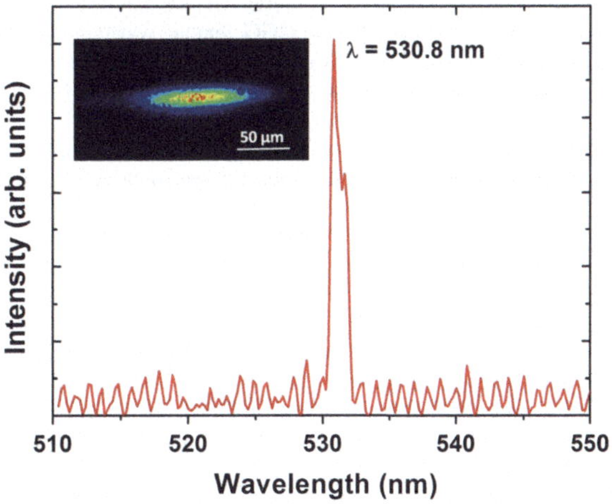

TM_0 mode. Further investigation shows that the maximum output power obtained from the intracavity SFD is ~36 μW under the pump power of ~65 mW, denoting an optical conversion efficiency of $\eta = 0.85\%W^{-1}$ from the 810 nm pump light to 531 nm SFD light. The Nd:YCOB SFD green waveguide laser is with relatively better performance in comparison to the early reported $Nd:LiNbO_3$ waveguides. It should be pointed that the doping of Nd^{3+} ions in YCOB crystals is a crucial factor for the SFD laser in waveguides. The YCOB crystals are with excellent nonlinear properties, whilst the Nd^{3+} doping is inhomogeneous in microscale volumes. This causes significant decrease of SFD laser efficiency. In addition, the further optimization of the self-frequency-doubling system of the waveguide, e.g., reduce the propagation loss of the waveguides, may increase the green laser generation to an acceptable level.

References

1. E. Snitzer, Optical maser action of Nd^{+3} in a barium crown glass. Phys. Rev. Lett. **7**, 444 (1961)
2. Y. Jia, F. Chen, Compact solid-state waveguide lasers operating in the pulsed regime: a review. Chin. Opt. Lett. **17**, 012302 (2019)
3. C. Grivas, Optically pumped planar waveguide lasers, Part I: fundamentals and fabrication techniques. Prog. Quantum Electron. **35**, 159 (2011)
4. C. Grivas, Optically pumped planar waveguide lasers: Part II: gain media, laser systems, and applications. Prog. Quantum Electron. **45–46**, 3 (2016)
5. S.J. Field, D.C. Hanna, A.C. Large, D.P. Shepherd, A.C. Tropper, P.J. Chandler, P.D. Townsend, L. Zhang, Low threshold ion-implanted Nd:YAG channel waveguide laser. Electron. Lett. **27**, 2375 (1991)
6. S.J. Field, D.C. Hanna, D.P. Shepherd, A.C. Tropper, P.J. Chandler, P.D. Townsend, L. Zhang, Ion-implanted $Nd:MgO:LiNbO_3$ planar waveguide laser. Opt. Lett. **16**, 481 (1991)
7. S.J. Field, D.C. Hanna, A.C. Large, D.P. Shepherd, A.C. Tropper, P.J. Chandler, P.D. Townsend, L. Zhang, Ion-implanted Nd:GGG channel waveguide laser. Opt. Lett. **17**, 52 (1992)
8. E. Flores-Romero, G.V. Vazquez, H. Marquez, R. Rangel-Rojo, J. Rickards, R. Trejo-Luna, Planar waveguide lasers by proton implantation in Nd:YAG crystals. Opt. Express **12**, 2264 (2004)
9. M. Domenech, G.V. Vázquez, E. Cantelar, G. Lifante, Continuous-wave laser action at $\lambda =$ 1064.3 nm in proton- and carbon-implanted Nd:YAG waveguides. Appl. Phys. Lett. **83**, 4110 (2003)
10. E. Flores-Romero, G.V. Vázquez, H. Márquez, R. Rangel-Rojo, J. Rickards, R. Trejo-Luna, Laser emission in proton-implanted Nd:YAG channel waveguides. Opt. Express **15**, 17874 (2007)
11. M. Domenech, G. V. Vázquez, E. Flores-Romero, E. Cantelar, G. Lifante, Continuous-wave laser oscillation at 1.3 μm in Nd:YAG proton-implanted planar waveguides. Appl. Phys. Lett. **86**, 151108 (2005)
12. Y.Y. Ren, N.N. Dong, F. Chen, A. Benayas, D. Jaque, F. Qiu, T. Narusawa, Swift heavy-ion irradiated active waveguides in Nd:YAG crystals: fabrication and laser generation. Opt. Lett. **35**, 3276 (2010)
13. Y. Ren, N. Dong, F. Chen, D. Jaque, Swift nitrogen ion irradiated waveguide lasers in Nd:YAG crystal. Opt. Express **19**, 5522 (2011)
14. Y.C. Yao, Y. Tan, N.N. Dong, F. Chen, A.A. Bettiol, Continuous wave Nd:YAG channel waveguide laser produced by focused proton beam writing. Opt. Express **18**, 24516 (2010)

15. Y. Tan, C. Zhang, F. Chen, F.Q. Liu, D. Jaque, Q.M. Lu, Room-temperature continuous wave laser oscillations in Nd:YAG ceramic waveguides produced by carbon ion implantation. Appl. Phys. B **103**, 837 (2011)
16. Y. Jia, N. Dong, F. Chen, J.R. Vazquez de Aldana, S. Akhmadaliev, S. Zhou, Continuous wave ridge waveguide lasers in femtosecond laser micromachined ion irradiated Nd:YAG single crystals. Opt. Mater. Express **2**, 657 (2012)
17. Y. Jia, C.E. Rueter, S. Akhmadaliev, S. Zhou, F. Chen, D. Kip, Ridge waveguide lasers in Nd:YAG crystals produced by combining swift heavy ion irradiation and precise diamond blade dicing. Opt. Mater. Express **3**, 433 (2013)
18. Y. Jia, Y. Tan, C. Cheng, J.R. Vazquez de Aldana, F. Chen, Efficient lasing in continuous wave and graphene Q-switched regimes from Nd:YAG ridge waveguides produced by combination of swift heavy ion irradiation and femtosecond laser ablation. Opt. Express **22**, 12900 (2014)
19. Y. Tan, S. Akhmadaliev, S. Zhou, S. Sun, F. Chen, Guided continuous-wave and graphene-based Q-switched lasers in carbon ion irradiated Nd:YAG ceramic channel waveguide. Opt. Express **22**, 3572 (2014)
20. Y. Yao, Y. Jia, F. Chen, S. Akhmadaliev, S. Zhou, Channel waveguide lasers at 1064 nm in Nd:YAG crystal produced by C^{5+} ion irradiation with shadow masking. Appl. Opt. **53**, 195 (2014)
21. J. Lv, Z. Shang, Y. Tan, J.R. Vázquez de Aldana, F. Chen, Cladding-like waveguide fabricated by cooperation of ultrafast laser writing and ion irradiation: characterization and laser generation. Opt. Express **25**, 19603 (2017)
22. Z. Shang, Y. Tan, S. Akhmadaliev, S. Zhou, F. Chen, Cladding-like waveguide structure in Nd: YAG crystal fabricated by multiple ion irradiation for enhanced waveguide lasing. Opt. Express **23**, 27612 (2015)
23. Y. Jia, J.R. Vazquez de Aldana, S. Akhmadaliev, S. Zhou, F. Chen, Femtosecond laser micromachined ridge waveguide lasers in Nd:YAG ceramics. Opt. Mater. **36**, 228 (2013)
24. Y. Tan, Z. Guo, L. Ma, H. Zhang, S. Akhmadaliev, S. Zhou, F. Chen, Q-switched waveguide laser based on two-dimensional semiconducting materials: tungsten disulfide and black phosphorous. Opt. Express **24**, 2858 (2016)
25. Y.Y. Ren, N.N. Dong, Y. Tan, J. Guan, F. Chen, Q.M. Lu, Continuous wave laser generation in proton implanted Nd:GGG planar waveguides. J. Lightwave Technol. **28**, 3578 (2010)
26. Y. Yao, N. Dong, F. Chen, S.K. Vanga, A.A. Bettiol, Proton beam writing of Nd:GGG crystals as new waveguide laser sources. Opt. Lett. **36**, 4173 (2011)
27. Y. Jia, N. Dong, F. Chen, J.R. Vazquez de Aldana, S. Akhmadaliev, S. Zhou, Ridge waveguide lasers in Nd:GGG crystals produced by swift carbon ion irradiation and femtosecond laser ablation. Opt. Express **20**, 9763 (2012)
28. M.E. Sánchez-Morales, G.V. Vázquez, E.B. Mejía, H. Márquez, J. Rickards, R. Trejo-Luna, Laser emission in Nd:YVO_4 channel waveguides at 1064 nm. Appl. Phys. B **94**, 215 (2009)
29. Y. Yao, N. Dong, F. Chen, L. Pang, Z. Wang, Q. Lu, Continuous wave waveguide lasers of swift argon ion irradiated Nd:YVO_4 waveguides. Opt. Express **19**, 24252 (2011)
30. C. Grivas, D.P. Shepherd, R.W. Eason, L. Laversenne, P. Moretti, C.N. Borca, M. Pollnau, Room-temperature continuous-wave operation of Ti:sapphire buried channel-waveguide lasers fabricated via proton implantation. Opt. Lett. **31**, 3450 (2006)
31. Y. Ren, N. Dong, Y. Jia, L. Pang, Z. Wang, Q. Lu, F. Chen, Efficient laser emissions at 1.06 μm of swift heavy ion irradiated Nd:YCOB waveguides. Opt. Lett. **36**, 4521 (2011)
32. S.J. Field, D.C. Hanna, D.P. Shepherd, A.C. Tropper, P.J. Chandler, P.D. Townsend, L. Zhang, Ion-implanted Nd:YAP planar waveguide laser. Electron. Lett. **26**, 1826 (1990)
33. L. Ma, Y. Tan, S. Wang, S. Akhmadaliev, S. Zhou, H. Yu, H. Zhang, F. Chen, Continuous-wave and Q-switched Yb:YSGG waveguide laser. J. Lightwave Technol. **35**, 2642 (2017)
34. D. Marcuse, Loss analysis of single-mode fiber splices. Bell Syst. Tech. J. **56**, 703 (1977)
35. C. Xu, F.W. Wise, Recent advances in fibre lasers for nonlinear microscopy. Nat. Photonics **7**, 875 (2013)
36. S. Hakobyan, V.J. Wittwer, P. Brochard, K. Gurel, S. Schilt, A.S. Mayer, U. Keller, T. Sudmeyer, Full stabilization and characterization of an optical frequency comb from a diode-pumped solid-state laser with GHz repetition rate. Opt. Express **25**, 20437 (2017)

37. H.M. Oubei, J.R. Duran, B. Janjua, H.-Y. Wang, C.-T. Tsai, Y.-C. Chi, T.K. Ng, H.-C. Kuo, J.-H. He, M.-S. Alouini, G.-R. Lin, B.S. Ooi, 4.8 Gbit/s 16-QAM-OFDM transmission based on compact 450 nm laser for underwater wireless optical communication. Opt. Express **23**, 23302 (2015)
38. S.H. Yun, S.J.J. Kwok, Light in diagnosis, therapy and surgery. Nat. Biomed. Eng. **1**, 0008 (2017)
39. H. Zhang, S.B. Lu, J. Zheng, J. Du, S.C. Wen, D.Y. Tang, K.P. Loh, Molybdenum disulfide (MoS_2) as a broadband saturable absorber for ultra-fast photonics. Opt. Express **22**, 7249 (2014)
40. G. Wang, K. Wang, B.M. Szydłowska, A.A. Baker-Murray, J.J. Wang, Y. Feng, X. Zhang, J. Wang, W.J. Blau, Ultrafast nonlinear optical properties of a graphene saturable mirror in the 2 μm wavelength region. Laser Photonics Rev. **11**, 1700166 (2017)
41. Q. Guo, Z. Qin, Z. Wang, Y.-X. Weng, X. Liu, G. Xie, J. Qiu, Broadly tunable plasmons in doped oxide nanoparticles for ultrafast and broadband mid-infrared all-optical switching. ACS Nano **12**, 12770 (2018)
42. X. Yu, P. Yu, D. Wu, B. Singh, Q. Zeng, H. Lin, W. Zhou, J. Lin, K. Suenaga, Z. Liu, Q.J. Wang, Atomically thin noble metal dichalcogenide: a broadband mid-infrared semiconductor. Nat. Commun. **9**, 1545 (2018)
43. L. Lu, Z.M. Liang, L.M. Wu, Y.X. Chen, Y.F. Song, S.C. Dhanabalan, J.S. Ponraj, B.Q. Dong, Y.J. Xiang, F. Xing, D.Y. Fan, H. Zhang, Few-layer bismuthene: sonochemical exfoliation, nonlinear optics and applications for ultrafast photonics with enhanced stability. Laser Photonics Rev. **12**, 1700221 (2018)
44. Z. Li, N. Dong, Y. Zhang, J. Wang, H. Yu, F. Chen, Mode-locked waveguide lasers modulated by rhenium diselenide as a new saturable absorber. APL Photonics **3**, 080802 (2018)
45. C. Pang, R. Li, Y. Zhang, Z. Li, N. Dong, L. Wu, H. Yu, J. Wang, F. Ren, F. Chen, Tailoring optical nonlinearities of $LiNbO_3$ crystals by plasmonic silver nanoparticles for broadband saturable absorbers. Opt. Express **26**, 31276 (2018)
46. L. Tao, X. Huang, J. He, Y. Lou, L. Zeng, Y. Li, H. Long, J. Li, L. Zhang, Y.H. Tsang, Vertically standing $PtSe_2$ film: a saturable absorber for a passively mode-locked Nd:$LuVO_4$ laser. Photonics Res. **6**, 750 (2018)
47. Y. Ye, Y. Xian, J. Cai, K. Lu, Z. Liu, T. Shi, J. Du, Y. Leng, R. Wei, W. Wang, X. Liu, G. Bi, J. Qiu, Linear and nonlinear optical properties of few-layer exfoliated SnSe nanosheets. Adv. Opt. Mater. **7**, 1800579 (2019)
48. G. Yanqi, Z. Zhengfeng, X. Yanhua, C. Yunxiang, C. Si, L. Zhiming, S. Yufeng, Z. Yousheng, Z. Haibo, X. Shixiang, Z. Han, F. Dianyuan, Broadband nonlinear photoresponse of 2D TiS_2 for ultrashort pulse generation and all-optical thresholding devices. Adv. Opt. Mater. **6**, 1701166 (2018)
49. G. Wang, G. Liang, A.A. Baker-Murray, K. Wang, J.J. Wang, X. Zhang, D. Bennett, J.-T. Luo, J. Wang, P. Fan, W.J. Blau, Nonlinear optical performance of few-layer molybdenum diselenide as a slow-saturable absorber. Photonics Res. **6**, 674 (2018)
50. B. Yan, B. Zhang, H. Nie, H. Wang, G. Li, X. Sun, R. Wang, N. Lin, J. He, High-power passively Q-switched 2.0 μm all-solid-state laser based on a $MoTe_2$ saturable absorber. Opt. Express **26**, 18505 (2018)
51. Z. Li, R. Li, C. Pang, N. Dong, J. Wang, H. Yu, F. Chen, 8.8 GHz Q-switched mode-locked waveguide lasers modulated by $PtSe_2$ saturable absorber, Opt. Express **27**, 8727 (2019)
52. Y.I. Jhon, J. Lee, M. Seo, J.H. Lee, Y.M. Jhon, van der waals layered tin selenide as highly nonlinear ultrafast saturable absorber. Adv. Opt. Mater. **7**, 1801745 (2019)
53. S. Yamashita, Nonlinear optics in carbon nanotube, graphene, and related 2D materials. APL Photonics **4**, 034301 (2019)
54. X. Dong, Z. Li, C. Pang, N. Dong, H. Jiang, J. Wang, F. Chen, Q-switched mode-locked Nd:GGG waveguide laser with tin disulfide as saturable absorber. Opt. Mater. **100**, 109702 (2020)
55. Y. Tan, C. Cheng, S. Akhmadaliev, S. Zhou, F. Chen, Nd:YAG waveguide laser Q-switched by evanescent-field interaction with graphene. Opt. Express **22**, 9101 (2014)

56. Y. Tan, H. Zhang, C. Zhao, S. Akhmadaliev, S. Zhou, F. Chen, Bi_2Se_3 Q-switched Nd:YAG ceramic waveguide laser. Opt. Lett. **40**, 637 (2015)
57. R. Li, C. Pang, Z. Li, N. Dong, J. Wang, F. Ren, S. Akhmadaliev, S. Zhou, F. Chen, Monolithic waveguide laser mode-locked by embedded Ag nanoparticles operating at 1 μm. Nanophotonics **8**, 859 (2019)
58. H. Jiang, Z. Li, X. Dong, C. Pang, S. Wang, F. Chen, WS_2-based Q-switched laser generation from Nd:YAG ridge waveguides fabricated by combination of swift heavy ion irradiation and laser ablation. Opt. Mater. **92**, 163 (2019)
59. L. Ma, Y. Tan, S. Akhmadaliev, S. Zhou, F. Chen, Electrically tunable Nd:YAG waveguide laser based on Graphene. Sci. Rep. **6**, 36785 (2016)
60. Z. Shang, Y. Tan, S. Zhou, F. Chen, Layer-to-layer compression and enhanced optical properties of few-layer graphene nanosheet induced by ion irradiation. Opt. Eng. **55**, 081303 (2016)
61. Y. Tan, X. Liu, Z. He, Y. Liu, M. Zhao, H. Zhang, F. Chen, Tuning of interlayer coupling in large-area graphene/WSe_2 van der Waals heterostructure via ion irradiation: optical evidences and photonic applications. ACS Photonics **4**, 1531 (2017)
62. Y. Tan, L. Chen, D. Wang, Y. Chen, S. Akhmadaliev, S. Zhou, M. Hong, F. Chen, Tunable picosecond laser pulses via the contrast of two reverse saturable absorption phases in a waveguide platform. Sci. Rep. **6**, 26176 (2016)
63. Y. Ren, Y. Jia, N. Dong, L. Pang, Z. Wang, Q. Lu, F. Chen, Guided-wave second harmonics in Nd:YCOB optical waveguides for integrated green lasers. Opt. Lett. **37**, 244 (2012)

Chapter 10
Tailoring of Optical Properties by Metallic Nanoparticles

Abstract Metallic nanoparticles (NPs) embedded into dielectrics can significantly enhance the optical absorption properties of the materials due to the strong localized surface plasmon resonance (LSPR) effect. In this chapter, optical properties of dielectrics with embedded metallic NPs by direct ion implantation are demonstrated, i.e., Sect. 10.1 for optical absorbance enhancement by LSPR, Sect. 10. 2 for giant enhancement of optical nonlinearities, and Sect. 10. 3 for nonlinear absorption tuning by metallic NPs. The applications of dielectric wafers embedded metallic NPs for efficient laser pulse generations in laser systems are introduced in Sect. 10. 4.

The optical absorption properties of the dielectric materials can be dramatically modified by the metallic NPs due to the strong LSPR effect. The dielectric materials include glass (e.g., fused silica) and crystals (e.g., Al_2O_3, Nd:YAG, $LiNbO_3$, $LiTaO_3$). By using ion implantation, NPs in spherical geometries can be synthesized in various dielectric matrix. The shapes of the NPs can be further transformed (e.g., elongated from nanospheres to nanorods or nano-ellipsoids) by swift heavy ion irradiation [1]. In Chaps. 4 and 5, the ion beam synthesis and reshaping of NPs embedded in dielectrics have been demonstrated in details. The physical mechanisms of the evolution of these NPs under thermal treatment are also indicated. Additional linear optical absorption of the dielectrics occurs due to the existence of NPs, which is mainly located in the visible wavelength band. In addition, the optical nonlinearities of dielectrics can be significantly modified by the embedded metallic NPs, resulting in effective tailoring of the nonlinear refractive index and related properties. Further investigation shows that the dielectrics with embedded metallic NPs may possess ultrafast saturable absorption features, which have been utilized to implementation of Q-switched or mode-locked lasing in visible and near infrared (NIR) 1 μm wavelength regimes.

F. Chen et al., *Ion Irradiation of Dielectrics for Photonic Applications*,
Springer Series in Optical Sciences 231, https://doi.org/10.1007/978-981-15-4607-5_10

10.1 Optical Absorbance Enhancement by Surface Plasmon Resonance

The metallic NP induced linear optical absorption is associated with the LSPR effect [2]. The obvious effect of LSPR is the color changes of the dielectrics as NPs are embedded into the matrix of materials. The early examples can be traced back to ancient Roman (e.g., the famous Lycurgus Cup during the ancient Roman period) and Middle Ages (e.g., colorful windows with stained glass containing metallic NPs in Christian churches) [2], see Chap. 3. There are a few parameters that determine the LSPR induced optical absorption, i.e., the species, sizes, shapes, and concentration of NPs, and the surrounding environment of the matrix [3]. The diameters of the ion implanted metallic NPs with spherical geometries are typically no larger than 50 nm, depending on the nature of dielectrics and implantation or post-implantation processing. In fact, the system of embedded NPs by ion implantation contains NPs with diverse diameters. As an example, Nie et al. reported on the synthesis Au NPs embedded into Nd:YAG crystal by Au ion implantation and following annealing at 800 °C for 1.5 h [4]. Figure 10.1a shows the cross-sectional overview HAADF-STEM images of Au NPs in Nd:YAG. Figure 10.1b, c depict the histogram of the Au NP size distribution and the element mapping, respectively. The average size of Au NPs embedded in Nd:YAG is calculated to be with diameter of 2.65 nm. In addition, the Au NPs as well as Nd:YAG matrix are shown to be crystalline.

According to the Mie's theory, for NPs with diameters smaller than 50 nm, the dipole approximation can be used, which generates the extinction coefficient γ describing as [5].

$$\gamma = \frac{18\pi p\varepsilon_d^{3/2}}{\lambda_0}\frac{\varepsilon''_m}{|\varepsilon_m + 2\varepsilon_d|^2} \tag{10.1}$$

where ε_m, ε_d and p denote the complex dielectric constants of the metal and insulator, as well as the volume fraction of the metal, respectively. ε''_m and λ_0 denote the imaginary part of ε_m and the wavelength of light in a vacuum, separately. Hereafter $'$ and $''$

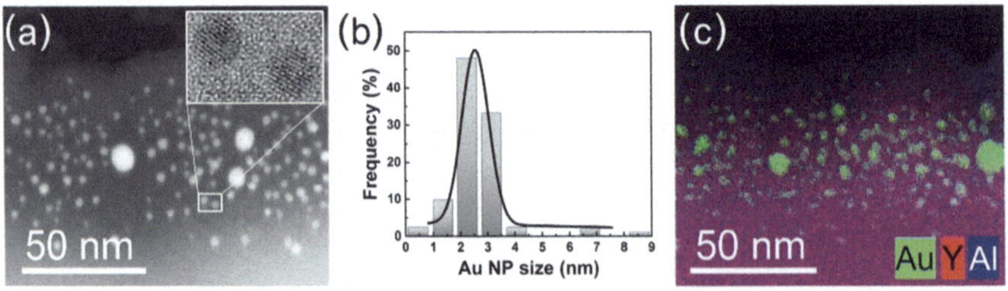

Fig. 10.1 **a** The HAADF-STEM image of Au NPs in Nd:YAG, **b** particle size distribution, and **c** superimposed Au (green), Y (red), and Al (blue) element distributions based on EDXS analysis. Reproduced with permission from [4], ©2018 Royal Society of Chemistry

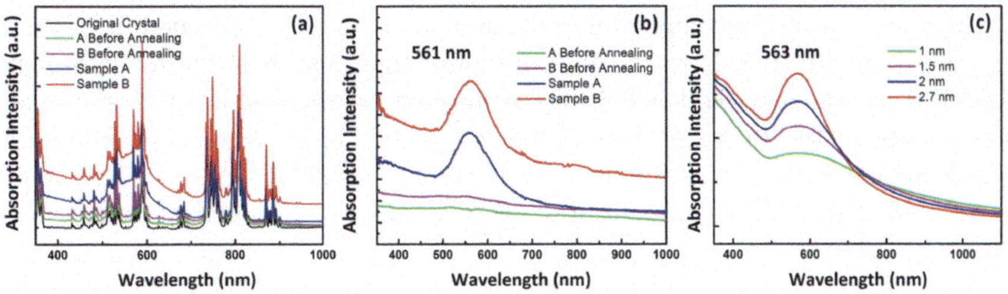

Fig. 10.2 **a** The measured absorption bands of the original crystal (Nd:YAG), sample A (1.5×10^{16} ions per cm^2) before (green line) and after annealing (blue line), sample B (2×10^{16} ions per cm^2) before (pink line) and after annealing (red line), respectively. **b** The measured absorption spectral differences in implanted and annealed samples after subtracting the original crystal absorption to exclude the absorption influence of Nd^{3+} ions. **c** The calculated absorption spectra using Mie theory of the Au sphere NPs with various diameters in the YAG crystal. Reproduced with permission from [4], ©2018 Royal Society of Chemistry

denote real and imaginary parts, respectively. Under the condition for an absorption maximum, it should also satisfy the following equation

$$\left\{\varepsilon_m'(\omega) + 2\varepsilon_d'(\omega)\right\}^2 + \left\{\varepsilon_m''(\omega)\right\}^2 = \text{minimum} \tag{10.2}$$

As a rough estimation, one could calculate the linear optical absorption by the Mie's theory. Let us apply it for the Au NPs in Nd:YAG system that the diameters of Au nanospheres is much less than 50 nm. Figure 10.2a shows the absorption of Au^+ ion implanted Nd:YAG samples before and after annealing, and Fig. 10.2b depicts the measured absorption spectral differences in implanted and annealed samples after subtracting the original crystal absorption to exclude the absorption influence of Nd^{3+} ions and the calculated absorption spectra using Mie's theory of the Au spherical NPs with various diameters in the Nd:YAG crystal [4]. Please note that in this case, the annealing is necessary to form Au NPs (see detailed mechanism in Chap. 4). For spherical Au NPs with diameters of 1, 1.5, and 2 nm and the measured statistical average diameter of 2.7 nm, it can be calculated that, based on the Mie's theory, all the LSPR bands of Au NPs develop consistently at a clear absorption peak ~564 nm, which is in good agreement with the experimental value of 561 nm. Moreover, as depicted in Fig. 10.2c, an increase in the absorption intensity with an increase in NP diameter is conspicuous, which well reproduces the measured results, manifesting that the greater ion fluence may give rise to the larger diameters of Au NPs. Please note that, although the main LSPR band is located at the visible band, the absorption at NIR 1 μm wavelength regime is not zero. This is because the diameters of Au NPs have almost Gaussian distribution from 0 to 9 nm, and the whole effect of LSPR extends to longer wavelength. The Mie's theory in simple cases cannot be used to calculate this weak absorption extending to NIR region. However, it seems to be of significant importance for this extended LSPR effect in nonlinear case. Pang et al. developed a new approach to overcome this shortcoming to obtain the absorbance at 1 μm wavelength, and applied it in the system of Au NPs embedded in $LiNbO_3$ crystal [6]. Figure 10.3 shows the high-resolution transmission electron microscopy (HRTEM) image of Au NPs in $LiNbO_3$. Note from the HRTEM image, some small NPs (diameters less than 2 nm) are in the vicinity of a large NP. This phenomenon containing one large NP surrounded by a number of small NPs exists widely in the system. Nevertheless, the amount of these small NPs is considerably large, which cannot be counted.

A system that consists of a 30 nm-diameter Au NP and surrounded by a few small-diameter NPs (e.g., 2 nm) is therefore constructed, which is termed by multidiameter model (as seen in Fig. 10.4a). In this model, the absorption of each NP γ_i is affected by influence factor a_i, and the total absorption γ_{sum} can be calculated as the sum of

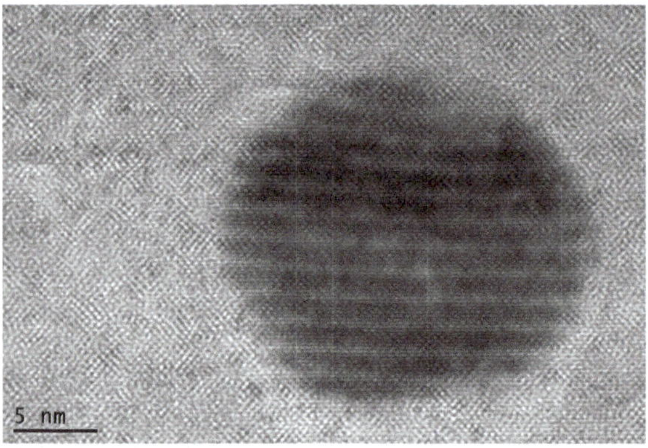

Fig. 10.3 The HRTEM image of Au NPs in $LiNbO_3$. Please note both the 30 nm diameter Au NP and $LiNbO_3$ matrix are crystalline

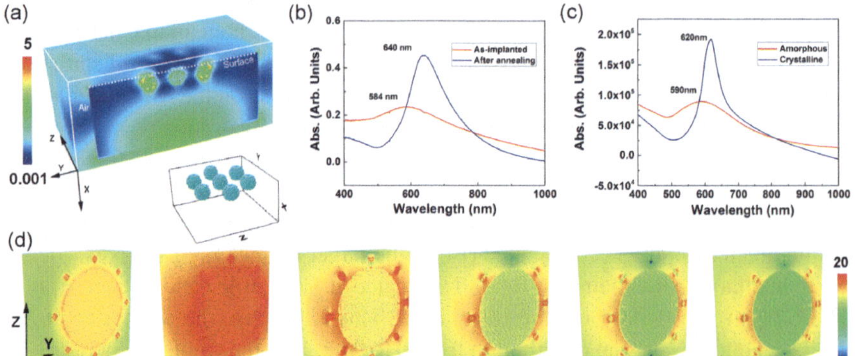

Fig. 10.4 **a** Near-field distribution of NPs with 30 nm diameter in the $LiNbO_3$ crystal. LSPR absorption of **b** experiments and **c** simulations. **d** Near-field distribution at different wavelengths. Reproduced with permission from [6], ©2018 Wiley

contribution of all NPs by Mie theory. The relationships among these parameters are as follows:

$$\gamma_i = \frac{18\pi p \varepsilon_d^{3/2}}{\lambda_0} \frac{\varepsilon_m''}{|\varepsilon_m + 2\varepsilon_d|^2} \tag{10.3}$$

$$\sum a_i \gamma_i = \gamma_{\text{sum}} \tag{10.4}$$

Figure 10.4b, c show the measured absorption spectrum and the calculated results, which indicates the optical absorption till near infrared wavelength region. It can also be confirmed by further calculation based on multidiameter model on the near-field distribution of the optical field in the system. Figure 10.4d shows the electric field distributions of NPs at different wavelengths (550, 640, 750, 850, 950, 1064 nm) for the system of one 30 nm NP surrounded by a few small NPs with diameters of 2 nm in the $LiNbO_3$ crystal. From the calculation results, it can be seen that the optical limitation of NPs depends on particle diameters as well as excitation wavelengths. The enhancement of near field is a synergetic effect of NPs of large and small scales. The large-scale NP plays a major role in contrast to small NPs at any wavelength. At short wavelengths (e.g., in visible light band), large NP exhibit stronger limitation of light field. The high electric field caused by the large NP covers surrounding small NPs, and consequently, the effect from the larger NP dominates in comparison to small NPs. In the longer wavelength band (e.g., near infrared, NIR), the effect of large NP becomes significantly weak, which enables smaller NPs to play a nonnegligible role on the limitation of light field. Basically, the applications of Au NPs are operating in visible light band, since the main LSPR effect mainly occurs at this band. For ion implanted Au NPs systems, the extension of LSPR effect of Au NPs to NIR band is the important fundamental for the applications in NIR bands.

Numerous works have been performed on ion implanted metallic NPs in glass, particularly in fused silica [7–9]. In dielectric crystals, noble metallic NPs, e.g., Au, Ag, and Cu, have been synthesized in Nd:YAG, $LiNbO_3$, $LiTaO_3$, BGO, Al_2O_3, $KTiOPO_4$, etc., and similar LSPR induced optical absorptions have been reported [10–15]. In addition to these metallic nanospheres, one can further reshape the NPs to nano-ellipsoids or nanorods, and change the optical absorption of the systems as well.

Li et al. reported on the fabrication of reshaped Ag NPs embedded in Nd:YAG crystal by combining Ag ion implantation and swift heavy ion (Xe^{18+}) irradiation [16]. The Ag^+ ion implantation is at energy of 200 keV and fluence of 5×10^{16} cm^{-2}. The Xe^{18+} ion irradiation is at an energy of 200 MeV and a fluence of 5×10^{13} cm^{-2}, which irradiates the sample at 45° incident angle. The LSPR effect is proved to be efficiently modulated according to the phenomenon of polarization-dependent absorption, which is closely correlated to the shape of the NPs. For the as-implanted Nd:YAG crystal with embedded Ag nanospheres, the LSPR induced absorption peak is located at 448 nm, whilst for the reshaped Ag NPs (i.e., nano-ellipsoids), there is a redshift or blueshift at 0° or 90° polarization, respectively (see Fig. 10.5 for details). This is because the 200 MeV Xe^{18+} ion irradiation reshapes the Ag nanospheres

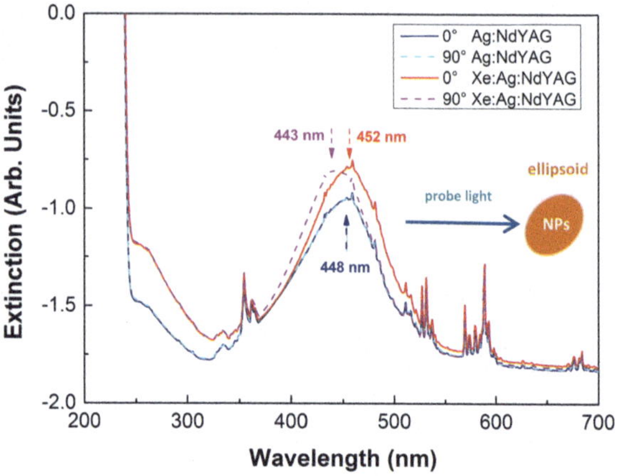

Fig. 10.5 Polarization-dependent optical extinction of Ag NPs in Nd:YAG before and after reshaping. Reproduced with permission from [16], ©2018 The Institute of Physics

to nano-ellipsoids with aspect ratio of 1.05:1. Although the reshaping of NPs is very slight, the obvious optical dichroism occurs under light excitation of different polarization. Figure 10.6 depicts the calculated results of LSPR induced absorption based on discrete dipole approximation (DDA), which is in good agreement with the experimental results shown in Fig. 10.5. Based on the near field intensity distribution, the interaction between reshaped NPs shows a non-ignorable effect on the optical absorption. Another interesting phenomenon is the polarization-dependence of photoluminescence (PL) due to the reshaping of Ag NPs. Further investigation by utilizing the confocal microscope system with an excitation of 532 nm, the polarization-dependence of PL intensity is found to be positively related to the modulated LSPR absorption, which suggests potential PL enhancement by reshaped Ag NPs.

Nano-ellipsoids of noble metals with larger aspect ratios in dielectrics brings out more obvious phenomena of polarization dependent optical absorption. Wolf et al. used high-energy Kr or Au ion irradiation to elongate the Ag NPs buried in $LiNbO_3$ [17]. The irradiation is carried out at room temperature and at an angle of 45° regarding the surface normal. By using 33 MeV Kr ions, the aspect ratio of the

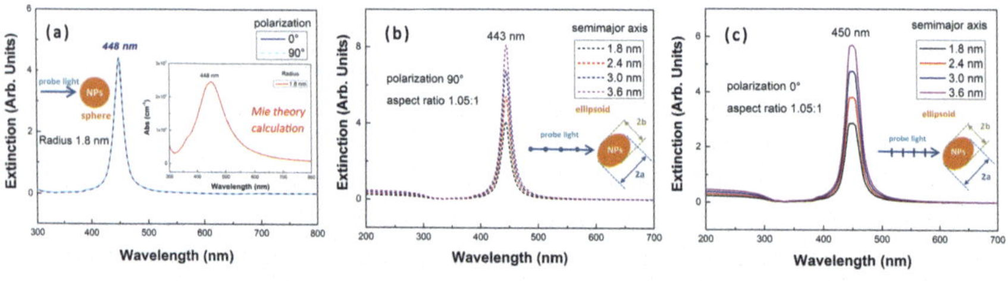

Fig. 10.6 **a** Simulated optical extinction spectra of spherical Ag NPs at 0° or 90° polarization; The insert is the calculated extinction spectrum by Mie theory. Simulated optical extinction spectra of ellipsoid Ag NPs with different semimajor axis at **b** 90° and **c** 0° polarization. Reproduced with permission from [16], ©2018 Institute of Physics

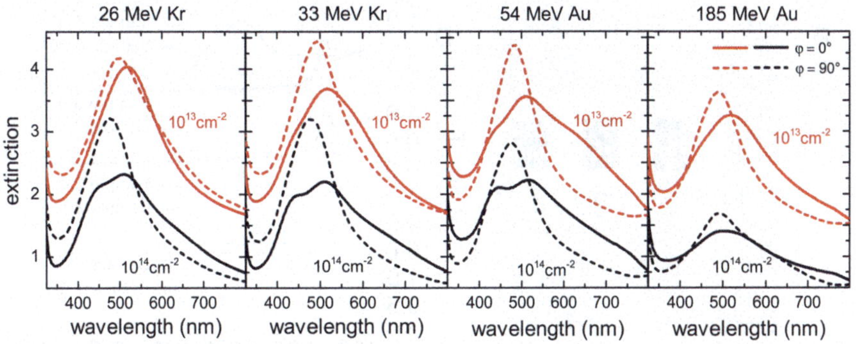

Fig. 10.7 UV-Vis spectra of $LiNbO_3$ with Ag nano-ellipsoids samples irradiated with different ion energies and fluences of 10^{13} and 10^{14} cm^{-2}. The polarization angle of the incoming light was 90° (dashed lines) or 0° (solid lines). Reproduced with permission from [17], ©2016 The Institute of Physics

Ag nano-ellipsoids is 1.6, whist for 185 MeV Au it increases to 2.4. Figure 10.7 shows the measured optical absorption curves for the Ag nano-ellipsoids produced by the swift heavy ion irradiation.

Another remarkable example is Cu NPs in $LiTaO_3$ crystal. Pang et al. synthesized Cu NPs in $LiTaO_3$ by Cu^+ ion implantation at fluence of 5×10^{16} or 1×10^{17} cm^{-2} [18]. The sample with higher-fluence implantation is further irradiated by 200 MeV Xe^{14+} ions at angle of 45°. Figure 10.8 shows the characterization of linear optical response of the $LiTaO_3$ samples: sample 1: pure crystal; sample 2: Ag NPs implanted by 5×10^{16} cm^{-2}; sample 3: Ag NPs implanted by 1×10^{17} cm^{-2}; and Ag nano-ellipsoids elongated by Xe^{14+} ions. The aspect ratio of the elongated Ag nano-ellipsoids is 1.63:1, compared to spherical NPs. From these figures, it is clearly seen that the elongated nano-ellipsoids possess polarization dependent absorption, whist the nanospheres do not feature this property. At 90° excitation the LSPR peak is located at 608 nm, and at 0° excitation it is located at 599 nm, in which difference of 9 nm is obtained for the nano-ellipsoids. The calculation fits the experimental results in a reasonable manner.

10.2 Giant Enhancement of Optical Nonlinearities

The optical nonlinearities of the dielectrics embedded with metallic NPs are significantly different from those without NPs. Usually the host materials do not possess strong third-order nonlinearity. The high-concentration doping of metallic NPs in dielectrics will enhance the optical nonlinearities of the hosts greatly. It was observed in early works on NPs embedded fused silica system [19–21]. The well-known z-scan arrangement can be used to measure the third-order optical nonlinearity of the materials. Two typical z-scan systems can be used for optical nonlinearity measurements, i.e., the closed-aperture and open-aperture. The closed-aperture z-scan technique is

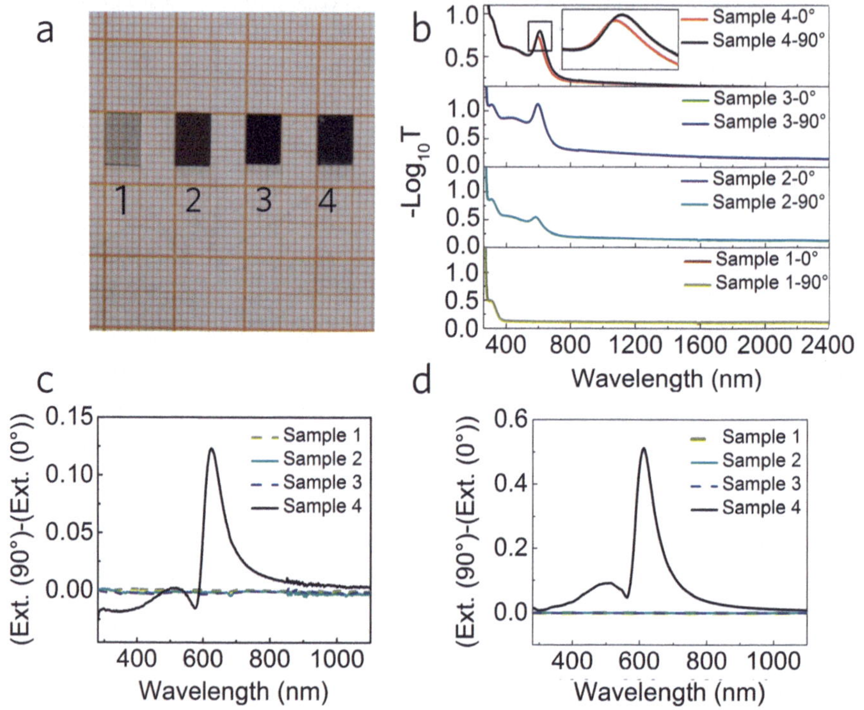

Fig. 10.8 **a** Photographs of Cu NPs embedded $LiTaO_3$ samples 1–4. **b** Polarization-dependent optical extinction of samples. **c** The difference between 90° polarization spectrum and 0° polarization spectrum of samples 1–4. **d** The simulated difference between 90° polarization spectrum and 0° polarization spectrum of samples 1–4, respectively. Reproduced with permission from [18], ©2019 American Chemical Society

used to measure the nonlinear index n_2 (Kerr nonlinearity), which gives the information on the materials: self-focusing or self-defocusing. The open-aperture technique is utilized to measure the nonlinear absorption coefficient, which also gives the information on the origins of nonlinear absorption: two-photon absorption or saturable absorption. For laser technology, the generation of laser pulses requires saturable absorption properties of the materials for Q-switching or mode-locking.

The z-scan data can be analyzed quantitatively according to the theory of nonlinear optics. The propagation equation in these samples can be written as:

$$\frac{dI}{dz'} = -\alpha_0 I - \alpha_{NL} I^2 \tag{10.5}$$

where I is the excitation intensity, z' is the propagation distance in the samples, α_0 is the linear absorption coefficient, and α_{NL} is the nonlinear absorption coefficient This equation can be solved for the z-scan method exactly and the normalized power transmission as a function of z is as follows [22]:

$$T_{Norm}(z) = \frac{\ln[1 + q_0(z)]}{q_0(z)} \tag{10.6}$$

$$q_0(z) = \alpha_{NL}(I_0 L_{\text{eff}})/(1 + z^2/z_0^2) \tag{10.7}$$

$$L_{eff} = [1 - e^{-\alpha_0 L}]/\alpha_0 \tag{10.8}$$

where L_{eff} and L are the effective and practical layer thickness of Au NPs, respectively. I_0 is the light intensity at the focus and z_0 is the diffraction length of beam.

For the closed-aperture z-scan measurement, the nonlinear refractive index n_2, which is a typical parameter for optical nonlinearity, can be determined. The doping of metallic NPs could significantly enhance the n_2 of the host materials. Figure 10.9 shows the normalized transmittance of the pure and Ag NPs embedded Nd:YAG crystal as function of z by fs-laser probing under pulse energies of 50 and 150 nJ [10]. A valley-peak type distribution is observed, indicating a self-focusing effect for both the host and the Ag NPs doped material.

By fitting the data by the following analytic formula [23], n_2 can be therefore deduced.

$$T_{Norm}(z) = 1/\{[1 - 4x\Delta\phi_0/(1 + x^2)^2] + [4\Delta\phi_0^2/(1 + x^2)^3]\} \tag{10.9}$$

$$\Delta\phi_0 = k \cdot \Delta n \cdot L_{\text{eff}} \tag{10.10}$$

$$z_0 = k\omega_0^2/2 \tag{10.11}$$

$$x = z/z_0 \tag{10.12}$$

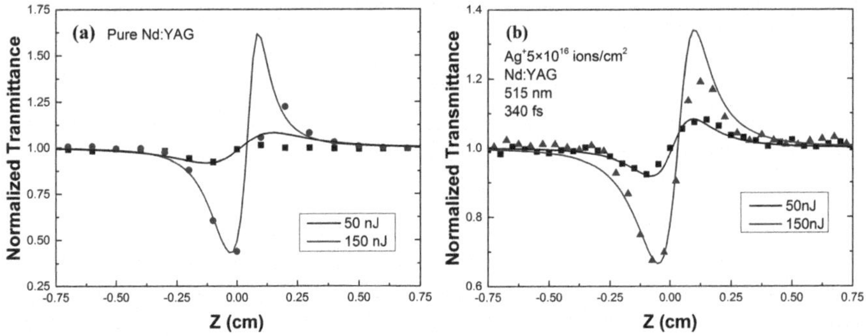

Fig. 10.9 The measured normalized transmittance of the **a** pure and **b** Ag NPs embedded Nd:YAG crystal as function of z by fs laser probing under pulse energies of 50 and 150 nJ. The used laser for excitation is at wavelength of 515 nm, with duration of 340 fs, and repetition rate of 100 Hz

Table 10.1 The reported nonlinear refractive indices of the dielectric systems embedded with various metallic NPs

Materials	Metallic NPs	n_2 (cm^2/W)	Refs.
Nd:YAG	Ag	2×10^{-12}	Li et al. [10]
$LiNbO_3$	Ag	9.38×10^{-16}	Pang et al. [24]
	Zn	-13.8×10^{-11}	Wang et al. [25]
	Cu	7.82×10^{-11}	
	Cu (annealed at 300 °C)	9.07×10^{-11}	
	Fe	11.7×10^{-16}	Wang et al. [26]
	Ni	3.4×10^{-16}	
SiO_2	Cu	1.66×10^{-11}	Tao et al. [27]
	Cu (annealed at 800 °C)	-7.52×10^{-11}	
Nd:YAG	Zn	-12.4×10^{-11}	Niu et al. [28]
	Ag	-17.3×10^{-11}	
	Ag (annealed at 700 °C)	-46×10^{-11}	

where $\Delta\phi_0$ is on-axis phase shift at the focus, Δn is nonlinear refractive-index change, L_{eff} is the effective thickness of sample, z_0 is the diffraction length of the beam, and k is wave number, w_0 is beam waist radius.

One can finally obtain giant enhancement of n_2 (4 order of magnitude) in Ag NPs embedded Nd:YAG crystal (1.8×10^{-12} cm^2/W at 50 nJ) over the pure crystal (1.2×10^{-16} cm^2/W at 50 nJ), which shows the significant roles of the LSPR effects from the NPs on the optical nonlinearities.

Similar enhancement of optical nonlinearities has been observed in other crystals embedded with metallic NPs. Table 10.1 shows the reported nonlinear refractive indices of the dielectric systems embedded with various metallic NPs. As it is indicated, in all systems the metallic NPs induce giant enhancement of n_2, reaching 4–6 orders of magnitude.

10.3 Nonlinear Absorption Tuning by Nanoparticles

By using embedded metallic NPs, the nonlinear absorption of the dielectrics can be modified significantly. Materials with saturable absorption features can be used in the modern laser technology to generate laser pulses through Q-switching or mode-locking mechanisms. The open-aperture z-scan technique is used to obtain the information of the origins of nonlinear absorption: two-photon absorption or saturable absorption, and to measure the nonlinear absorption coefficients of the modified dielectric systems as well. Some materials possess two-photon absorption properties, which cannot be used for laser pulses generation. However, the embedded metallic NPs may serve as a crucial factor to change or even inverse the nonlinear absorption from two-photon absorption to saturable absorption. Nanomaterials such as zero-dimensional quantum dots, one-dimensional single-wall carbon tubes, and two-dimensional layered materials (e.g., graphene, black phosphorous, MoS_2 etc.), are well-known saturable absorbers (SAs) for passive Q-switching or mode-locking [29–34]. Some materials doped by NPs by growth through chemical techniques are also used for ultrafast laser generation. By using ion implantation to form a 100 nm thick layer of metallic NPs, the nonlinear absorbance of the dielectric wafers could be modified and tailored. Basically, the nonlinear optical responses could be considered as the synergistic effect of contributions from metallic NPs and dielectric materials. In addition, the concentration of the metallic NPs plays important roles for the total effect. Figure 10.10 shows the open-aperture z-scan results for the Ag NPs embedded $LiNbO_3$ produced by different fluences of Ag^+ ion implantation under excitation at 515 and 1030 nm, respectively [24]. From Fig. 10.10a, one can obtain that, for Ag^+ ion implantation at fluence of 1×10^{16} and 3×10^{16} cm^{-2}, the $LiNbO_3$ embedded with Ag NPs possesses two-photon absorption property, which is as same as pure $LiNbO_3$ crystal. This means that the synergy of Ag NPs and bulk is that the bulk effect is still dominant over the NPs. However, for $LiNbO_3$ implanted by Ag^+ ions at 3×10^{16} cm^{-2}, the feature of two-photon absorption is degraded in comparison to that implanted at 1×10^{16} cm^{-2}. As the concentration of Ag NPs increases, which is performed by higher fluence Ag^+ ion implantation, the $LiNbO_3$ with embedded Ag NPs starts to have saturable absorption properties, which means that the contribution from the NPs becomes more obvious. This offers a unique solution to tailor the nonlinear absorptions of $LiNbO_3$ crystals by embedded metallic NPs. It should be also pointed out that the data from Fig. 10.10a are obtained under excitation at 515 nm, which is in the main LSPR effect band (visible light wavelength regime). Since the Ag NPs is with diverse diameters (the size distribution is quasi-Gaussian), the LSPR effect extends to longer wavelength region (till NIR wavelength of 1 μm). This NIR absorption is usually very weak, and only at high concentration of NPs the NPs play non-negligible effect on the nonlinearities of the system. As one can see from Fig. 10.10b, under excitation at 1030 nm, the $LiNbO_3$ with embedded Ag NPs possesses obvious saturable absorption feature that could be used for generation of laser pulses.

Some crystals like BGO are difficult to change the nonlinear absorption from two-photon to saturable absorption by NPs. However, it is still possible to engineer the nonlinearities of the crystal to some extent. Li et al. used 200 keV Ag^+ ion implantation at fluence of 1×10^{17} cm^{-2}, and measured the nonlinear absorption coefficient [13]. In case of multi-photon absorption effect, traditional z-scan theory should be corrected accordingly. The propagation equation in these samples can be written as:

$$dI/dz' = -\alpha_0 I - \sum_{m=2} \alpha_m I^m - \sigma_a N_{e-h} I \tag{10.13}$$

where I is the excitation intensity, z' is the propagation distance in the samples, σ_a is the cross term of absorption and refraction by the photo-excited charge carriers,

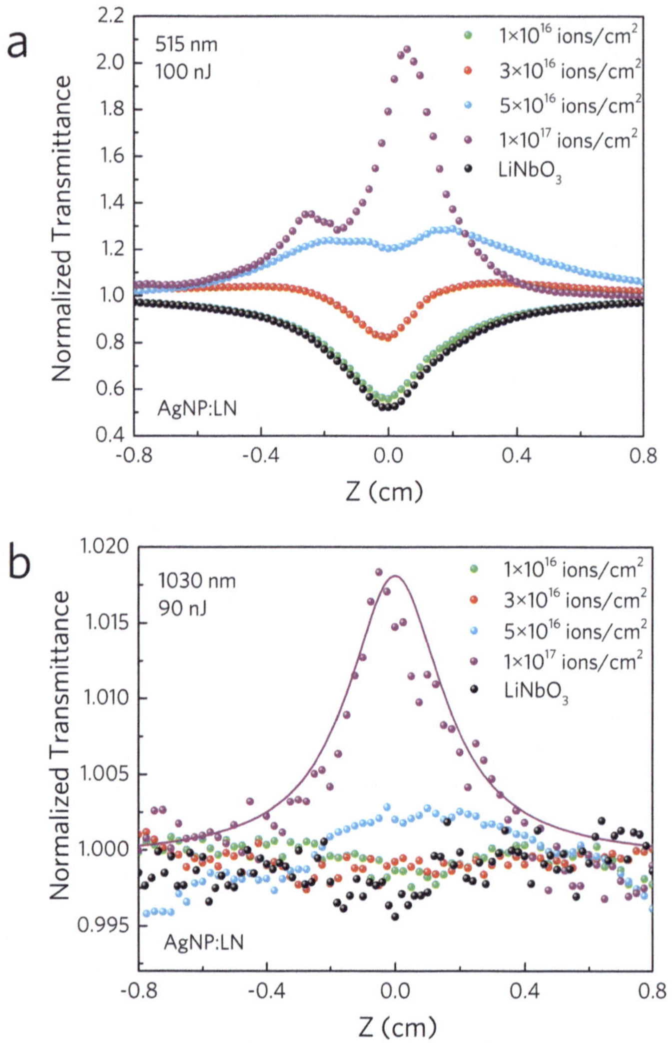

Fig. 10.10 The open-aperture z-scan results for the Ag NPs embedded $LiNbO_3$ produced by different fluences of Ag^+ ion implantation under excitation at **a** 515 and **b** 1030 nm

N_{e-h} is the number density of the photo-excited charge carriers, α_0 is the linear absorption coefficient and α_m is the nonlinear absorption coefficient. This equation can be solved as follows [35],

$$T_{Norm}(z) = \frac{1}{\pi^{1/2} q_0} \int \ln[1 + q_0 \exp(-x^2)] dx \tag{10.14}$$

$$T'_{Norm}(z) = \frac{1}{\pi^{1/2} p_0} \int \ln\left\{[1 + p_0^2 \exp(-2x^2)]^{1/2} + p_0 \exp(-x^2)\right\} dx \tag{10.15}$$

where $q_0 = \alpha_2(I_0 L_{\mathrm{eff}})/(1 + z^2/z_0^2)$, $p_0 = (2\alpha_2 L'_{\mathrm{eff}})^{1/2}(I_0)/(1 + z^2/z_0^2)$, $L_{\mathrm{eff}} = [1 - e^{-\alpha_0 L}]/\alpha_0$, $L'_{\mathrm{eff}} = [1 - e^{-2\alpha_0 L}]/2\alpha_0$. L_{eff} and L'_{eff} is the effective thickness of the Ag NPs layer at the condition of two-photon and three-photon absorption respectively. L is the thickness of the Ag NPs layer, I_0 is the light intensity at the focus and z_0 is the beam's diffraction length. Further approximation enables Eqs. 10.14 and 10.15 to be rewritten as [36]:

$$\ln(1 - T_{Norm}(z)) = \ln(I(z)) + C \tag{10.16}$$

$$\ln(1 - T'_{Norm}(z)) = 2\ln(I(z)) + C' \tag{10.17}$$

where C and C' are the constants. Equations 10.14 and 10.16 are appropriate for the two-photon absorption while the Eqs. 10.15 and 10.17 are suitable for the three-photon absorption condition, from which we can easily distinguish the order of the nonlinear optical response by $\ln(1 - T(z))$ versus $\ln(I(z))$. Figure 10.11 shows the $\ln(1 - T(z))$ versus $\ln(I(z))$ curves for the pure BGO and BGO with embedded Ag NPs, which can be used to determine the optical nonlinearity of the samples.

From the obtained normalized transmittance versus position (z) curves, one can calculate the absorption coefficient. The nonlinear absorption coefficients are measured to be ~3.1×10^{-9} cm/W (two-photon absorption coefficient) and ~8.9×10^{-14} cm^3/W^2 (effective three-photon absorption coefficient) for pure BGO crystal and the sample embedded with Ag NPs, respectively. The results indicate that the nonlinear optical response of the Ag NPs is responsible for the three-photon absorption.

For the dielectrics with weak optical nonlinearities, the contribution of the nonlinear absorption is mainly from the metallic NPs. For example, the Nd:YAG exhibits neither two-photon absorption nor saturable absorption. With embedded Au NPs, the systems are offered with saturable absorption properties [4]. For materials already

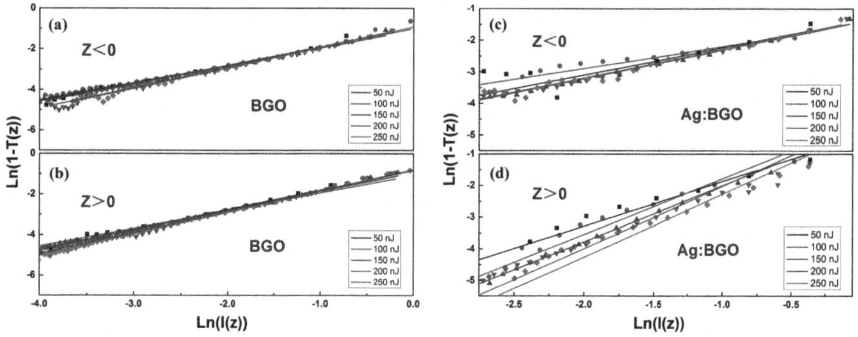

Fig. 10.11 Plots of $\ln(1 - T(z))$ versus $\ln(I(z))$ of BGO when **a** $z < 0$ and **b** $z > 0$; plots of $\ln(1 - T(z))$ versus $\ln(I(z))$ of Ag implanted BGO as **c** $z < 0$ and **d** $z > 0$. Reproduced with permission from [13], ©2018 Springer-Nature

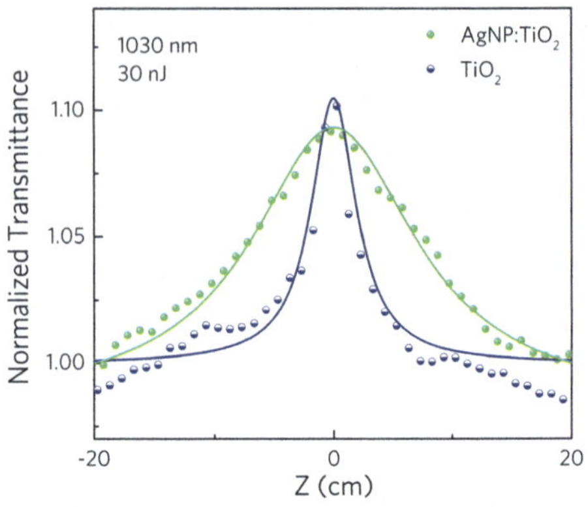

Fig. 10.12 Z-scan results of TiO_2 and TiO_2 with embedded Ag NPs under the excitation light of 340 fs pulses at 1030 nm

possessing saturable absorption property, the embedded metallic NPs further modify feature, such as the saturation intensity, modulation depth, enabling alternative solutions for laser systems (Fig. 10.12) [37].

10.4 Pulsed Waveguide Lasers Based on Nanoparticles as Saturable Absorbers

Saturable absorption is a property of materials where the absorption of light decreases with increasing light intensity. Metallic NPs with saturable absorption properties may be used for pulsed laser generation through the configurations of passive Q-switching or mode-locking. The dielectric materials with embedded metallic NPs may also be utilized as SAs since the optical nonlinearities of the host have been significantly enhanced. One of the intriguing advantages of the NPs doped dielectric systems is the possible combination of other dielectric materials in a stable way, which is particularly useful for compact devices.

The dielectrics with embedded metallic NPs are considered as "fast" SAs. A fast SA is one with a recovery time well below the pulse duration, whilst a slow absorber is one with a recovery time well above the pulse duration. Figure 10.13 shows the pump-probe result of $LiNbO_3$ crystal with embedded Ag NPs under a femtosecond laser system (1040 nm, 340 fs, 1 kHz). The pump energy is measured to be 2 μJ and the probe energy is 35 nJ. Corresponding relaxation time can be obtained by single-exponential model fitting:

$$g(t) = A\exp\left(-\frac{t}{\tau}\right) \tag{10.18}$$

where A denotes amplitude, τ represents the lifetime of the sample, and t is the time delay. The obtained relaxation time is 261 fs, which is limited by the temporal resolution of the setup (~340 fs) and the dispersion of the optics in the setup. In

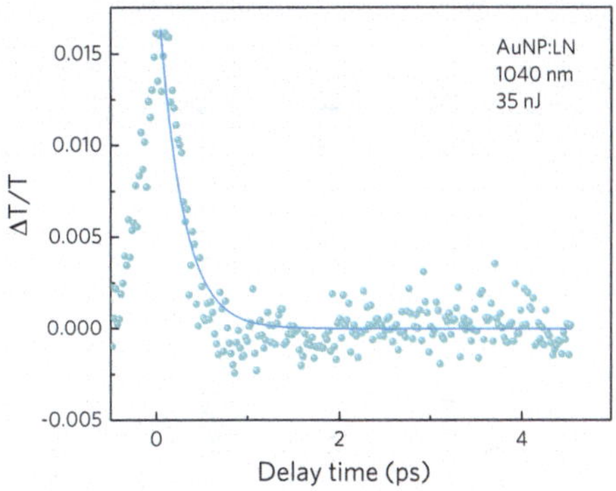

Fig. 10.13 Pump-probe results of $LiNbO_3$ and $LiNbO_3$ with embedded Au NPs under the excitation light of 340 fs pulses at 1040 nm

addition, from the obtained curve of the pump-probe test, the Ag NPs have indeed modified $LiNbO_3$ crystal as a fast SA material. The short relaxation time suggests potential applications of the $LiNbO_3$ crystal with embedded Ag NPs could be used in ultrafast laser systems.

Table 10.2 summarizes the obtained results for Q-switched or mode-locked lasers based on dielectrics with embedded NPs as SAs. The main LSPR band is located at the visible light regime in most dielectric systems with embedded metallic NPs, considering the species, size, shapes, and dielectric surroundings (refractive index) [38–42]. The metallic NPs can tailor the nonlinear absorption to saturable absorption, as a result, the modified dielectrics could serve as SAs in visible light regime. The lasing wavelength is in accordance with operation wavelength of the SAs, which is also located at visible light band. The first pulsed laser based on SA of ion beam synthesized metallic NPs in dielectrics was realized by Nie et al. [4], in which a Pr^{3+}:$LuLiF_4$ crystal is used as laser gain medium owing to the energy level system of Pr^{3+} ions (transition band of $^3P_0 \rightarrow {}^3F_2$ related to Pr^{3+} ions) for lasing in visible light band. The Q-switched lasing operates at 639 nm. The obtained minimum pulse duration is 330 ns, and the maximum repetition rate reaches 240 kHz. These parameters are generally comparable to those from the 2D materials-based laser systems. Also taking Pr^{3+}:$LuLiF_4$ crystal as gain medium, Pang et al. used $LiNbO_3$ with embedded Ag NPs as SA, and realized passively Q-switched laser at 639 nm [24]. The shortest pulse duration is 963 ns, and the repetition rate is 137 kHz.

The extension of LSPR effects to NIR band enables wider applications. The Nd^{3+} or Yb^{3+} ion doped crystals are favorite gain media for 1 μm wavelength lasing. Chen's group in Shandong University deployed various dielectrics with embedded NPs to develop pulsed laser systems. By using waveguide platforms, efficient Q-switching or mode-locking has been achieved. The systems are similar to those based on 2D materials as SAs. Because of the ultrafast saturable absorption property, pulse duration in picosecond level and repetition rate up to 10 GHz has been obtained in Nd^{3+} ion-doped waveguide laser systems (see Table 10.2). The first mode-locked pulses were realized by Pang et al., in which a $LiNbO_3$ crystal with embedded Au NPs was used as SA [6]. With Nd:YVO_4 waveguide as laser cavity, the 1064 nm laser pulses with repetition rate of 6.4 GHz and the pulse duration of 74.1 ps were obtained.

Table 10.2 Summary of the obtained results for Q-switched or mode-locked lasers based on dielectrics with embedded NPs as SAs

NPs	Materials	Operation regimes	Laser types	Lasing Wavelength	Performance	Ref.
Au	Nd:YAG	Q-switched	Bulk	639 nm	RR_{max}: 240 kHz PD_{min}: 330 ns	Nie et al. [4]
Au	$LiNbO_3$	Q-switched mode-locked	Waveguide	1064 nm	RR_{max}: 6.4 GHz PD_{min}: 74.1 ps PE_{max}: 8 nJ	Pang et al. [6]
Ag	SiO_2	Q-switched mode-locked	Waveguide	1064 nm	RR_{max}: 6.5 GHz PD_{min}: 27 ps PE_{max}: 28.2 nJ	Pang et al. [43]
Ag	$LiNbO_3$	Q-switched	Waveguide	1064 nm	RR_{max}: 4.95 MHz PD_{min}: 41.5 ns PE_{max}: 58.3 nJ	Pang et al. [24]
			Bulk	639 nm	RR_{max}: 137 kHz PD_{min}: 963 ns PE_{max}: 47.3 nJ	
Ag	$LiTaO_3$	Q-switched mode-locked	Waveguide	1064 nm	RR_{max}: 8.74 GHz PD_{min}: 35 ps PE_{max}: 29 nJ	Pang et al. [12]
Ag	TiO_2	Q-switched mode-locked	Waveguide	1064 nm	RR_{max}: 26.4 GHz PD_{min}: 16.8 ps PE_{max}: 10.8 nJ	Pang et al. [37]
Ag	Nd:YAG	Q-switched mode-locked	Waveguide	1064 nm	RR_{max}: 10.5 GHz PD_{min}: 29.5 ps PE_{max}: 44.5 nJ	Li et al. [44]
Cu	$LiTaO_3$	Q-switched mode-locked	Waveguide	1064 nm	RR_{max}: 8.6 GHz PD_{min}: 23 ps	Pang et al. [18]
Cu	$LiNbO_3$	Q-switched mode-locked	Waveguide	1064 nm	RR_{max}: 8.6 GHz PD_{min}: 55 ps	Wang et al. [11]

RRF_{max}: repetition rate; PD_{min}: pulse duration; PE_{max}: pulse energy

Later other systems with embedded NPs have been utilized as efficient SAs to achieve lasing of ultrashort pulse durations and multi-GHz repetition rates. Figure 10.14 shows the obtained results for Q-switched mode-locked Nd:YAG waveguide lasers operating at 1064 nm based on $LiTaO_3$ embedded Ag NPs as SA [12]. The pulse duration as short as 35 ps, the repetition rate of 8.74 GHz, and signal-to-noise ratio (SNR) of 47.2 dB indicate the Q-switched mode-locked laser operates in a relatively stable system.

Since the metallic NPs are embedded in the dielectrics, it is intriguing to integrate the NPs with optical waveguides monolithically, i.e., one can use ion beam techniques to fabricate waveguides and synthesize NPs in a single chip. Li et al. proposed an integrated chip in Nd:YAG crystal, containing ridge waveguide structure with Ag NPs embedded in top 100 nm region [44]. The Ag NPs were firstly synthesized by Ag^+ ion implantation onto the Nd:YAG. And then the ridge waveguide was produced by 12 MeV O^{5+} ion implantation and followed by diamond saw dicing. In this case, the monolithic chip was formed. Under 808 nm pump, the lasers interact with Ag NPs through the evanescent field absorption, achieving mode-locking. Figure 10.15

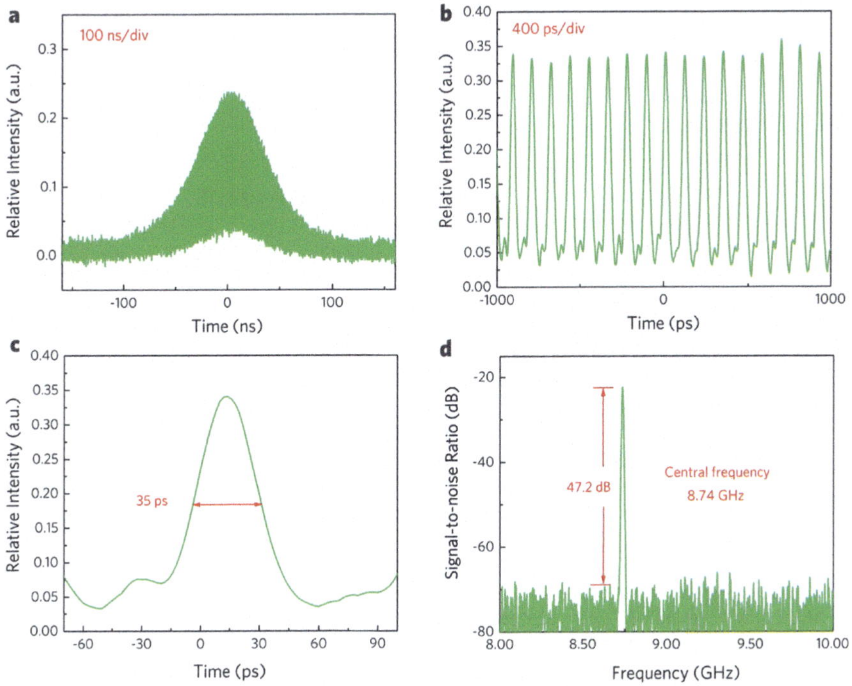

Fig. 10.14 Q-switched mode-locked Nd:YAG waveguide lasers operating at 1064 nm based on $LiTaO_3$ embedded Ag NPs as SA: **a** Q-switched pulse envelopes; **b** mode-locked pulse trains on picosecond time scale; **c** single mode-locked pulse train; and **d** radio frequency spectrum. The length of Nd:YAG waveguide is ~1 cm. Reproduced with permission from [12], ©2019 The Institute of Physics

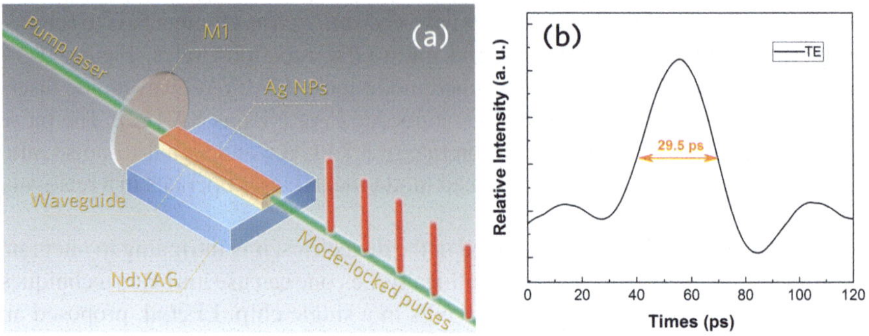

Fig. 10.15 **a** The schematic diagram of the Q-switched mode-locked laser generation, and **b** the obtained single pulse profile of mode-locking at TE polarization

shows (a) the schematic diagram of the Q-switched mode-locked laser generation and (b) the obtained single pulse profile of mode-locking at TE polarization. The measured fundamental repetition rate is 10.53 GHz with signal-to-noise ratio of 44.6 dB, and obtained pulse duration is as short as 29 ps.

References

1. W. Wesch, E. Wendeler, *Ion Beam Modifications of Solids* (Springer, 2016)
2. M. Quinten, *Optical Properties of Nanoparticle Systems: Mie and Beyond* (Wiely-VCH, 2011)
3. A.L. Stepanov, X. Xiao, F. Ren, T. Kavetskyy, N.Y. Osin, Catalytic and biological sensitivity of TiO_2 matrices with silver nanoparticles created by ion implantation: a review. Rev. Adv. Mater. Sci. **34**, 107 (2013)
4. W.J. Nie, Y.X. Zhang, H.H. Yu, R. Li, R.Y. He, N.N. Dong, J. Wang, R. Hübner, R. Böttger, S.Q. Zhou, H. Amekura, F. Chen, Plasmonic nanoparticles embedded in single crystals synthesized by gold ion implantation for enhanced optical nonlinearity and efficient Q-switched lasing. Nanoscale **10**, 4228 (2018)
5. G. Mie, Contributions to the optics of turbid media, particularly of colloidal metal solutions. Ann. Phys. **25**, 377 (1908)
6. C. Pang, R. Li, Z.Q. Li, N.N. Dong, C. Cheng, W.J. Nie, R. Bottger, S.Q. Zhou, J. Wang, F. Chen, Lithium niobate crystal with embedded au nanoparticles: a new saturable absorber for efficient mode-locking of ultrafast laser pulses at 1 μm. Adv. Opt. Mater. **6**, 1800357 (2018)
7. M.C. Ridgway, R. Giulian, D.J. Sprouster, P. Kluth, L.L. Araujo, D.J. Llewellyn, A.P. Byrne, F. Kremer, P.F.P. Fichtner, G. Rizza, H. Amekura, M. Toulemonde, Role of thermodynamics in the shape transformation of embedded metal nanoparticles induced by swift heavy-ion irradiation. Phys. Rev. Lett. **106**, 095505 (2011)
8. J. Guzman, S.J. Shin, C.Y. Liao, C.W. Yuan, P.R. Stone, O.D. Dubon, K.M. Yu, J.M. Beeman, M. Watanabe, J.W. Ager, Photoluminescence enhancement of Er-doped silica containing Ge nanoclusters. Appl. Phys. Lett. **95**, 201904 (2009)
9. H. Amekura, H. Kitazawa, N. Umeda, N. Umeda, Y. Takeda, N. Kishimoto, Nickel nanoparticles in silica glass fabricated by 60 keV negative-ion implantation. Nucl. Instrum. Meth. Phys. Res. B **222**, 114 (2004)

10. R. Li, N.N. Dong, C. Cheng, F. Ren, R. Huebner, J. Wang, S.Q. Zhou, F. Chen, Giant enhancement of nonlinear optical response in Nd:YAG single crystals by embedded silver nanoparticles. ACS Omega **2**, 1279 (2017)
11. S.X. Wang, C. Pang, Z.Q. Li, R. Li, N.N. Dong, Q.M. Lu, F. Ren, J. Wang, F. Chen, 8.6 GHz Q-switched mode-locked waveguide lasing based on $LiNbO_3$ crystal embedded Cu nanoparticles. Opt. Mater. Express **9**, 3808 (2019)
12. C. Pang, R. Li, Z.Q. Li, N.N. Dong, J. Wang, F. Ren, F. Chen, Plasmonic Ag nanoparticles embedded in lithium tantalate crystal for ultrafast laser generation. Nanotechnology **30**, 334001 (2019)
13. R. Li, N.N. Dong, F. Ren, H. Amekura, J. Wang, F. Chen, Nonlinear absorption response correlated to embedded ag nanoparticles in BGO single crystal: from two-photon to three-photon absorption. Sci. Rep. **8**, 1977 (2018)
14. M. Nogami, L.X. Hung, H.V. Tuyen, V.X. Quang, Formation of Ni nanoparticles in Al_2O_3–SiO_2 glass by reacting with hydrogen gas. J. Mater. Sci. **54**, 13883 (2019)
15. R.Y. He, F. Ren, F. Chen, Embedded silver nanoparticles in KTP crystal produced by ion implantation. Mater. Lett. **193**, 158 (2017)
16. R. Li, C. Pang, H. Amekura, F. Ren, R. Hubner, S.Q. Zhou, N. Ishikawa, N. Okubo, F. Chen, Ag nanoparticles embedded in Nd:YAG crystals irradiated with tilted beam of 200 MeV Xe ions: optical dichroism correlated to particle reshaping. Nanotechnology **29**, 424001 (2018)
17. S. Wolf, J. Rensberg, A. Johannes, R. Thomae, F. Smit, R. Neveling, M. Moodley, T. Bierschenk, M. Rodriguez, B. Afra, S. Bin Hasan, C. Rockstuhl, M. Ridgway, K. Bharuth-Ram, C. Ronning, Shape manipulation of ion irradiated Ag nanoparticles embedded in lithium niobate, Nanotechnology **27**, 145202 (2016)
18. C. Pang, R. Li, Z. Li, N. Dong, H. Amekura, S. Wang, H. Yu, J. Wang, F. Ren, N. Ishikawa, N. Okubo, F. Chen, Copper nanoparticles embedded in lithium tantalate crystals for multi-GHz lasers. ACS Appl. Nano Mater. **2**, 5871 (2019)
19. J. Bornacelli, C. Torres-Torres, H.G. Silva-Pereyra, L. Rodríguez-Fernández, M. Avalos-Borja, J.C. Cheang-Wong, A. Oliver, Nanoscale influence on photoluminescence and third order nonlinear susceptibility exhibited by ion-implanted Pt nanoparticles in silica. Methods Appl. Fluores. **5**, 025001 (2017)
20. M.H. Mezher, W.Y. Chong, R. Zakaria, Third-order optical nonlinearity studies of bilayer Au/Ag metallic films. Laser Phys. **26**, 055401 (2016)
21. G. Battaglin, P. Calvelli, E. Cattaruzza, F. Gonella, P. Mazzoldi, Z-scan study on the nonlinear refractive index of copper nanocluster composite silica glass. Appl. Phys. Lett. **78**, 3953 (2001)
22. M. Sheik-Bahae, A.A. Said, T.H. Wei, D.J. Hagan, E.W. Van Stryland, Sensitive measurement of optical nonlinearities using a single beam. IEEE J. Quantum Elect. **26**, 760 (1990)
23. C.H. Kwak, Y.L. Lee, S.G. Kim, Analysis of asymmetric Z-scan measurement for large optical nonlinearities in an amorphous As_2S_3 thin film. J. Opt. Soc. Am. B **16**, 600 (1999)
24. C. Pang, R. Li, Y. Zhang, Z. Li, N. Dong, L. Wu, H. Yu, J. Wang, F. Ren, F. Chen, Tailoring optical nonlinearities of $LiNbO_3$ crystals by plasmonic silver nanoparticles for broadband saturable absorbers. Opt. Express **26**, 31276 (2018)
25. Y. Wang, Y. Niu, G. Wang, Y. Sun, C. Liu, Enhanced nonlinear optical properties of $LiNbO_3$ crystal embedded with CuZn alloy nanoparticles by ion implantation. J. Alloy. Compd. **778**, 691 (2019)
26. Y. Wang, X. Yu, F. Liu, Y. Wang, Nonlinear refraction of lithium niobate crystal doped with different metal nanoparticles. Mater. Lett. **123**, 35 (2014)
27. C. Tao, G. Jia, X. Mu, H. Dai, C. Liu, Synthesis and optical properties of Cu Core/Ti-related shell nanoparticles in silica sequentially implanted with Ti and Cu ions. Plasmonics **10**, 1869 (2015)
28. Y. Niu, Y. Wang, G. Wang, T. Zhang, C. Liu, Enhanced optical linearity and nonlinearity of Nd:YAG crystal embedded with Ag nanoparticles by prior Zn ion implantation. Opt. Mater. Express **8**, 3666 (2018)
29. W.B. Cho, J.H. Yim, S.Y. Choi, S. Lee, A. Schmidt, G. Steinmeyer, U. Griebner, V. Petrov, D.I. Yeom, K. Kim, Boosting the nonlinear optical response of carbon nanotube saturable absorbers for broadband mode-locking of bulk lasers. Adv. Funct. Mater. **20**, 1937 (2010)

30. H.R. Chen, Y.G. Wang, C.Y. Tsai, K.H. Lin, T.Y. Chang, J. Tang, W.F. Hsieh, High-power, passively mode-locked Nd:$GdVO_4$ laser using single-walled carbon nanotubes as saturable absorber. Opt. Lett. **36**, 1284 (2011)
31. S. Wang, H. Yu, H. Zhang, A. Wang, M. Zhao, Y. Chen, L. Mei, J. Wang, broadband few-layer MoS_2 saturable absorbers. Adv. Mater. **26**, 3538 (2014)
32. E.U. Rafailov, M.A. Cataluna, W. Sibbett, Mode-locked quantum-dot lasers. Nat. Photonics **1**, 39 (2007)
33. P.T. Guerreiro, S. Ten, N.F. Borrelli, J. Butty, N. Peyghambarian, PbS quantum-dot doped grasses as saturable absorbers for mode locking of a Cr:forsterite laser. Appl. Phys. Lett. **71**, 1595 (1997)
34. Y. Chen, G. Jiang, S. Chen, Z. Guo, X. Yu, C. Zhao, H. Zhang, Q. Bao, S. Wen, D. Tang, D. Fan, Mechanically exfoliated black phosphorus as a new saturable absorber for both Q-switching and mode-locking laser operation. Opt. Express **23**, 12823 (2015)
35. S. Ijaz, A. Mahendru, D. Sanderson, Sensitive measurement of optical nonlinearities using a single beam. IEEE J. Quant. Electron. **26**, 760 (2002)
36. H. Li, J. He, J. Mi, W. Ji, Y. Qu, Three-photon absorption in ZnO and ZnS crystals. Opt. Express **13**, 9235 (2005)
37. C. Pang, R. Li, J. Huang, Z. Li, N. Dong, H. Yu, J. Wang, L. Wu, F. Ren, Z. Fang, F. Chen, Ultrafast nonlinear excitation dynamics in titanium dioxide by silver nanoparticles modulation for enhanced nonlinear optical responses, unpublished (2019)
38. J.Z. Dong, X.L. Zhang, Y.A. Cao, W.S. Yang, J.G. Tian, Shape dependence of nonlinear optical behaviors of gold nanoparticles. Mater. Lett. **65**, 2665 (2011)
39. Y. Hua, K. Chandra, D.H.M. Dam, G.P. Wiederrecht, T.W. Odom, Shape-dependent nonlinear optical properties of anisotropic gold nanoparticles. J. Phys. Chem. Lett. **6**, 4904 (2015)
40. K.H. Kim, A. Husakou, J. Herrmann, Linear and nonlinear optical characteristics of composites containing metal nanoparticles with different sizes and shapes. Opt. Express **18**, 7488 (2010)
41. Z.P. Zeng, C.S. Garoufalis, A.F. Terzis, S. Baskoutas, Linear and nonlinear optical properties of ZnO/ZnS and ZnS/ZnO core shell quantum dots: Effects of shell thickness, impurity, and dielectric environment. J. Appl. Phys. **114**, 023510 (2013)
42. R. Sato, M. Ohnuma, K. Oyoshi, Y. Takeda, Experimental investigation of nonlinear optical properties of Ag nanoparticles: effects of size quantization. Phys. Rev. B **90**, 125417 (2014)
43. R. Li, C. Pang, Z. Li, M. Yang, H. Amekura, N. Dong, J. Wang, F. Ren, Q. Wu, F. Chen, Fused silica with embedded 2D-like Ag nanoparticle monolayer: Tunable saturable absorbers by interparticle spacing manipulation, unpublished (2019)
44. R. Li, C. Pang, Z. Li, N. Dong, J. Wang, F. Ren, S. Akhmadaliev, S. Zhou, F. Chen, Monolithic waveguide laser mode-locked by embedded Ag nanoparticles operating at 1 μm. Nanophotonics **8**, 859 (2019)

Chapter 11
Summary and Outlook

Abstract In this chapter, the summary of the ion irradiation of dielectrics for photonic applications will be given. A brief outlook in this research field is presented on the new materials and new applications.

11.1 Summary

In summary, this book overviews the state of the art in the intensely developing area of ion beam processed dielectrics for photonic applications. The fundamentals of the ion beam techniques and facilities, the basic theories of the ion-materials interactions, and the demonstrations of the waveguides, nanoscale particles, and thin films produced by ion beams in dielectrics have been presented. The various properties of the ion beam processed dielectric materials have been introduced, together with the following overviews on their applications as photonic devices. The research on ion beam modifications of dielectrics receives considerable attentions in multidisciplinary community of physics, material sciences, and other areas, facilitating the emergence of practical photonic material platforms produced by ion beam technology to be commercially available. Nevertheless, there are a number of topics requiring special attentions, which will be introduced in the following subsection.

11.2 Outlook

The dielectrics contain various materials that could be used for photonic applications. By using diverse ion beam processing, those materials can be tailored to possess a number of new properties or be with on-demand geometries as functional devices for intriguing applications. The following topics may be future research highlights in this area.

F. Chen et al., *Ion Irradiation of Dielectrics for Photonic Applications*, Springer Series in Optical Sciences 231, https://doi.org/10.1007/978-981-15-4607-5_11

11.2.1 New Materials

First, the new crystalline thin-film materials may serve as unique platforms for on-chip photonic devices. By using ion implantation and wafer bonding, high-quality $LiNbO_3$ thin film wafers have been produced successfully. Although remarkable achievements have been obtained on high-speed electrooptic (EO) modulation and high-efficiency nonlinear optical frequency conversion, the investigations on the doped $LiNbO_3$ thin films are almost blank. The doping of special ions into the $LiNbO_3$ crystals offers additional features that could be used in many areas. For examples, $LiNbO_3$ crystals doped with rare-earth ions (e.g., Nd^{3+}, Er^{3+}, or Tm^{3+}) can be utilized as gain media for amplifiers in telecommunications, monolithic solid-state lasers, or quantum memory devices. These active $LiNbO_3$ thin films are expected to exhibit potential in new applications in the LNOI based photonic integrated circuits. Doped by MgO, the ability for anti-optical-damage of $LiNbO_3$ is enhanced, whilst doped by Fe ions, $LiNbO_3$ becomes an excellent photorefractive material. The doping of selected ions into $LiNbO_3$ bulks could be performed by modified crystal growth techniques or thermal doping (in-diffusion). In addition, the technique containing ion implantation and wafer bonding enables possibility of potential extension to other suitable dielectrics, such as $LiTaO_3$, $KTiOPO_4$ etc. These wafer-size thin films may be utilized to meet the requirements of diverse applications, such as EO modulation or nonlinear optical wavelength conversion.

Second, the nanoparticles (NPs) embedded in the dielectrics offer new properties for a number of potential applications. The major research in this area was focused on the mechanisms of the nanoparticle synthesis and reshaping. The bulk materials are mainly fused silica and other glasses. The nonlinear optical properties of the nanoparticles buried inside the dielectrics have been investigated in detail very recently. Nevertheless, there are much more works to be done for the ion beam synthesized NPs in dielectrics. For example, the synergistic effect of embedded NPs and dielectrics requires in-depth study on the physical mechanisms to fulfill the gaps between fundamental research and applications. Novel optical phenomenon induced by the unique effects in nanostructures is also expected in the future work. Considering the universality and versatility of dielectric crystals, compact nanostructures with on-demand design by ion beams in crystalline dielectrics are of great significance to the photonic applications.

Third, new hybrid materials fabricated by ion beam techniques will play significant roles in the future applications. Two-dimensional (2D) materials belong to a newly emerging family of ultrathin films, since the great success on graphene and other layered materials has been achieved. It has been proved that, by using suitable ion beam processing, the properties of the layered 2D materials can be flexibly modified, which brings out new features with potential novel applications. For example, the ion-beam induced defects can form intermediate states in the electron band structure of the 2D materials, which may extend the optical effects of the 2D materials from the visible to the near infrared band. The integration of these 2D materials with dielectrics could be used for a number of new applications in photodetectors, sensors, lasers, and other passive/active devices in photonics. For example, the ion beam processed graphene may possess enhanced coupling effect with the dielectric substrates, which improves the performance of the devices.

11.2.2 Novel Devices for Micro-nano Photonics

The new materials bring out novel applications. Based on the LNOI platforms, the micro-resonators or whispering-gallery resonators (WGRs) enable compact and wavelength-sensitive optical devices, finding a wide spectrum of applications in photonics. Chip-integrated WGRs are highly promising based on LNOI platform since LN itself is a multi-functional ferroelectric crystal, opening up exciting opportunities and possibilities for integrated photonics. In addition, the doped LNOI offers light sources of the LNOI-based chips, which, together the fast EO modulators and nonlinear optical components, constructs the unique LNOI-based photonic chip devices. The applications of the LNOI in quantum technology and information processing are also very intriguing for future photonic networks. For example, Er^{3+} or Tm^{3+} ion doped LNOI platforms may be excellent candidates for quantum memories. Combined with the high-quality nonlinear optical and EO components, compact quantum chips may be expected on LNOI platforms.

NPs offer new properties of the bulks. The combination of the NPs with the dielectrics are potential platforms for diverse applications. For example, the gain media with embedded metallic NPs can be used as monolithic gain chips for all-solid-state lasers to generate ultrafast pulses. This design avoids the additional saturable absorbers and therefore improve the stability of the laser systems. To use 2D materials, some interesting devices may be fabricated by taking both advantages of the 2D materials and ion beam produced waveguides, including the photodetectors, new laser sources, and beam tailoring elements (e.g., polarizers).

Index

F. Chen et al., *Ion Irradiation of Dielectrics for Photonic Applications*, Springer Series in Optical Sciences 231, https://doi.org/10.1007/978-981-15-4607-5

The manufacturer's authorised representative in the EU is Springer Nature Customer Service Centre GmbH, Europaplatz 3, 69115 Heidelberg, Germany. If you have any concerns regarding our products, please contact ProductSafety@springernature.com

Printed and bound by CPI Group (UK) Ltd, Croydon, CR0 4YY
05/10/2026
02228123-0011